高等学校应用型新工科创新人才培养计划系列教材

高等学校计算机类专业课改系列教材

Android 程序设计及实践

(第二版)

青岛英谷教育科技股份有限公司
青岛农业大学
编著

西安电子科技大学出版社

内 容 简 介

Android是基于Linux的自由及开放源代码的操作系统，广泛应用于各种移动设备。

本书分为理论篇与实践篇两部分。理论篇共11章，深入讲解了Android开发的基础知识，其内容包括Android概述、活动(Activity)、用户界面、意图(Intent)、广播(Broadcast)、服务(Service)、数据存储、碎片(Fragment)、网络通信、消息处理机制以及Android特色开发等。

实践篇侧重于项目实战，通过"餐饮点餐系统"Android客户端的实现，并结合知识拓展内容，使读者能够循序渐进地理解Android理论知识，并提高项目开发实战能力。另外，实践篇还介绍了Android开发环境的搭建，以及使用LogCat对Android程序进行调试与监视等。

本书适用面广，可作为计算机科学与技术、软件工程、网络工程、计算机软件、计算机信息管理、电子商务和经济管理等专业程序设计课程的教材，也适合Android爱好者和Android应用开发人员使用。

图书在版编目(CIP)数据

Android程序设计及实践 / 青岛英谷教育科技股份有限公司，青岛农业大学编著.
—2版. —西安：西安电子科技大学出版社，2019.7(2022.9重印)
ISBN 978-7-5606-5351-8

Ⅰ. ①A… Ⅱ. ①青… ②青… Ⅲ. ①移动终端—应用程序—程序设计
Ⅳ. ①TN929.53

中国版本图书馆CIP数据核字(2019)第117273号

策　　划　毛红兵
责任编辑　刘小莉
出版发行　西安电子科技大学出版社(西安市太白南路2号)
电　　话　(029)88202421　88201467　　邮　　编　710071
网　　址　www.xduph.com　　电子邮箱　xdupfxb001@163.com
经　　销　新华书店
印刷单位　陕西天意印务有限责任公司
版　　次　2019年7月第2版　　2022年9月第8次印刷
开　　本　787毫米×1092毫米　1/16　印　张　34
字　　数　809千字
印　　数　21 001～24 000册
定　　价　84.00元
ISBN　978-7-5606-5351-8/TN

XDUP 5653002-8

高等学校计算机类专业
课改系列教材编委会

❖❖❖ 前　　言 ❖❖❖

随着我国计算机技术的迅猛发展，社会对具备计算机基本能力的人才需求急剧增加，“全面贴近企业需求，无缝打造专业实用人才”是目前高校计算机专业教育的革新方向。为了适应高等教育体制改革的新形势，积极探索适应 21 世纪人才培养的教学模式，编委会组织编写了高等学校计算机类专业系列课改规划教材。

该系列教材面向高校计算机类专业应用型新工科人才的培养，强调产学研结合，经过了充分的调研和论证，并参考多所高校一线专家的意见，具有系统性、实用性等特点，旨在帮助读者系统掌握软件开发知识，同时着重培养其综合应用能力和解决问题的能力。

该系列教材具有如下几个特点。

1. 以培养应用型人才为目标

本系列教材以培养应用型软件人才为目标，并在原有体制教育的基础上对课程进行了改革，强化了“应用型”技术的学习，从而使读者在经过系统、完整的学习后能够掌握如下技能：

- ✧ 掌握软件开发所需的理论和技术体系以及软件开发过程的规范体系。
- ✧ 能够熟练地进行设计和编码工作，并具备良好的自学能力。
- ✧ 具备一定的项目经验，包括代码的调试、文档编写、软件测试等内容。
- ✧ 达到软件企业的用人标准，做到学校学习与企业的无缝对接。

2. 以新颖的教材架构来引导学习

本系列教材采用的教材架构打破了传统的以知识为标准编写教材的方法，采用理论篇与实践篇相结合的组织模式，引导读者在学习理论知识的同时，加强实践动手能力的训练。

- ✧ 理论篇：学习内容的选取遵循“二八原则”，即，重点内容由企业中常用的20%的技术组成。每个章节设有本章目标，明确本章的学习重点和难点。章节内容结合示例代码，引导读者循序渐进地理解和掌握这些知识和技能，培养学生的逻辑思维能力，掌握软件开发的必备知识和技巧。
- ✧ 实践篇：集多点于一线，任务驱动，以完整的具体案例贯穿始终，力求使学生在动手实践的过程中加深对课程内容的理解，培养学生独立分析和解决问题的能力，并配备相关知识的拓展讲解和拓展练习，以拓宽学生的知识面。

另外，本系列教材借鉴了软件开发中的“低耦合，高内聚”的设计理念，并在组织结构上遵循软件开发中的 MVC 理念，即在保证最小教学集的前提下可以根据自身的实际情况对整个课程体系进行横向或纵向裁剪。

3. 提供全面的教辅产品来辅助教学实施

为充分体现“实境耦合”的教学模式，方便教学实施，该系列教材配备可配套使用的项目实训教材和全套教辅产品。

- ✧ 实训教材：集多线于一面，以辅助教材的形式，提供适应当前课程及先行课程的综合项目，遵循软件开发过程，进行讲解、分析、设计、指导，注重工作过程的系统性，培养读者解决实际问题的能力，是实施“实境”教学的关键环节。
- ✧ 立体配套：为适应教学模式和教学方法的改革，本系列教材提供完备的教辅产品，主要包括教学指导、实验指导、电子课件、习题集、实践案例等内容，并配以相应的网络教学资源。教学实施方面，本系列教材提供全方位的解决方案(课程体系解决方案、实训解决方案、教师培训解决方案和就业指导解决方案等)，以适应软件开发教学过程的特殊性。

本书还在第一版基础上进行了内容更新：改用 Google 官方开发工具 Android Studio，在理论篇新增加了广播(Broadcast)、碎片(Fragment)以及消息处理机制三章，实践篇新增加了相对应的实践和第三方框架，并在原有章节中增加了若干全新的知识点，使全书内容更加丰富，紧跟技术趋势，符合学生和企业的需求。

本书由青岛英谷教育科技股份有限公司和青岛农业大学编写，参与本书编写工作的有张坤、何莉娟、刘江林、王振芳、王万琦、王友君、刘立彬、孟洁、金成学、王燕等。本书在编写期间得到了各合作院校专家及一线教师的大力支持与协作，在此，衷心感谢每一位老师与同事为本书出版所付出的努力。

教材问题反馈

由于水平有限，书中难免有不足之处，欢迎大家批评指正。读者在阅读过程中如发现问题，可通过邮箱(yinggu@121ugrow.com)或扫描右侧二维码进行反馈，以期进一步完善。

本书编委会

2019 年 6 月

❖❖❖ 目　　录 ❖❖❖

理　论　篇

实 践 篇

理论篇

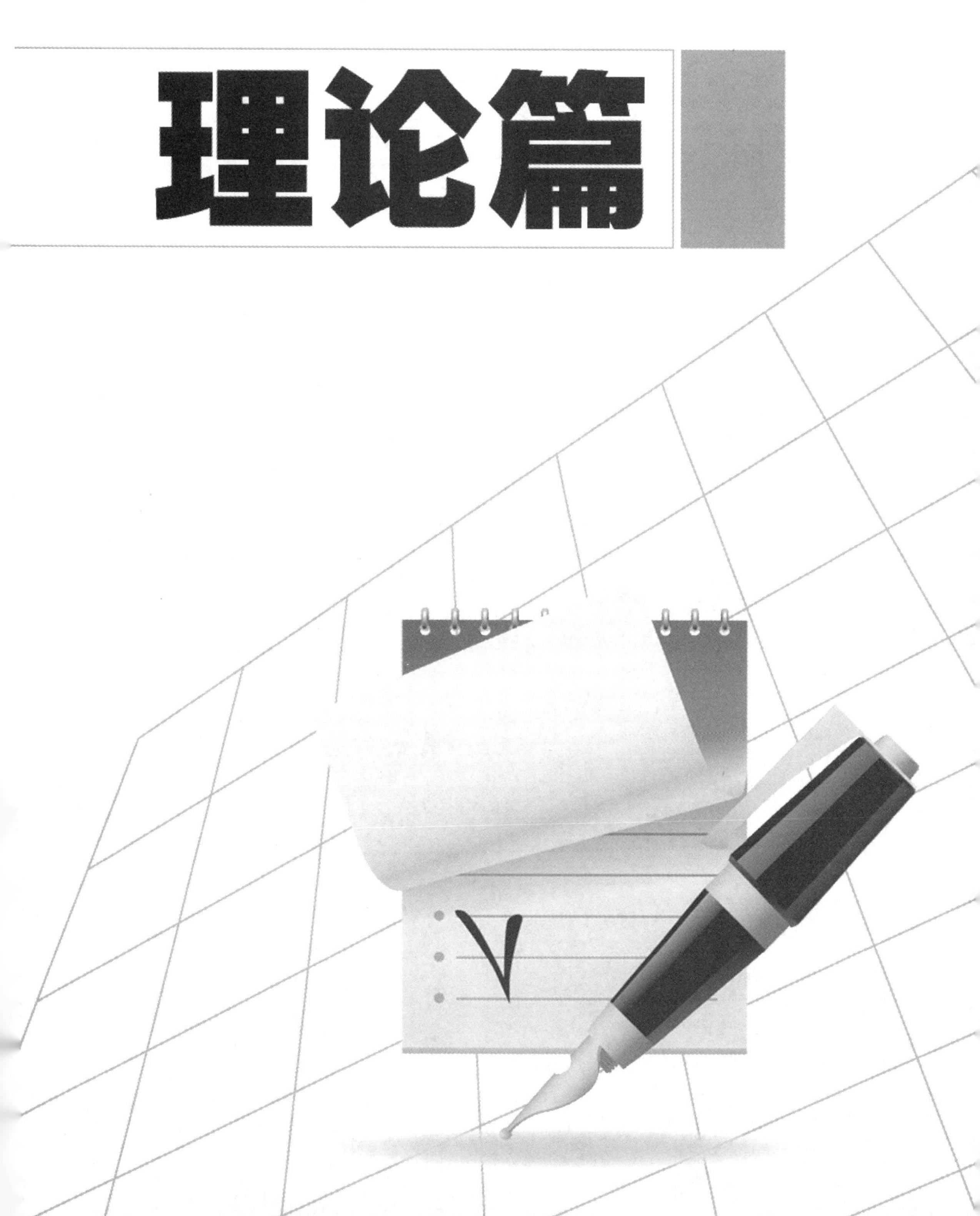

第1章　Android 概述

本章目标

- 了解移动信息设备分类
- 了解 Android 的历史和优缺点
- 熟悉 Android 平台的体系架构
- 掌握 Android 应用程序结构
- 掌握 Android 应用程序的编写

1.1 移动设备开发平台

目前主要有两大主流移动端开发平台：Android 与 iOS。这两大平台的用户体验各不相同，各有特色。iOS 依靠稳定的平台技术和面向高端的移动设备，显得势如猛虎；Android 则依靠开源的平台系统和不断完善的用户体验，发展势如破竹。

1.1.1 移动信息设备系统

随着计算机技术和无线通信技术的发展，移动信息设备正在深刻地改变着人们的生活，以手机、平板电脑等为代表的移动信息设备已经渗透到生活中的各个角落。一方面，新的移动设备与移动应用不断涌现；另一方面，人们从网络信息服务中受益，并正以前所未有的主动性去创造信息、共享信息。这些事实必将带来大量移动设备上的应用程序需求，因此，移动信息设备编程将成为今后计算机软件开发的热点之一。

但与 PC 不同的是，移动信息设备存在多种操作系统。从全球市场占有率看，PC 中的 Windows 系列占了 90%以上的市场，而移动信息设备中的操作系统却呈现出群雄割据的局面，常用的操作系统有：iOS、Windows Phone、BlackBerry OS、Linux(含 Android、Maemo 和 WebOS)。这些系统之间的应用软件互不兼容，因此移动信息设备中的应用程序需要根据不同的操作系统进行专门的开发。

1. iOS

iOS 是苹果公司开发的移动操作系统，目前用于苹果公司生产的 iPhone、iPod touch、iPad 以及 Apple TV 等产品上，甚至用到了车载设备上。iOS 凭借着系统的安全性、高度的稳定性、简单易用的页面、令人惊叹的内置功能和硬件设备的独特设计赢得了众多用户的支持，iPhone 手机如图 1-1 所示。

图 1-1　iPhone 手机

2. Android

Android 是网络巨头 Google 公司发布的基于 Linux 平台的开源操作系统，主要应用于移动设备，但其凭借着强大的开源机制，目前不仅用于手机、平板电脑等终端，而且应用到了如智能电视、车载导航、智能可穿戴设备和物联网设备中。正因 Android 采用的编码语言是 Java，之前从事 Java 的开发人员可以很容易地过渡到 Android 开发上来，因此吸引了很大一批 Java 开发人员投入到 Android 的开发中。Android 手机如图 1-2 所示。

图 1-2　Android 手机

图 1-3 所示为 2018 年各种移动开发平台市场份额的调查，从图中可看出 Android 已经成为市场份额最大的开发平台。

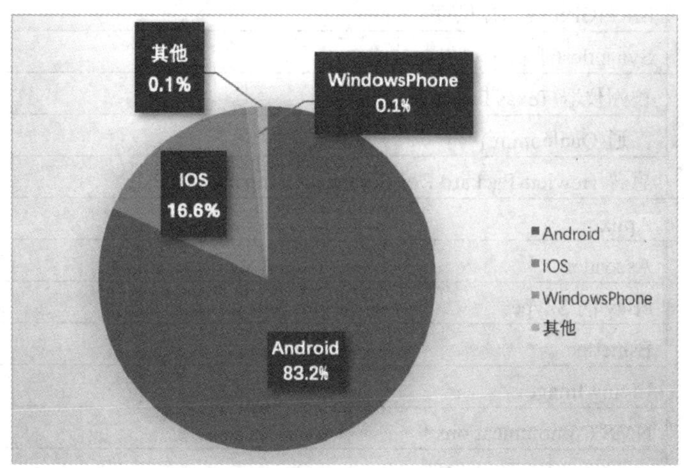

图 1-3　2014 年各种移动开发平台受开发者欢迎程度的调查

1.1.2　开放手机联盟

开放手机联盟(Open Handset Alliance，OHA)由一群共同致力于构建更好的手持移动信息设备的公司组成，该组织由 Google 领导，包括移动运营商、手持设备制造商、零部件制造商、软件解决方案和平台提供商以及市场营销公司。Android 平台就是开放手机联盟的成果。目前，联盟成员包括 Google、中国移动、T-Mobile、宏达电子(HTC)、摩托罗拉

等，这些领军企业将通过开放手机联盟携手开发 Android 及其应用程序。开放手机联盟成员如表 1-1 所示。

表 1-1 开放手机联盟成员

成员分类	成　员
手机制造商	华为技术有限公司——中国
	摩托罗拉(美国最大的手机制造商)
	韩国三星电子
	韩国 LG 电子
移动运营商	中国移动(全球最大的移动运营商，8 亿用户)
	日本 KDDI(2900 万用户)
	日本 NTT DoCoMo(5200 万用户)
	美国 Sprint Nextel(美国第三大移动运营商，5400 万用户)
	意大利电信(意大利主要的移动运营商，3400 万用户)
	西班牙 Telefónica(在欧洲和拉美有 1.5 亿用户)
	T-Mobile(德意志电信旗下公司，在美国和欧洲有 1.1 亿用户)
半导体公司	Audience Corp(声音处理器公司)
	Broadcom Corp(无线半导体主要提供商)
	英特尔(Intel)
	Marvell Technology Group
	Nvidia(图形处理器公司)
	SiRF(GPS 技术提供商)
	Synaptics(手机用户界面技术)
	德州仪器(Texas Instruments)
	高通(Qualcomm)
	惠普(Hewlett-Packard Development Company，L.P)
软件公司	Aplix
	Ascender
	eBay 的 Skype
	Esmertec
	Living Image
	NMS Communications
	Noser Engineering AG
	Nuance Communications
	PacketVideo
	SkyPop
	Sonix Network
	TAT-The Astonishing Tribe
	Wind River Systems

随着 Android 平台的发展，越来越多的相关企业加入开放手机联盟，最新的开放手机联盟成员名单可以在其官网(http://www.openhandsetalliance.com/oha_members.html)中查看到，但此网站在国内可能无法正常打开。像我国的电信、移动、联通这三大运营商以及华为、中兴等通信设备制造商都已经加入开放手机联盟。

开放手机联盟旨在开发多种技术，以及大幅削减移动设备及其服务的开发和推广成本。因为开放手机联盟中的厂商都将基于 Android 平台开发手机的新型业务，所以应用之间的通用性和互联性将在最大程度上得到保持。开放手机联盟表示，Android 平台可以促进移动设备的创新，让用户体验到最优越的移动服务，同时，开发商也将得到一个新的开放级别，更方便进行协同合作，从而保障新型移动设备的研发速度。随着越来越多的移动运营商和手机厂商推出 Android 手机，Android 平台的发展已进入一个全新的快速发展的阶段。

1.2　Android 简介

Android 一词本意是指“机器人”，是 Google 公司推出的一款基于 Linux 内核的开源的移动操作系统。Android 作为 Google 移动互联网战略的重要组成部分，其目的是为了推进“随时随地为每个人提供信息”这一企业目标的实现，并完善企业移动发展战略：通过与全球各地的手机制造商和移动运营商成为合作伙伴，开发既实用又有吸引力的移动服务，并推广这些产品。

1.2.1　Android 的历史

Android 目前已成为移动信息设备操作系统中的重量级成员，正吸引越来越多的追随者加入，包括开发者、设备生产商、软件开发商等。通过表 1-2 所示的 Android 发展历程大事记，可以看到 Android 市场占有率正飞速攀升，带来的周边效益也越来越被从事相关产品开发的业界人士所关注和重视。

表 1-2　Android 发展历程大事记

时　间	事　　件
2007 年 11 月 5 日	Google 公司宣布组建一个全球性的开放手机联盟，这一联盟将会支持 Google 发布的手机操作系统或者应用软件，共同开发名为 Android 的开放源代码的移动系统。开放手机联盟包括手机制造商、手机芯片厂商和移动运营商等。创建时，联盟成员数量已经达到了 34 家
2008 年 9 月 22 日	美国运营商 T-Mobile 在纽约正式发布第一款 Google 手机——T-Mobile G1，该款手机为 HTC 代工制造，是世界上第一部使用 Android 操作系统的手机，支持 WCDMA/HSPA 网络，理论下载速率 7.2 Mb/s，并支持 WiFi
2009 年 1 月 1 日	Google 的 Android 应用程序市场(App Market)在 2009 年初开始出售 Android 付费应用程序，这标志着 Android Market 营收的开始

续表一

时 间	事 件
2009 年 11 月 25 日	AdMob 的调查显示，在美国，10 月份使用苹果 iOS 操作系统所浏览的智能手机广告量占美国市场的 55%；第二位的是 Android 系统的 20%。至于全球市场，10 月份透过 iOS 系统浏览的广告量，以市场占有率 50%居冠；其次是 Symbian 操作系统的 25%；接着是 Android 系统的 11%，居于第三位。作为一个智能手机平台的新成员来说，Android 系统的受欢迎程度正在快步上升
2009 年 12 月 9 日	HTC 将逐渐放弃 Windows Mobile 系统，继而转向 Android 系统
2009 年 12 月 23 日	Google 在中国大陆推出中文版 Android Market，国内已经有开发者推出针对国内用户的 Android Market，易联致远公司已经推出名为 eoeMarket 的专门针对国内用户的第三方 Android Market
2010 年 1 月 6 日	Google 正式发布首款自有品牌手机 Nexus One，该机采用 Android 2.1 操作系统
2010 年 2 月 24 日	全球瞩目的世界移动大会(Mobile World Congress 2010)召开，华为公司在此次大会上展出了 5 款 Android 终端，并创造性地把 Android 平台运用到家庭互联网终端上，首次发布了 SmaKit S7 Tablet
2010 年 3 月 3 日	运营商 AT&T 宣布即将推出首款 Android 手机
2010 年 3 月 10 日	网络分析公司 Quantcast 最新报告显示，2 月份，Google 和 RIM 移动互联网流量份额增长，而苹果 iOS 份额则下滑。报告指出，Android 份额在过去一年中几乎翻番，RIM 份额增长 7.5%，iOS 份额同期下滑 10.2%。但苹果仍是移动互联网流量份额的遥遥领先者，2 月份份额近 64%；其次是 Android，份额约 15%；RIM 份额约 9%
2010 年 5 月 20 日	Android 2.2 Froyo 发布
2010 年 12 月 7 日	Android 2.3 Ginerbread 发布
2011 年 2 月 2 日	Android 3.0 Honeycomb 发布，优化针对平板；全新设计的 UI 增强了网页浏览功能
2011 年 5 月 7 日	Android 的市场份额已占有 43.7%，在智能手机中已位居第一
2011 年 5 月 11 日	Android 3.1 Honeycomb 发布，经过优化的 Gmail 电子邮箱；全面支持 Google Maps ；将 Android 手机系统跟平板系统再次合并从而方便开发者；任务管理器可滚动，支持 USB 输入设备(键盘、鼠标等)；支持 Google TV；支持 XBOX 360 无线手柄；Widget 支持的变化，能更加容易地定制屏幕 widget 插件
2011 年 7 月 13 日	Android 3.2 Honeycomb 发布，支持 7 英寸设备；引入了应用显示缩放功能
2011 年 10 月 19 日	Android 4.0 Ice Cream Sandwich 发布，全新的 UI；增强了截图功能；更强大的照片和编辑功能等
2012 年 10 月 9 日	Android 4.1.2 JellyBean(果冻豆)发布，加入主画面的旋转功能，并对一些错误进行修正，提升了性能及稳定性
2013 年 7 月 25 日	Android 4.3 发布，最重要的更新是开始支持 OpenGL ES3.0
2013 年 9 月 4 日	Android 4.4 KitKat(奇巧巧克力)发布，RAM、新图标、锁屏、启动动画和配色方案等得到了进一步优化
2014 年 10 月 15 日	Android 5.0L(棒棒糖)发布，其界面采用新的 Material Design 设计风格，加入了整合碎片化，支持 64 位处理器以及使用 ART 虚拟机等新功能

续表二

时　间	事　　件
2015 年 9 月 30 日	Android 6.0 Marshmallow(棉花糖)发布，支持指纹识别、更完整的应用权限管理等功能
2016 年 5 月 18 日	Android 7.0 N(牛轧糖)发布，支持分屏任务、号码拦截、更便捷的通知中心等功能
2017 年 8 月 22 日	谷歌正式发布了 Android 8.0 的正式版，其正式名称为 Android Oreo(奥利奥)
2018 年 8 月 7 日	谷歌正式发布 Android 9.0 系统，其正式名称为 Pie(开心果冰淇淋)

1.2.2 Android 的优缺点

Android 作为一个出现不久的移动信息设备开发平台，因为具有一些巨大的先天优势，使其具有良好的发展前景。

Android 的优势主要体现在以下几个方面：

(1) 系统的开放性和免费性。

Android 最震撼人心之处在于 Android 手机系统的开放性和服务免费。Android 是一个对第三方软件完全开放的平台，开发者在为其开发程序时拥有更大的自由度，而且 Android 操作系统免费向开发人员提供。这一点对开发者、厂商来说是最大的诱惑。

(2) 移动互联网的发展。

Android 采用 WebKit 浏览器引擎，具备触摸屏、高级图形显示和上网功能，用户能够在手机上查看电子邮件、搜索网址和观看视频节目等，比 iPhone 等其他手机更强调搜索功能，界面更强大，是一种融入全部 Web 应用的互联网络平台，顺应了移动互联网大潮流，有助于 Android 的推广及应用。

(3) 相关厂商的大力支持。

Android 目前正在从移动运营商、手机制造厂商、开发者和消费者那里获得大力支持。从组建开放手机联盟开始，Google 一直向服务提供商、芯片厂商和手机销售商提供 Android 平台的技术支持。

但是 Android 也并不是一个完美的系统，同样面临着以下挑战：

(1) 用户体验不一致，安全性有待提高。

由于 Android 的开放性，导致许多生产厂商开发不同的 Android 系统和不同规格的硬件产品，因此会出现体验不一致的问题，与此同时，因发展非常迅速的第三方 Android 应用市场鱼龙混杂，缺乏统一的管理，使之安全性遭到比较大的威胁。

(2) 技术的进一步完善。

目前，Android 系统在技术上仍有许多不足，例如：不支持桌面同步功能，还有自身系统的一些 Bug。这些都是 Android 需要继续完善的地方。

(3) 开放手机联盟模式的挑战。

Android 由开放手机联盟去开发、维护、完善。很多人会担心，最终的结局是否会像当年的 Linux 和 Windows 操作系统之争那样？这种开放式联盟的模式对 Android 未来的发展、定位是否存在阻碍作用？这些未知的隐忧也会影响到一些开发者的信心。

1.2.3 Android 平台的技术架构

Android 平台采用了软件栈(Software Stack)，又名软件叠层的架构，如图 1-4 所示，由低到高分为四部分：

(1) Linux 内核层：该层是基础，包含各种驱动，并提供操作系统的基本功能。

(2) 中间层：该层包括程序库(Libraries)和 Android 运行时环境。

(3) 应用程序框架：该层是编写核心应用所使用的 API 框架，开发者可以使用这些框架来开发自己的应用，但必须遵守该框架的开发原则。

(4) 应用层：该层是各种应用软件，包括通话、短信、日历、地图、浏览器等核心应用程序，这些应用程序都是使用 Java 编写的。

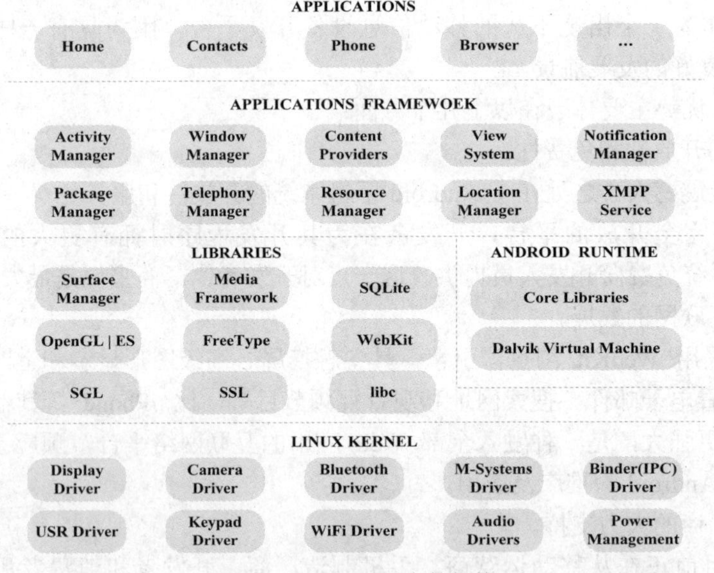

图 1-4 Android 平台的架构

1. Linux 内核

Android 核心系统服务依赖于 Linux 内核，包括安全性、内存管理、进程管理、网络协议栈和驱动模型等。Linux 内核也同时作为硬件和软件栈之间的抽象层。

2. 程序库

Android 包含一个能被 Android 系统中各种不同组件所使用的 C/C++ 库，该库通过 Android 应用程序框架为开发者提供服务，主要包括：

(1) 系统 C 库：一个从 BSD 继承的标准 C 系统函数库，是专门为基于嵌入式 Linux 设备定制的。

(2) 媒体库：基于 PacketVideo 的 OpenCORE，该库支持多种常用的音频、视频格式文件的回放和录制，同时支持静态图像文件，编码格式包括 MPEG4、H.264、MP3、AAC、AMR、JPG 和 PNG 等。

(3) Surface Manager：管理显示子系统，并且为多个应用程序提供 2D 和 3D 图层的无缝融合。

(4) SGL：底层的 2D 图形引擎。

(5) 3D 库：基于 OpenGL ES 1.0 API 实现，该库可以使用 3D 硬件加速或者使用高度优化的 3D 软件加速。

(6) FreeType：用于位图和矢量字体显示。

(7) WebKit：Web 浏览器引擎。

(8) SQLite 库：一个用于本地存储的、轻型关系型数据库引擎。

3. Android 运行时环境

Android 运行时环境由一个核心库和 Dalvik 虚拟机组成。核心库提供 Java 编程语言核心库的大多数功能。每一个 Android 应用程序都在自己的进程中运行，都拥有一个独立的 Dalvik 虚拟机实例。Dalvik 在一个设备中可以同时高效运行多个虚拟系统，它依赖于 Linux 内核的一些功能，例如，线程机制和底层内存管理机制等。Dalvik 虚拟机执行 .dex 类型的 Dalvik 可执行文件，该格式文件针对小内存的使用进行了优化，同时虚拟机是基于寄存器的。所有的类由 Java 编译器编译，然后通过 SDK 中的“dx”工具转化成.dex 格式，最后由 Dalvik 虚拟机执行。

4. 应用程序框架

开发者可以访问 Android 应用程序框架中的 API，该应用程序架构简化了组件的重用，任何一个应用程序都可以发布它的功能块，并且任何其他的应用程序都可以使用这些发布的功能块。同样，该应用程序的重用机制也使用户可以方便地替换程序组件。

Android 提供一系列的服务和管理器，其中包括：

(1) 丰富而又可扩展的视图(Views)：包括列表(Lists)、网格(Grids)、文本框(Text Boxes)、按钮(Buttons)，甚至包括可嵌入的 Web 浏览器，这些视图可以用来构建应用程序。

(2) 内容提供器(Content Providers)：使得应用程序可以访问另一个应用程序的数据，例如，联系人数据库，或者可以共享它们自己的数据。

(3) 资源管理器(Resource Manager)：提供非代码资源的访问，例如，本地字符串、图形和布局文件(Layout Files)等。

(4) 通知管理器(Notification Manager)：使得应用程序可以在状态栏中显示自定义的提示信息。

(5) 活动管理器(Activity Manager)：用来管理应用程序生命周期，并且提供常用的导航回退功能。

5. 应用程序

Android 会附带一系列核心应用程序包，这些应用程序包包括 E-mail 客户端、SMS 短信程序、日历、地图、浏览器、联系人管理程序等。Android 中所有的应用程序都是使用 Java 语言编写的。

1.3 Android 应用程序构成

一个 Android 应用程序通常由以下 4 个组件构成：活动(Activity)、广播接收者(BroadcastReceiver)、服务(Service)和内容提供者(ContenProvider)。这 4 个组件是构成

Android 应用程序的基础，但并不是每个 Android 应用程序都必须包含这 4 个组件，除了 Activity 是必要部分之外，其余组件都是可选的，某些应用程序可能只需要其中部分组件即可。

1.3.1 活动(Activity)

活动(Activity)是最基本的 Android 应用程序组件。在应用程序中，一个活动通常就是一个单独的屏幕。每个活动都通过继承活动基类而被实现为一个独立的活动类。活动类将会显示由视图控件组成的用户接口，并对事件做出响应。

大多数的应用程序都是由多个屏幕显示组成的。例如，一个发送邮件的应用，第一个屏幕用来显示发送邮件的联系人列表，第二个屏幕用来写邮件内容和选择收件人，第三个屏幕可以查看历史邮件或者邮件设置操作等。这里每个屏幕都是一个活动，当打开一个新屏幕时，之前的屏幕会被置为停止状态并且压入历史堆栈中。用户可以通过回退操作退回到之前打开过的屏幕，也可以选择性地移除一些没有必要保留的屏幕。

1.3.2 广播接收者(BroadcastReceiver)

广播接受者(BroadcastReceiver)是四大 Android 应用程序组件之一，是 Android 广播机制的基本要素，在 Android 中有着广泛的应用。

在 Android 中，有一些操作完成以后会发送广播，比如发送一条短信，或者打出一个电话，如果某个程序接收了这个广播，就会进行相应的处理。

Android 中的广播和传统意义上的电台广播有相似之处，因为发送方并不关心接收方是否接受数据，也不关心接收方如何处理数据。广播可以被一个或者多个应用程序所接收，也可能不被任何应用程序所接收。

Android 中的广播是非常灵活的，因为 Android 中的每个应用程序都可以对自己感兴趣的广播进行注册，这样该程序就只会接收到自己关心的广播内容，这些广播可能是来自于系统的，也可能是来自于其他应用程序的。Android 提供了一套完整的 API，允许应用程序自由地发送和接收广播。

Android 广播机制包含三个基本要素：广播(Broadcast)——用于发送广播；广播接收者(BroadcastReceiver)——用于接收广播；意图内容(Intent)——用于保存广播相关信息的媒介。

1.3.3 服务(Service)

服务(Service)是 Android 应用程序中具有较长的生命周期但是没有用户界面的程序。它在后台运行，并且可以与其他程序进行交互。Service 跟 Activity 的级别差不多，但是不能独立运行，需要通过某一个 Activity 来调用。

Android 应用程序的生命周期是由 Android 系统来决定的，不是由具体的应用程序线程来控制的。如果应用程序要求在没有界面显示的情况还能正常运行(要求有后台线程，而且直到线程结束，后台线程才会被系统回收)，此时就需要用到 Service。

Service 的典型例子是一个具有播放列表功能的媒体播放器。在媒体播放器应用中，

可能会有一个或多个活动，让使用者可以选择并播放音乐。然而活动本身并不处理音乐播放功能，因为用户期望在切换到其他屏幕后，音乐应该还在后台继续播放，因此使用 Service 是合适的方案。

1.3.4　内容提供者(ContentProvider)

Android 应用程序可以使用文件或 SQLite 数据库来存储数据。内容提供者(ContentProvider)提供了一种多应用间数据共享的方式。当某个应用程序的数据需要与其他应用程序共享时，内容提供者就会发挥作用。一个 ContentProvider 类实现一组标准的方法，能够让其他的应用保存或读取此内容提供者处理的各种数据类型，即一个应用程序可以通过实现一个 ContentProvider 的抽象接口将自己的数据暴露出去。外界根本看不到，也不用看到该应用程序暴露的数据是如何存储的，但是外界可以通过这一套标准及统一的接口和应用程序里的数据打交道，可以读取应用程序的数据，也可以删除应用程序的数据。

1.4　第一个 Android 应用

Android 应用程序的编程语言是 Java/Kotlin 语言(Kotlin 语言暂未大规模使用，本书以 Java 语言讲解)，在正式编写 Android 应用时，需要搭建开发环境(详见实践 1)。在 2013 年的 I/O 大会上，Google 谷歌针对 Android 开发推出了一款新的官方 IDE 工具 Android Studio，相比之前的 Eclipse，Android Studio 具有编译速度更快、UI 更漂亮、整合 gradle 构建工具等优点，可以更便捷地创建一个 Android 项目。本书中所有的代码都将在 Android Studio 上进行编写。

1.4.1　创建一个新的 Android 项目

启动 Android Studio，弹出如图 1-5 所示的欢迎窗口，单击其中的【Start a new Android Studio project】，可以新建一个 Android 项目。

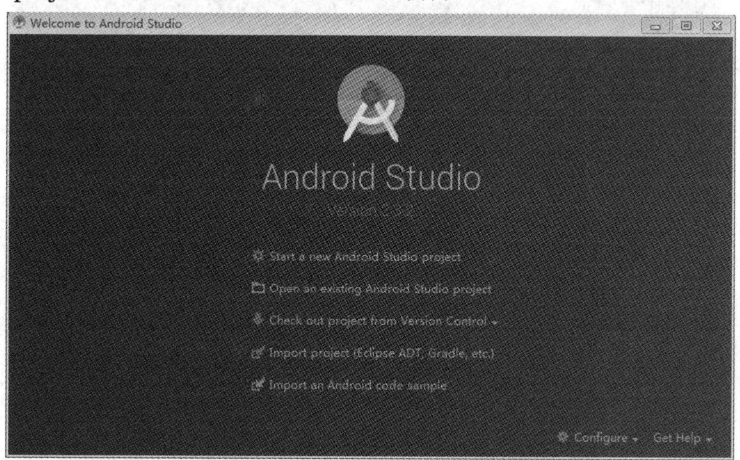

图 1-5　新建 Android 项目

在弹出的设置窗口【Create New Project】中，可以对新建项目的名称、公司域名、包名、工作空间进行设置，如图 1-6 所示。

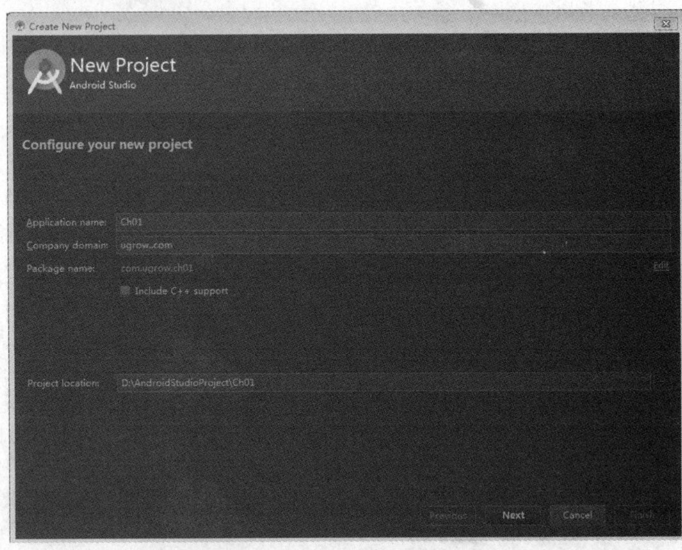

图 1-6　设置新建 Android 项目的基本信息

具体设置项目如下：

(1) Application Name：应用程序在手机中显示的名称(项目名称)。

(2) Company domain：公司域名。

(3) Package Name：项目包名。

(4) Project location：项目存储位置。

设置完毕，单击界面中的【Next】按钮，将出现【Target Android Devices】界面，在其中可以设置项目程序兼容的最低 Android 系统版本，设置完毕，单击【Next】按钮，如图 1-7 所示。

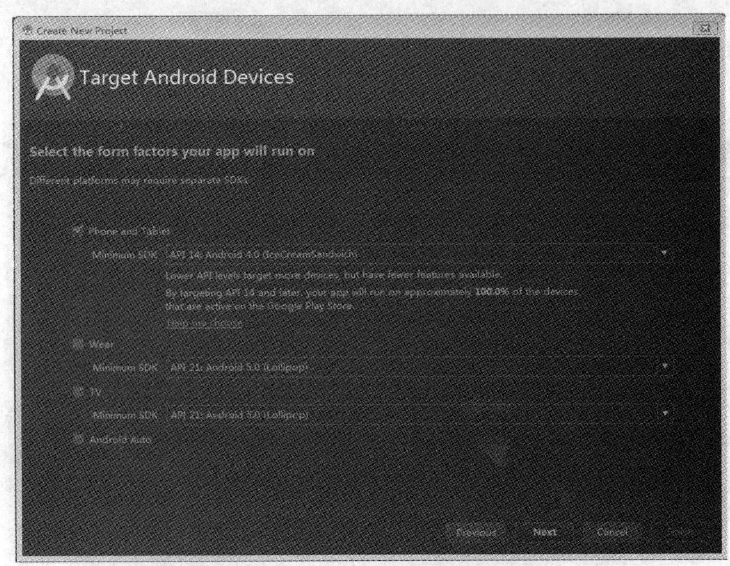

图 1-7　设置新建 Android 项目兼容的最低 Android 系统版本

本书均采用 Android 4.0 版本，对应最低 SDK 版本 14。

在出现的【Customize the Activity】界面中的【Activity Name】后面设置第一个 Activity 的名称，在【Layout Name】后面设置对应的布局文件的名称。注意：修改前者时，后者也会自动进行相应修改，如图 1-8 所示。

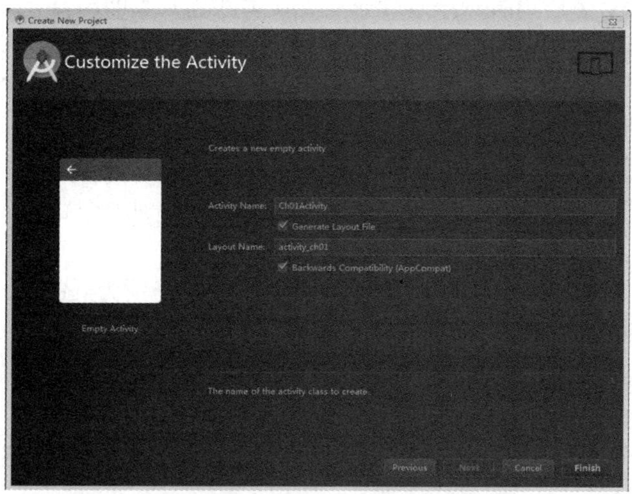

图 1-8　设置新建 Android 项目的第一个 Activity

设置完毕，单击【Finish】按钮，即可完成 Android 项目的创建。

在 Android Studio 的 Project 模式下，可以查看新创建的 Android 项目的架构，如图 1-9 所示。

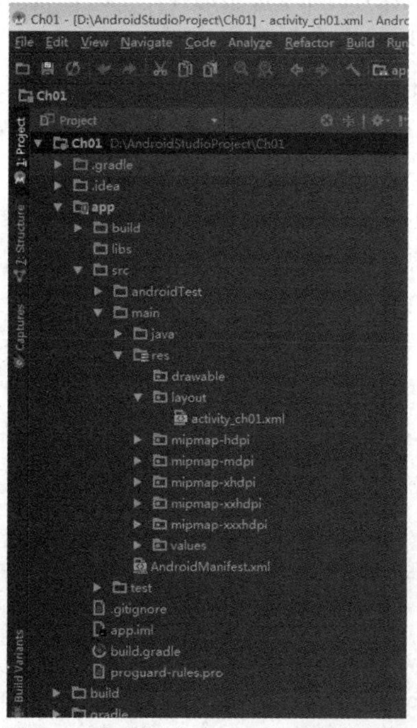

图 1-9　查看新建 Android 项目的架构

可以看到，项目的根目录 Ch01 中包含了一些自动生成的文件夹和文件，其主要功能及作用如下：

(1) .gradle 文件夹：是 gradle 运行以后生成的缓存文件夹，可以查看当前项目使用的 gradle 版本。

(2) .idea 文件夹：Android Studio 工程启动后生成的工作环境配置文件夹，包括一些与复制版权、编译、编码语言、运行环境、工作空间等有关的配置。

(3) app 文件夹：该文件夹中包含源代码文件夹 src、build 文件夹以及 libs 文件夹。

- ✧ src 文件夹：该文件夹下的 main 文件夹包含三个部分。第一个是 java 文件夹，用来存放 Java 代码，包括一些 Activity、Fragment 的代码；第二个是 res 文件夹，用来存放资源，包括布局文件(Layout)、图片资源(mipmap)、其他资源(Values)等；第三个是 AndroidMainfest.xml 清单文件，用来配置项目的权限及注册信息等。
- ✧ build 文件夹：该文件夹包含四个子文件夹，分别为 generated、intermediates、outputs 和 tmp，其中 outputs 文件夹用于存放生成的 APK 包，generated 文件夹用于存放某些 AIDL(Android 接口定义语言)生成的 JAVA 文件。
- ✧ libs 文件夹：该文件夹用于存放项目所需的 jar 包。

(4) build 文件夹：编译时的缓存文件夹，会在每次运行时生成，并在执行 gradle clean 任务后被删除掉。

(5) gradle 文件夹：该文件夹中包含一个 wrapper 文件夹，其中有一个 gradle-wrapper.jar 与一个 gradle-wrapper.properties 文件，配置文件 gradle-wrapper.properties 中的属性 distributionUrl 指定了项目所用的 gradle 版本。

(6) .gitignore 文件：git 版本控制的忽略清单。

(7) buidle.gradle 文件：项目的全局配置文件。

(8) gradle.properties 文件：Java 虚拟机内存大小的配置文件。

(9) local.properties 文件：个人电脑环境的配置文件，通常配置的是 SDK 在电脑中的存放位置。

(10) settings.gradle：子项目的配置文件，为 Project 配置子项目。默认内容为 include ':app'，如果指定相应的子项目(module)能在主项目中使用，则需要将子项目添加进去，如：include ':app', ':example'(其中 example 为子项目的名称)。

(11) External Libraries：此文件夹下存放了项目加载的一些第三方库。

下面对项目部分主要文件的内容进行介绍：

(1) 字符串引用文件 res/values/strings.xml，内容如下：

```
<?xml version="1.0" encoding="utf-8"?>
<resources>
<string name="app_name">Ch01</string>
</resources>
```

(2) 界面布局文件 res/layout/activity_ch01.xml，内容如下：

```
<RelativeLayoutxmlns:android="http://schemas.android.com/apk/res/android"
xmlns:tools="http://schemas.android.com/tools"
```

```
android:layout_width="match_parent"
android:layout_height="match_parent"
tools:context="com.ugrow.ch01.Ch01Activity" >
        <TextView
                android:layout_width="wrap_content"
                android:layout_height="wrap_content"
                android:text="Hello World" />
</RelativeLayout>
```

上述代码设置了一个相对布局(RelativeLayout)，在布局中还添加了一个文本标签控件(TextView)，可使用该控件的 android:text 属性设置控件显示的文本内容。

关于布局及控件的详细设置参见第 3 章用户界面相关内容。

(3) 资源引用文件 app/build/generated/source/r/debug/com.dh.ch01/R.java，内容如下：

```
public final class R {
    public static final class anim {
        public static final int abc_fade_in=0x7f050000;
        public static final int abc_fade_out=0x7f050001;
        public static final int abc_grow_fade_in_from_bottom=0x7f050002;
        public static final int abc_popup_enter=0x7f050003;
        public static final int abc_popup_exit=0x7f050004;
        public static final int abc_shrink_fade_out_from_bottom=0x7f050005;
        public static final int abc_slide_in_bottom=0x7f050006;
        public static final int abc_slide_in_top=0x7f050007;
        public static final int abc_slide_out_bottom=0x7f050008;
        public static final int abc_slide_out_top=0x7f050009;
    }
        public static final class dimen {
        public static final int abc_action_bar_content_inset_material=0x7f07000c;
        public static final int abc_action_bar_elevation_material=0x7f070015;
        }
       public static final class layout {
        public static final int activity_ch01=0x7f04001b;
        }

        public static final class drawable {
        public static final int abc_ab_share_pack_mtrl_alpha=0x7f020000;
                public static final int abc_btn_borderless_material=0x7f020002;
                public static final int abc_btn_check_material=0x7f020003;
                public static final int abc_btn_check_to_on_mtrl_000=0x7f020004;
```

```
            public static final int abc_btn_check_to_on_mtrl_015=0x7f020005;
            public static final int abc_btn_colored_material=0x7f020006;
            public static final int abc_btn_default_mtrl_shape=0x7f020007;
            public static final int abc_btn_radio_material=0x7f020008;
        }
}
```

上述代码中，针对不同资源定义了多个静态的内部类，在这些静态类中又定义了不同资源的指针，例如，属性值 activity_ch01 指向的是对 activity_ch01.xml 布局文件的引用，这些引用可以在其他文件中调用，也可以在 Java 代码中调用。

(4) JAVA 源文件 src/ main/java/com.ugrow.ch01/Ch01Activity.java，内容如下：

```
public class Ch01Activity extends AppCompatActivity {
    @Override
    protected void onCreate(Bundle savedInstanceState) {
        super.onCreate(savedInstanceState);
        setContentView(R.layout.activity_ch01);
    }
}
```

上述代码中的 Ch01Activity 类继承了 Android 提供的 v7 包中的 AppCompatActivity 类，并重写了 onCreate()方法，在该方法中先调用父类中的 onCreate()方法，然后调用 setContentView()方法设置要使用的布局文件，其中的“R.layout.activity_ch01”即表示使用布局文件 activiity_ch01.xml。

注 意

Activity 中的 onCreate()方法包含了一个 Bundle 类型的参数，可用于不同 Activity 之间的消息传递。本章只定义了一个 Activity，在第 2 章中会详细介绍 Activity 之间的消息传递。

(5) 项目配置清单文件 AndroidManifest.xml，内容如下：

```
<?xml version="1.0" encoding="utf-8"?>
<manifest xmlns:android="http://schemas.android.com/apk/res/android"
    package="com.ugrow.ch01"
        <application
            android:allowBackup="true"
            android:icon="@mipmap/ic_launcher"
            android:label="@string/app_name"
            android:theme="@style/AppTheme">
        <activity
            android:name=".Ch01Activity"
            >
            <intent-filter>
                <action android:name="android.intent.action.MAIN" />
                <category android:name="android.intent.category.LAUNCHER" />
```

```
            </intent-filter>
        </activity>
    </application>
</manifest>
```

对上述代码中的配置元素简介如下：

✧ <manifest>是根元素，在其中指定项目的命名空间、包等信息。

✧ <application>元素用于设置项目相关信息，其中的两个属性分别对应项目的图标和标题。

✧ <activity>元素用于设置 Activity 信息，其中的属性 android:name 指定的是 Activity 的类名；并通过<intent-filter>(意图过滤器)元素的子元素<action>和<category>将该 Activity 指定为程序的入口。

Android 应用程序没有类似于 C 语言的 main()函数入口，而是会以指定的 Activity 作为程序执行的入口点。

1.4.2 运行 Android 应用程序

至此，一个 Android 应用已经创建完成，接下来需要运行这个应用：单击【Android Studio】菜单栏上的运行按钮，在弹出的菜单中选择要使用的模拟器，然后单击【OK】按钮运行，如图 1-10 所示。如果没有启动 AVD(Android 模拟器)，则需要创建模拟器。

Android 应用程序的运行结果如图 1-11 所示。

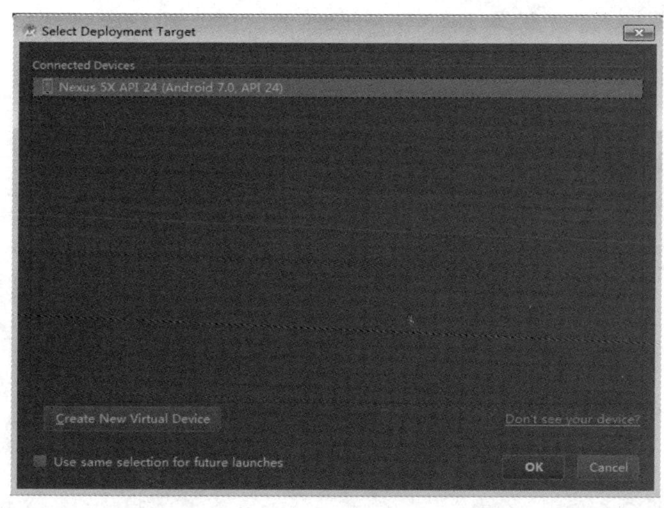

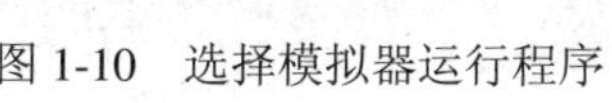

图 1-10　选择模拟器运行程序

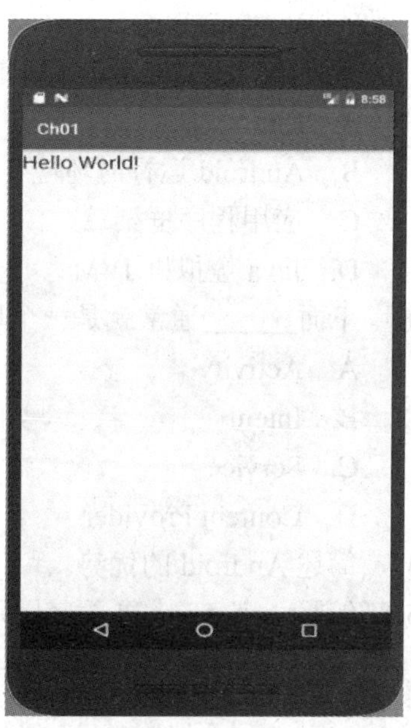

图 1-11　运行新建的 Android 项目

创建 AVD(Android 模拟器)的具体操作会在本章实践部分介绍。

本 章 小 结

通过本章的学习，读者应当了解：

✧ Android 是 Google 公司推出的开源手机操作系统。

✧ Android 上的应用程序开发使用 Java 语言。

✧ Android 平台采用软件栈的架构，主要分为活动、广播接收者、服务、内容提供者四个部分。

✧ 活动(Activity)是最基本的 Android 应用程序组件。

✧ 广播接收者(BroadcastReceiver)是用来接收广播实现特定功能的 Android 应用程序组件。

✧ 服务(Service)是具有较长的生命周期但没有用户界面的 Android 应用程序组件。

✧ 内容提供者(ContentProvider)提供了一种在多个 Android 应用间共享数据的方式。

本 章 练 习

1. 下列不属于移动设备操作系统的是______。

   A. Windows Phone

   B. iOS

   C. Android

   D. Windows 7

2. 下列不属于 Android 平台的技术架构的是______。

   A. Linux 内核

   B. Android 运行时环境

   C. 应用程序框架

   D. Java 虚拟机 JVM

3. 下面______通常就是一个单独的屏幕。

   A. Activity

   B. Intent

   C. Service

   D. Content Provider

4. 简述 Android 的优势。

5. 编写一个 Android 的应用，并在屏幕中显示字符串“欢迎来到 Android 世界！”

第2章　活动(Activity)

本章目标

- 熟悉 Android 程序框架
- 理解 Activity 的生命周期及方法
- 掌握 Activity 的启动模式及跳转方法
- 熟悉 Android 中各种资源的使用

2.1 Activity 简介

Activity 是 Android 应用程序中最基本的组成单位。Activity 主要负责创建显示窗口，一个 Activity 对象通常就代表了一个单独的屏幕。Activity 是用户唯一可以看得到的组件，所以几乎所有的 Activity 都是用来与用户进行交互的。对于熟悉 Windows 或者 Java ME 编程的读者来说，可以将 Activity 理解为 Windows 编程中的 WinForm 窗口类或者 Java ME 编程中的 Display 类。

在 Android 应用中，如果需要有显示的界面，则在应用中至少要包含一个 Activity 类。Activity 用于提供可视化的用户界面，是 Android 应用中使用频率最高的组件。在具体实现时，每个 Activity 都被定义为一个独立的类，并且继承 Android 提供的 android.app.Activity，例如：

```
import android.app.Activity;
import android.os.Bundle;
public class MyActivity extends Activity {
......
}
```

在这些 Activity 类中将使用 setContentView(View)方法来显示由视图控件组成的用户界面，并对用户通过这些视图控件所触发的事件做出响应。

大多数的应用程序根据功能的需要都是由多个屏幕显示组成的，因此必须包含多个 Activity 类。这些 Activity 可以通过一个 Activity 栈进行管理，当一个新的 Activity 启动的时候，它首先会被放置在 Activity 栈顶部并标记为运行状态的 Activity，而之前正在运行的 Activity 也在 Activity 栈中，但是它将被保存在这个新的 Activity 下边，只有当这个新的 Activity 退出以后，之前的 Activity 才能重新回到前台界面。

2.1.1 Activity 生命周期

Activity 具有生命周期，在生命周期的过程中共有四种状态：

(1) 激活或者运行状态：此时 Activity 运行在屏幕的前台。

(2) 暂停状态：此时 Activity 失去了焦点但是仍然对用户可见，例如在该 Activity 上遮挡了一个透明的或者非全屏的 Activity。

(3) 停止状态：此时 Activity 被其他 Activity 完全覆盖。

(4) 终止状态：此时 Activity 将会被系统清理出内存。

注意 处于暂停状态和停止状态的 Activity 仍然保存了其所有的状态和成员信息，直到被系统终止。当被系统终止的 Activity 需要重新显示的时候，它必须重新启动并且将关闭之前的状态全部恢复回来。

Activity 从一个状态运行到另一个状态，状态改变时会执行相应的生命周期方法。Android 中的 android.app.Activity 类定义了 Activity 生命周期中所包含的全部方法，其具体

定义代码如下：

```
public class Activity extends ApplicationContext{
    protected void onCreate(Bundle SavedInstanceState);
    protected void onStart();
    protected void onRestart();
    protected void onResume();
    protected void onPause();
    protected void onStop();
    protected void onDestroy();
}
```

Activity 类中的这 7 个方法定义了 Activity 完整的生命周期，这些方法在生命周期中的作用以及相互之间的转换关系如表 2-1 所示。

表 2-1　Activity 类中的方法

方法	功 能 描 述	下一个方法
onCreate()	Activity 初次创建时被调用，在该方法中一般进行一些静态设置，如创建 View 视图、进行数据绑定等。如果 Activity 是首次创建，本方法执行完后将会调用 onStart()，如果 Activity 是停止后重新显示则调用 onRestart()	onStart()
onStart()	当 Activity 对用户即将可见的时候调用	onRestart()或 onResume()
onRestart()	当 Activity 从停止状态重新启动时调用	onResume()
onResume()	当 Activity 将要与用户交互时调用此方法，此时 Activity 在 Activity 栈的栈顶，用户输入的信息可以传递给它。如果其他的 Activity 在它的上方恢复显示，则调用 onPause()	onPause()
onPause()	当系统要启动一个其他的 Activity 时(其他的 Activity 显示之前)，这个方法将被调用，用于提交持久数据的改变、停止动画等	onResume()或 onStop()
onStop()	当另外一个 Activity 恢复并遮盖住当前的 Activity，导致其对用户不再可见时，这个方法将被调用	onStart()或 onDestroy()
onDestroy()	在 Activity 被销毁前所调用的最后一个方法	无

在如图 2-1 所示的 Activity 生命周期状态转换图中，椭圆形表示的是 Activity 所处的状态，矩形框代表了生命周期中的回调方法，开发者可以重载这些方法从而使自定义的 Activity 在状态改变时执行所期望的操作。当然这些方法不是要求必须都被实现的，一般情况下，所有 Activity 都应该实现自己的 onCreate()方法来进行初始化设置，大部分还应该实现 onPause()方法来准备终止与用户的交互，至于其他的方法则可以在需要时实现。

Activity 的生命周期还可以根据不同的标准分为完整生命周期、可见生命周期和前台生命周期。

(1) 完整生命期：从 Activity 最初调用 onCreate()方法到最终调用 onDestroy()方法的这个过程称为完整生命周期。Activity 会在 onCreate()方法中进行所有全局状态的设置，在 onDestroy()方法中释放其占据的所有资源。

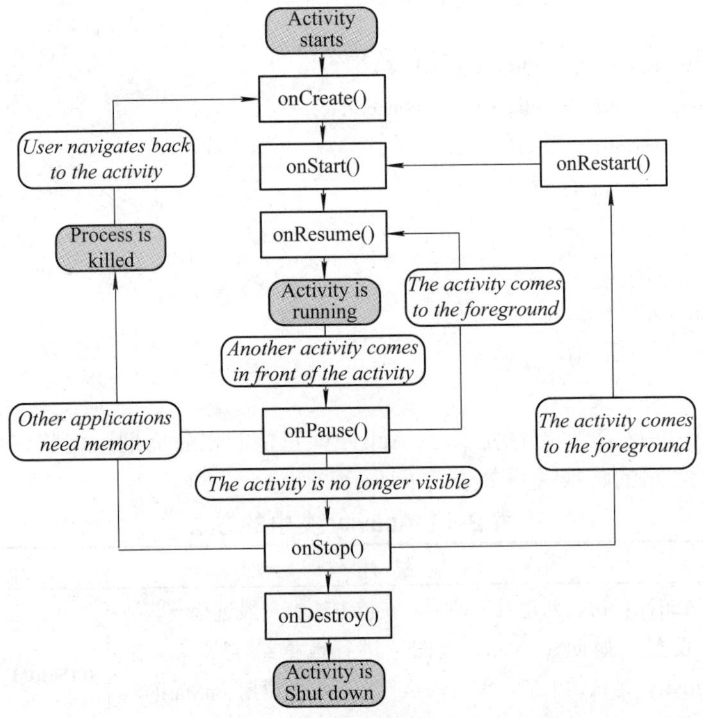

图 2-1　Activity 生命周期状态转换图

(2) 可见生命周期：从 Activity 调用 onStart()方法开始，到调用对应的 onStop()方法为止的这个过程称为可见生命周期。在这段时间内，用户可以在屏幕上看到这个 Activity，尽管并不一定是在前台显示，也不一定可以与其交互。在这两个方法之间，开发者可以维护 Activity 在显示时所需的资源。因为每当 Activity 显示或者隐藏时都会调用相应的方法，所以 onStart()方法和 onStop()方法在整个生命周期中可以多次被调用。

(3) 前台生命周期：从 Activity 调用 onResume()方法开始，到调用对应的 onPause()方法为止的这个过程称为前台生命周期，这段时间当前的 Activity 处于其他所有 Activity 的前面，且可以与用户交互。

【示例 2.1】 重写 Activity 的各个生命周期的回调方法，测试生命周期事件。

在 Android Studio 中新建 Android 项目 ch02_2D1，在 MainActivity 中编写以下代码：

```
public class MainActivity extends Activity {
    @Override
    public void onCreate(Bundle savedInstanceState) {
        super.onCreate(savedInstanceState);
        setContentView(R.layout. activity_main);
        Log.d("TAG", "onCreate");
    }
    @Override
    protected void onStart() {
        super.onStart();
        Log.d("TAG", "onStart");
```

```
    }
    @Override
    protected void onRestart() {
            super.onRestart();
            Log.d("TAG", "onRestart");
    }
    @Override
    protected void onResume() {
            super.onResume();
            Log.d("TAG", "onResume");
    }
    @Override
    protected void onPause() {
            super.onPause();
            Log.d("TAG", "onPause");
    }
    @Override
    protected void onStop() {
            super.onStop();
            Log.d("TAG", "onStop");
    }
    @Override
    protected void onDestroy() {
            super.onDestroy();
            Log.d("TAG", "onDestroy");
    }
}
```

上述代码中，MainActivity 继承了 Activity，并重写了 onCreate()、onStart()、onRestart()、onResume()、onPause()、onStop()、onDestroy()方法，其中调用 Log 类的静态方法 d()输出调试日志，日志信息为当前方法名。

android.util.Log 类提供了日志功能，使用 Log 类的下列静态方法可以输出各种级别的日志信息，且对应的级别依次升高，如表 2-2 所示。

表 2-2　Log 类常用的静态方法

静态方法	级别分类	功能说明
v()	verbose，对应 LogCat 视图中的 Verbose	最低级别，所有信息
d()	debug，对应 LogCat 视图中的 Debug	调试信息
i()	info，对应 LogCat 视图中的 Info	一般信息
w()	warn，对应 LogCat 视图中的 Warn	警告信息
e()	error，对应 LogCat 视图中的 Error	错误信息
Wtf()	assert，对应 LogCat 视图中的 Assert	错误信息

上述方法都至少有以下两种重载形式(以 d()为例)：

(1) public static int d (String tag, String msg)。

(2) public static int d (String tag, String msg, Throwable tr)。

其中，tag 为日志标记；msg 为日志信息；tr 为异常信息。

启动 Android 模拟器，在 Android Studio 窗口底部单击【Android Monitor】选项卡，打开 Android LogCat 视图，会看到其中显示了大量日志信息，如图 2-2 所示。

图 2-2 LogCat 视图显示的日志信息

在窗口右上方的日志类型选择下拉菜单中，可以选择要显示的日志信息的级别，默认为【Verbose】。

单击日志类型选择菜单右侧的下拉菜单，选择【Edit Filter Configuration】，创建过滤器，在弹出的界面【Create New Logcat Filter】左侧单击【+】号，在右侧的【Filter Name】和【Log Tag】后面输入“TAG”，然后单击【OK】按钮，即可在 LogCat 视图中只显示标记为“TAG”的日志信息，如图 2-3 所示。重复同样方法，可创建多个过滤器。

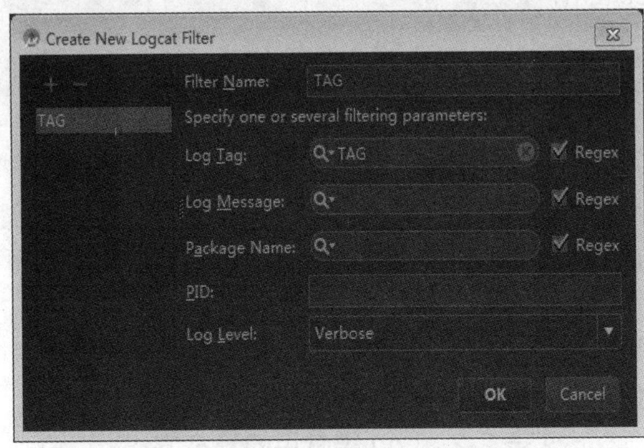

图 2-3 创建过滤器

设置完毕，运行项目 ch02_2D1，在 LogCat 视图中将显示以下日志信息：

```
06-28 02:39:13.946 19228-19228/com.ugrow.ch02_2d1 D/TAG: onCreate
06-28 02:39:13.947 19228-19228/com.ugrow.ch02_2d1 D/TAG: onStart
06-28 02:39:13.948 19228-19228/com.ugrow.ch02_2d1 D/TAG: onResume
```

由上述信息可知，当 Activity 启动时，调用了 onCreate()、onStart()、onResume()方法。

单击 Android 模拟器上的 Home 键，返回 Android 桌面，如图 2-4 所示。

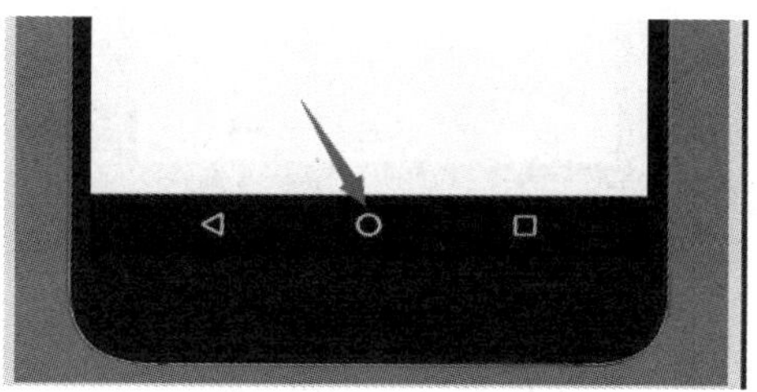

图 2-4　单击 Home 键

此时在 LogCat 视图中新增的日志信息如下所示：

```
06-28 02:42:57.859 19228-19228/com.ugrow.ch02_2d1 D/TAG: onPause
06-28 02:42:58.151 19228-19228/com.ugrow.ch02_2d1 D/TAG: onStop
```

上述信息表明，此时调用了 onPause()、onStop()方法，应用程序已停止。

在应用程序列表中找到 ch02_2D1，单击图标运行应用程序，如图 2-5 所示。

图 2-5　单击图标运行 Android 应用程序

此时在 LogCat 视图中新增的日志信息如下所示：

```
06-28 02:46:49.295 19228-19228/com.ugrow.ch02_2d1 D/TAG: onRestart
06-28 02:46:49.296 19228-19228/com.ugrow.ch02_2d1 D/TAG: onStart
06-28 02:46:49.299 19228-19228/com.ugrow.ch02_2d1 D/TAG: onResume
```

上述信息表明，此时程序执行了 onRestart()、onStart()、onResume()方法。

单击模拟器上的返回键，如图 2-6 所示。

图 2-6　单击返回键

此时在 LogCat 视图中新增的日志信息如下所示：

```
06-28 02:50:01.677 19228-19228/com.ugrow.ch02_2d1 D/TAG: onPause
06-28 02:50:02.628 19228-19228/com.ugrow.ch02_2d1 D/TAG: onStop
06-28 02:50:02.628 19228-19228/com.ugrow.ch02_2d1 D/TAG: onDestroy
```

上述信息表明，此时程序执行了 onPause()、onStop()、onDestroy()方法，Activity 已被系统销毁。

2.1.2　Activity 创建和注册

前面对 Activity 的生命周期进行了一个基本的介绍，但在实际开发过程中，Activity 的生命周期方法并不会全部被调用，而是根据项目操作流程的需要确定。

下面在 Android Studio 中新建一个 Android 项目 Ch02_2D2，演示创建并配置 Activity 的过程。

1．Activity 的创建

在 Android Studio 中创建 Activity，需要在图 1-9 所示的 src 文件夹下，右键单击新建的项目包名，在弹出的菜单中选择【new】/【Activity】/【Empty Activity】，在弹出的图 1-8 所示界面中，设置 Activity 与对应的 Layout 的名称，最后单击【Finish】按钮，即可创建一个 Activity。

新建 Activity 的代码如下：

```
public void SecondActivity extends  AppCompatActivity {
        @Override
        protected void onCreate(Bundle SaveInstanceState){
                super.onCreate(saveInstanceState);
                setContentView(R.layout.activity_sencond);
        }

}
```

2．Activity 的注册

当一个新的 Activity 创建完成后，必须要在清单文件 AndroidManifest.xml 中注册这个 Activity，如果没有注册就在程序中启动了该 Activity，程序就会崩溃，并会在日志平台中看到程序抛出异常(ActivityNotFoundException)。

注意：如果按照上述方式创建 Activity，Android Studio 会自动把相应的 Activity 在清

单文件中进行注册，无需手动注册。但并不是所有的创建方式都具备此项功能，因此新建 Activity 以后，一定要在清单文件中查看该 Activity 是否被注册。

在清单文件中注册 SecondActivity 的代码如下：

```
<?xml version="1.0" encoding="utf-8"?>
<manifest xmlns:android="http://schemas.android.com/apk/res/android"
    package="com.ugrow.ch02_2d2">

    <application
        android:allowBackup="true"
        android:icon="@mipmap/ic_launcher"
        android:label="@string/app_name"
        android:roundIcon="@mipmap/ic_launcher_round"
        android:supportsRtl="true"
        android:theme="@style/AppTheme">
        <activity android:name=".MainActivity">
            <intent-filter>
                <action android:name="android.intent.action.MAIN" />
                <category android:name="android.intent.category.LAUNCHER" />
            </intent-filter>
        </activity>
        <activity android:name=".SecondActivity"></activity>
    </application>
</manifest>
```

2.1.3　Activity 启动模式

在上一节中介绍了 Activity 的创建和注册，而一个完整的应用程序会包括多个 Activity，为了更好地管理 Activity，合理利用手机的内存空间，需要进一步了解 Activity 的启动模式。

1. 任务栈

一个任务(Task)就是应用程序在执行某项工作时与用户进行交互的 Activity 的集合，这些 Activity 会按照被打开的顺序依次被安排在一个堆栈中，此时，该堆栈被称为任务栈。

设备的主屏是大多数任务的启动位置，当用户触摸一个应用程序的启动图标(或者 App 快捷图标)时，应用程序的任务就会在前台显示，如果相关应用程序的任务不存在，那么就会有一个新的任务被创建，且会将应用程序打开的“主 Activity”作为任务的根 Activity。

2. 回退栈

在应用程序当前的 Activity 下启动了另一个 Activity 时，这个新的 Activity 就会被放到堆栈的栈顶，并且带有焦点，而前一个 Activity 并没有消失，而是保存在堆栈中并处于停止状态。如果在应用程序中打开了多个 Activity，当用户按下回退按钮时，当前的

Activity 会被从该堆栈的顶部弹出(这个 Activity 即被销毁)，而前一个 Activity 被恢复，此时，该堆栈被称为回退栈；当用户继续按回退按钮时，处于回退栈顶部的(当前显示的)Activity 会被依次弹出，同时前一个 Activity 被恢复，直到用户返回主屏(或者返回到任务开始运行的那个 Activity)；而当所有的 Activity 从回退栈中被删除时，这个任务就不存在了。

堆栈中的 Activity 不会被重新排列，因此，回退栈的操作顺序和后进先出的栈对象结构是一样的。图 2-7 展示了堆栈中的 Activity 在每个时间点上的处理过程。

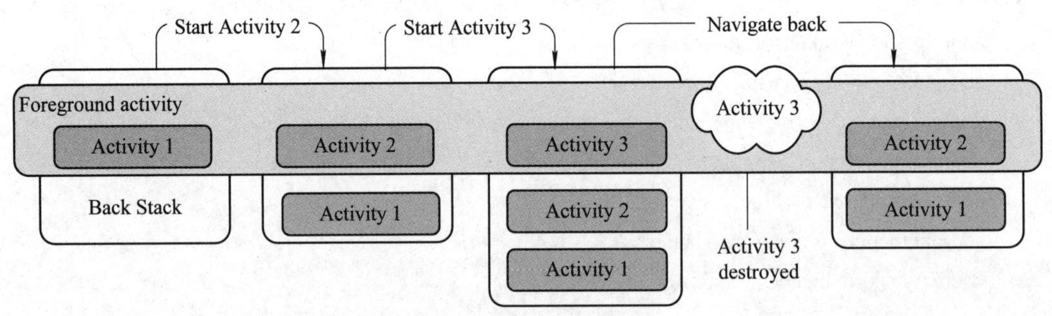

图 2-7　堆栈中的 Activity 处理过程

3. 启动模式

Activity 的启动模式决定了 Activity 的启动运行方式。Activity 有四种启动模式：

(1) standard：标准启动模式。系统默认的启动模式，每次激活 Activity 都会创建一个新的 Activity，并放入任务栈中，每个窗体的 getTaskId()保持不变，但 this.hashCode()会发生改变。

(2) singleTop：栈顶复用模式。如果在任务的栈顶正好存在某个 Activity 的实例，就重用该实例，而不会创建新的 Activity 对象；如果栈顶不存在某个 Activity 实例(即使栈中已经存在该 Activity 实例，但不在栈顶)，就会创建新的实例并放入栈顶。

(3) singleTask：栈内复用模式。如果在栈中已经有了某个 Activity 的实例，就会重用该 Activity 实例，重用时，会让该 Activity 实例回到栈顶，而在它上面的 Activity 实例会被移除；如果栈中不存在某个 Activity 实例，则会创建新的实例放入栈中。

singleTask 模式与 singleTop 模式在名字上就可以看出区别，即 singleTop 模式每次只检测当前栈顶的 Activity 实例是不是需要创建的，而 singleTask 则会从上至下检测栈中全部的 Activity 实例，如果检测到所请求的 Activity 实例，则会移除在其上面的全部 Activity 实例，直接把检测到所需要的 Activity 实例置为栈顶。

(4) singleInstance：单实例模式。在此模式下，系统会启动一个新的堆栈来管理 Activity 实例，配置该模式的 Activity 在整个系统中只有一个，无论从哪个任务栈中启动该 Activity 实例，都会将该 Activity 所在的任务栈转移到前台，从而使该 Activity 显示。此模式会节省大量的系统资源，因为它能保证所调用的 Activity 对象在当前的栈中只存在一个，可以实现在同一程序中共享一个 Activity 实例。

总而言之，在开发 Android 项目时，巧妙地设置 Activity 的启动模式可以节省系统开销，并提高程序的运行效率。在项目的清单文件 AndroidManifest.xml 中找到对应的

Activity，修改其中的属性 android:launchMode 的值，即可指定该 Activity 的启动模式，代码如下：

```
<activity android:name=".SecondActivity"
            android:launchMode="singleTop"></activity>
```

2.1.4 Activity 跳转方式

在 Android 中，每一个界面都是一个 Activity，切换界面操作其实是多个不同 Activity 之间相互跳转。Activity 的跳转方式基本可分为以下两种。

1. startActivity()

该跳转方式是从起始 Activity 跳转到目标 Activity，可以只是单纯的跳转，也可以向目标 Activity 传递数据，当目标 Activity 执行 finish()方法后，不会向起始 Activity 传递数据。

【示例 2.2】 在项目 ch02_2D2 中，使用 startActivity()方法实现 Activity 的跳转：单击 MainActivity 中的按钮，即跳转到 SecondActivity。

在 activity_main.xml 中编写代码如下：

```
<?xml version="1.0" encoding="utf-8"?>
<LinearLayout xmlns:android="http://schemas.android.com/apk/res/android"
    xmlns:tools="http://schemas.android.com/tools"
    android:layout_width="match_parent"
    android:layout_height="match_parent"
    tools:context="com.ugrow.ch02_2d2.MainActivity">

    <Button
        android:id="@+id/btn_one"
        android:layout_width="wrap_content"
        android:layout_height="wrap_content"
        android:text="跳转"/>

</LinearLayout>
```

上述代码中的父布局是一个线性布局(LinearLayout)，子控件是一个按钮(Button)，使用属性 android:id=""给 Button 设置 ID，使用属性 android:text=""设置 Button 的显示内容。

在 MainActivity 中编写代码如下：

```
public class MainActivity extends AppCompatActivity {

    private Button   mBtnOne;

    @Override
    protected void onCreate(Bundle savedInstanceState) {
```

```
        super.onCreate(savedInstanceState);
        setContentView(R.layout.activity_main);
        mBtnOne = (Button) findViewById(R.id.btn_one);
        mBtnOne.setOnClickListener(new View.OnClickListener() {
            @Override
            public void onClick(View view) {
                Intent intent = new Intent(MainActivity.this,SecondActivity.class);
                startActivity(intent);
            }
        });
    }
}
```

上述代码在 onCreate()方法中，通过 findViewById()方法初始化 Button，然后给 Button 设置监听事件；在 onClick()方法中，使用意图 Intent 指明起始 Activity 和目标 Activity，并通过 startActivity(intent)方法实现跳转。

在 activity_second.xml 中编写代码如下：

```
<?xml version="1.0" encoding="utf-8"?>
<LinearLayout xmlns:android="http://schemas.android.com/apk/res/android"
    xmlns:tools="http://schemas.android.com/tools"
    android:layout_width="match_parent"
    android:layout_height="match_parent"
    tools:context="com.ugrow.ch02_2d2.SecondActivity">
    <TextView
        android:layout_width="match_parent"
        android:layout_height="wrap_content"
        android:textSize="16sp"
        android:text="这是SecondActivity"/>

</LinearLayout>
```

上述代码中的父布局是 LinearLayout，布局中只有一个 TextView 控件，用于设置在 SecondActivity 中显示的内容。

在 SecondActivity 中编写代码如下：

```
public class SecondActivity extends AppCompatActivity {

    @Override
    protected void onCreate(Bundle savedInstanceState) {
        super.onCreate(savedInstanceState);
        setContentView(R.layout.activity_second);
    }
}
```

上述代码通过 onCreate()方法中的 setContentView(R.layout.activity_second)方法加载指定的布局，并显示相应的内容。

单击 Android Studio 的运行按钮，运行项目，可以看到在起始界面 MainActivity 中只有一个跳转按钮，如图 2-8 所示；单击该【跳转】按钮，可以跳转到目标界面 SecondActivity，同时界面中显示“这是 SecondActivity”，如图 2-9 所示。

图 2-8　起始界面 MainActivity

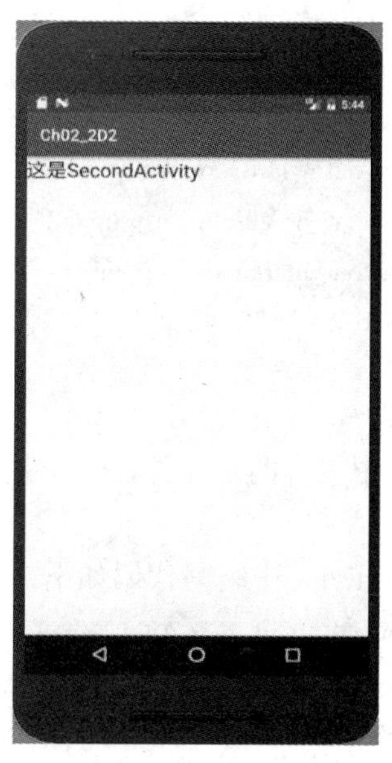

图 2-9　目标界面 SecondActivity

2．startActivityForResult()

此跳转方式可以理解为带有返回值的跳转方式，即：将当前活动界面的 Activity 实例看作一个父窗体，将目标 Activity 实例看作子窗体，当子窗体关闭时，父窗体就会执行 onActivityResult()方法，并能获取子窗体的返回值。

【示例 2.3】 在 Android Studio 中新建项目 Ch02_2D3，使用 startActivityForResult()方法实现跳转功能：在 MainActivity 中单击【跳转】按钮，会跳转到 SecondActivity；在 SecondActivity 中的输入框 EditText 中输入内容，单击【返回】按钮，会返回到 MainActivity 中，同时更新 TextView 显示的内容。

在 activity_main.xml 文件中编写代码如下：

```
<?xml version="1.0" encoding="utf-8"?>
<LinearLayout xmlns:android="http://schemas.android.com/apk/res/android"
    xmlns:tools="http://schemas.android.com/tools"
    android:layout_width="match_parent"
    android:layout_height="match_parent"
    android:orientation="vertical"
    tools:context="com.ugrow.ch02_2d3.MainActivity">


    <TextView
```

```
        android:id="@+id/tv_main"
        android:layout_width="match_parent"
        android:layout_height="wrap_content"
        android:layout_marginTop="16dp"
        android:text="Hello World!"
        android:textSize="20sp"/>
    <Button
        android:id="@+id/btn_main"
        android:layout_width="match_parent"
        android:layout_height="wrap_content"
        android:layout_marginTop="60dp"
        android:text="跳转"/>

</LinearLayout>
```

上述代码中包含两个控件，一个是 TextView，用来显示内容；另一个是 Button，用来实现单击操作。

在 MainActivity 中编写代码如下：

```
public class MainActivity extends AppCompatActivity {

    private TextView mTvMain;
    private Button mBtnMain;
    private int REQUEST_CODE = 0;

    @Override
    protected void onCreate(Bundle savedInstanceState) {
        super.onCreate(savedInstanceState);
        setContentView(R.layout.activity_main);
        initView();
    }

    private void initView() {
        mTvMain = (TextView) findViewById(R.id.tv_main);
        mBtnMain = (Button) findViewById(R.id.btn_main);
        mBtnMain.setOnClickListener(new View.OnClickListener() {
            @Override
            public void onClick(View view) {
                Intent intent = new Intent(MainActivity.this, SecondActivity.class);
                startActivityForResult(intent, REQUEST_CODE);
            }
        });
```

```
    }

    @Override
    protected void onActivityResult(int requestCode, int resultCode, Intent data) {
        super.onActivityResult(requestCode, resultCode, data);
        if (requestCode == REQUEST_CODE) {
            switch (resultCode) {
                case 1:
                    String content = data.getStringExtra("content");
                    mTvMain.setText(content);
                    break;
                default:
                    break;
            }
        }
    }
}
```

上述代码在 onCreate()方法中声明了一个 initView()方法，在该方法中通过 findViewById()方法初始化 TextView 和 Button 两个控件；然后为 Button 控件设置单击监听事件，重写 onClick()方法，在该方法中使用 startActivityForRcsult()方法实现跳转，该方法中的第一个参数是 Intent，第二个参数 REQUEST_CODE 是一个 int 类型的请求码；当接收回传数据时，需要在 MainActivity 中重写 onActivityResult()方法，其中三个参数分别代表请求码(requestCode)、返回码(resultCode)、返回数据(data)，需要判断 requestCode 是否与参数 REQUEST_CODE 定义的请求码相同，如果相同，则从 data 中读取数据，然后通过 setText()方法更新 TextView 的显示内容。

在 activity_second.xml 文件中编写代码如下：

```
<?xml version="1.0" encoding="utf-8"?>
<LinearLayout xmlns:android="http://schemas.android.com/apk/res/android"
    xmlns:tools="http://schemas.android.com/tools"
    android:layout_width="match_parent"
    android:layout_height="match_parent"
    android:orientation="vertical"
    tools:context="com.ugrow.ch02_2d3.SecondActivity">

    <EditText
        android:id="@+id/et_second"
        android:layout_width="match_parent"
        android:layout_height="60dp"
        android:hint="请输入内容!"
        android:textSize="16sp"/>
```

```
    <Button
        android:id="@+id/btn_second"
        android:layout_width="match_parent"
        android:layout_height="wrap_content"
        android:layout_marginTop="60dp"
        android:text="返回"/>

</LinearLayout>
```

上述代码中包含两个控件：一个是输入框 EditText，实现单击输入内容的功能；另一个是按钮 Button，实现单击返回的功能。

在 SecondActivity 中编写代码如下：

```
public class SecondActivity extends AppCompatActivity {

    private EditText mEtSec;
    private Button mBtnSec;
    private int RESULT_CODE = 1;

    @Override
    protected void onCreate(Bundle savedInstanceState) {
        super.onCreate(savedInstanceState);
        setContentView(R.layout.activity_second);
        initView();
    }

    private void initView() {
        mEtSec = (EditText) findViewById(R.id.et_second);
        mBtnSec = (Button) findViewById(R.id.btn_second);
        mBtnSec.setOnClickListener(new View.OnClickListener() {
            @Override
            public void onClick(View view) {
                String content = mEtSec.getText().toString();
                Intent intent = new Intent();
                intent.putExtra("content",content);
                setResult(RESULT_CODE,intent);
                finish();
            }
        });
    }
}
```

在其 onCreate()方法中，声明一个 initView()方法，在方法中通过 findViewById()找到 EditText 和 Button 两个控件，给 Button 设置监听事件，在 onClick()方法中，通过 mEtSec.getText().toString()方法获取输入框输入的内容，把输入的内容通过 Intent 传递，使用 setResult()方法，第一个参数是定义的 int 类型的返回码，第二个参数是 intent，最后，要调用 finish()方法，将本 Activity 销毁掉。

单击 Android Studio 上方的运行按钮，运行项目，可以看到 TextView 默认显示的内容为“Hello World!”，如图 2-10 所示；单击界面中的【跳转】按钮，在出现的输入框内输入“Android”，如图 2-11 所示，然后单击【返回】按钮，TextView 显示的内容就会改变为“Android”，如图 2-12 所示。

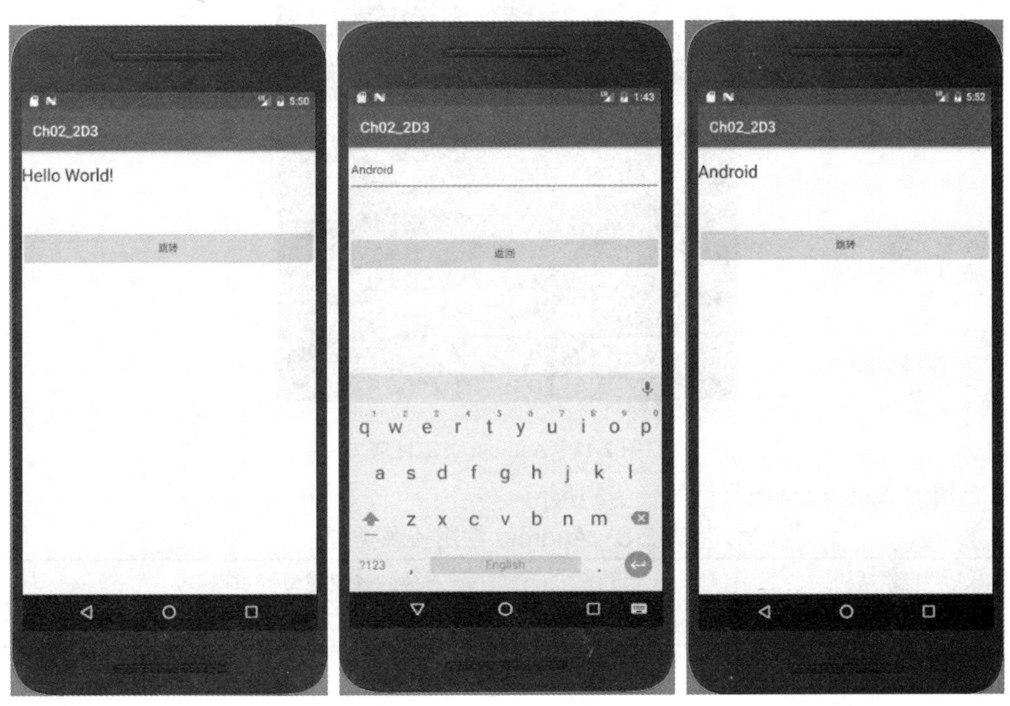

图 2-10　项目启动界面　　图 2-11　内容输入界面　　图 2-12　启动界面更新

2.2　Android 中的资源使用

Android 中的资源是指非代码部分，是代码中使用的外部文件，如图片、音频、动画、字符串等，作为应用程序的一部分，这些文件将被编译到应用程序中。将资源与代码分离能够提高程序的可维护性，例如通过字符串资源文件可以轻松实现程序的国际化，而无需修改代码。

如图 2-13 所示，在 Android 项目中，资源文件分别存放在 res 和 assets 两个文件夹中：

(1) res 目录存放 Android 程序能通过 R 资源类直接访问的资源。

(2) assets 目录存放 Android 程序不能直接访问的资源(原生文件)，如 MP3 文件，必须通过 AssetManager 类以二进制流的形式读取。

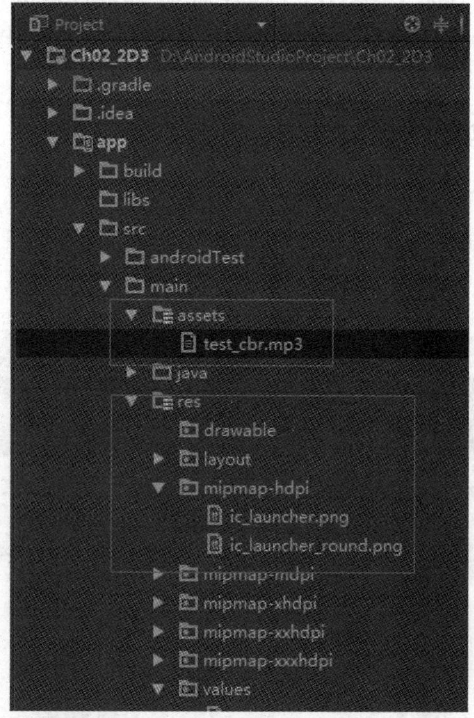

图 2-13 Android 资源目录

常用的 Android 资源类型如表 2-3 所示。

表 2-3 Android 常用资源类型

目录结构	存放的资源类型
res/anim	动画文件
res/mipmap	图标文件
res/layout	布局文件
res/xml	任意的 XML 文件
res/raw	直接复制到设备中的原生文件
res/menu	菜单文件
res/values	各种 XML 资源文件： ● strings.xml：字符串文件 ● arrays.xml：数组文件 ● colors.xml：颜色文件 ● dimens.xml：尺寸文件 ● styles.xml：样式文件

当在项目中加入新的资源时，资源引用文件 R.java 中会自动生成对新资源的引用。

2.2.1　字符串资源

字符串是最简单的一种资源，程序用到的字符串资源需要在 res/values/strings.xml 文件中定义，在其他的资源文件中或代码中都可以访问字符串资源。

在其他资源文件中采用“@string/资源名称”的形式访问，例如：

```
<TextView
    ......
    android:text="@string/hello_world" />
```

上述代码中，TextView 控件的 text 属性值为“@string/hello_world”，即名称为“hello”的字符串资源的值。

在代码中可通过“R.string.资源名称”的形式访问字符串资源，例如：

```
TextView tv = ...... // 初始化 TextView 控件
tv.setText(R.string.hello_world);
```

上述代码中，调用了 TextView 控件的 setText()方法，并用其引用了名称为“hello”的字符串资源。

【示例 2.4】 在代码和布局文件中使用字符串资源。

新建 Android 项目 Ch02_2D4，Android Studio 会自动生成 strings.xml 文件，内容如下：

```
<?xml version="1.0" encoding="utf-8"?>
<resources>
    <string name="app_name">ch02_2D4</string>
</resources>
```

上述文件中定义了一个字符串资源，名称为“app_name”。仿照此格式，可以修改或添加新的字符串资源，示例代码如下：

```
<?xml version="1.0" encoding="utf-8"?>
<resources>
    <string name="app_name">ch02_2D42</string>
    <string name="hello_world">Hello World, MyActivity!</string>
</resources>
```

新建项目后，Android Studio 会自动生成布局文件 res/layout/activity_main.xml，代码如下：

```
<?xml version="1.0" encoding="utf-8"?>
<LinearLayout xmlns:android="http://schemas.android.com/apk/res/android"
    android:orientation="vertical"
    android:layout_width="match_parent"
    android:layout_height="match_parent">
    <TextView
        android:layout_width="match_parent"
        android:layout_height="wrap_content"
```

```
                android:text="@string/hello_world" />
</LinearLayout>
```

实际上，布局文件 activity_main.xml 中已使用了字符串资源“hello”，其中 TextView 控件的 android:text 属性值为“@string/hello_world”，即代表使用字符串资源“hello_world”。

在 activity_main.xml 中添加一个新的 TextView 控件，并指定其 ID，代码如下：

```
<?xml version="1.0" encoding="utf-8"?>
<LinearLayout xmlns:android="http://schemas.android.com/apk/res/android"
        android:orientation="vertical"
        android:layout_width="match_parent"
        android:layout_height="match_parent">
        <TextView
                android:layout_width="match_parent"
                android:layout_height="wrap_content"
                android:textSize="25sp"
                android:text="@string/hello_world" />
        <TextView
                android:id="@+id/tv"
                android:layout_width="match_parent"
                android:layout_height="wrap_content"
                android:textSize="25sp"/>
</LinearLayout>
```

上述代码中添加了一个新的 TextView 控件，并增加了属性 android:id，该属性用于指定控件的唯一标识 ID，“@+id/tv”指明在资源引用 R.id 中增加一个 ID 为“tv”；同时，为了显示清晰，两个 TextView 的字体大小都通过 android:textSize 属性指定为“25sp”。

在静态资源引用文件 R.java 中，其 id 内部类会自动增加一个 tv 引用，代码如下：

```
public final class R {
......
    public static final class id {
public static final int tv=0x7f0b005e;
    }
}
```

注 意

Android 中，用于描述字体大小的单位用“sp”表示；关于布局文件的详细介绍请见本书第 3 章。

编写 Activity，代码如下：

```
public class MainActivity extends Activity {
        @Override
        public void onCreate(Bundle savedInstanceState) {
```

```
            super.onCreate(savedInstanceState);
            setContentView(R.layout.activity_main);
            TextView tv = (TextView)findViewById(R.id.tv);
            tv.setText(R.string.app_name);
    }
}
```

图 2-14　显示字符串资源

上述代码在 onCreate()方法中通过调用 findViewById()方法来标识 ID，以获取按钮 TextView 控件，并通过调用其 setText()方法将显示的文字设置为 app_name 代表的字符串资源的内容。

设置完毕，运行程序，效果如图 2-14 所示。

2.2.2　图片资源

图片资源的使用与字符串资源非常类似，程序用到的图片资源需要存放在 res 文件夹中的 mipmap 资源目录下，在其他的资源文件或代码中都可以访问其中的图片资源。

访问图片资源的方式与访问字符串资源也是类似的，在其他资源文件中可采用“@mipmap/资源名称”的形式访问，在代码中可通过“R. mipmap.资源名称”的形式访问。

注 意　res/mipmap-hdpi、res/mipmap-mdpi、res/mipmap-ldpi 三个目录分别用于存放高、中、低三种分辨率下的图片文件，Android 程序运行时，会根据当前分辨率自动到对应的目录下查找图片。

【示例 2.5】 使用图片资源设置 Activity 的背景。

新建 Android 项目 Ch02_2D5，复制需要使用的图片 td.jpg 到 res/mipmap-mdpi 目录，编辑布局文件 res/layout/activity_main.xml，代码如下：

```
<?xml version="1.0" encoding="utf-8"?>
<LinearLayout xmlns:android="http://schemas.android.com/apk/res/android"
    android:orientation="vertical"
    android:layout_width="match_parent"
    android:layout_height="match_parent"
    android:background="@mipmap/td">
    <TextView
        android:layout_width="match_parent"
        android:layout_height="wrap_content"
        android:text="@string/hello" />
</LinearLayout>
```

图 2-15　显示图片资源

上述代码在最外层的 LinearLayout 布局中，将 android:background 属性的值设置为“@mipmap/td”，即使用名称为 td 的图片资源作为程序背景。

设置完毕，运行程序，显示结果如图 2-15 所示。

本 章 小 结

通过本章的学习，读者应当了解：

✧ Activity 是 Android 应用程序中最基本的组成单位。

✧ 大部分的 Android 应用中包含多个 Activity 类。

✧ Activity 共有四种状态：激活或者运行状态、暂停状态、停止状态、终止状态。

✧ Activity 有四种启动模式：standard、singleTop、singleTask 和 singleInstance。

✧ Activity 可使用两种跳转方式：startActivity()和 startActivityForResult()。

✧ 每个 Activity 类在定义时都必须继承 android.app.Activity。

✧ android.app.Activity 类中的方法定义了 Activity 完整的生命周期。

✧ Android 中的资源是指非代码部分，是代码中使用的外部资源。

✧ 对于字符串资源，在其他资源文件中使用“@string/资源名称”的形式访问，在代码中可通过“R.string.资源名称”的形式访问。

✧ 对于图片资源，在其他资源文件中使用“@mipmap/资源名称”的形式访问，在代码中可通过“R.mipmap.资源名称”的形式访问。

本 章 练 习

1. Activity 生命周期中的______方法在 Activity 初次创建时被调用。

   A．OnStart()　　B．OnCreate()

   C．OnPause()　　D．OnResume()

2. ______状态下的 Activity 失去了焦点但是仍然对用户可见。

   A．激活状态　　B．停止状态

   C．运行状态　　D．暂停状态

3. Activity 的可见生命周期是______。

   A．从 onCreate()方法到 onDestroy()方法的这个过程

   B．从 onResume()方法到 onPause()方法的这个过程

   C．从 onStart()方法到 onStop()方法的这个过程

   D．从 onCreate()方法到 onPause()方法的这个过程

4. Android 程序不能直接访问的资源(原生文件)存放在______目录下。

   A．src 目录　　B．res 目录

   C．assets 目录　　D．Bres/raw 目录

5. 简述 Activity 的生命周期中的各个方法。

6. 简述 Activity 的启动模式和跳转方式。

7. 编写一个 Activity，在其中显示一张图片。

第 3 章　用户界面

本章目标

- 熟悉基本的 Android 界面组件
- 掌握 UI 的事件驱动机制
- 掌握常用的 Layout
- 掌握对话框以及 Toast 组件的使用
- 掌握常用的 Widget 组件
- 掌握菜单组件的使用
- 掌握 ActionBar 的使用
- 掌握适配器 Adapter 的使用

3.1 用户界面元素分类

Android 系统提供了丰富的可视化用户界面(UI)组件，包括菜单、对话框、按钮、文本框等。Android 借用了 Java 中的 UI 设计思想，以及事件响应机制和布局管理。Android 中的界面元素主要由以下几个部分构成：

(1) 视图组件(View)。

(2) 视图容器(ViewGroup)。

(3) 布局管理(Layout)。

一个复杂的 Android 界面设计往往需要组合不同的组件才能实现，本节将简要介绍 Android 各主要界面组件的特点及其功能。

3.1.1 视图组件(View)

View 是用户界面的基础元素，View 对象存储了 Andiord 屏幕上一个特定的矩形区域的布局和内容属性的数据体。View 对象可实现对布局、绘图、焦点变换、滚动条、屏幕区域的按键、用户交互等功能。

Android 的窗体功能是通过 Widget(窗体部件)类实现的，而 View 类是 Widget 类的基类。View 类的常见子类及功能如表 3-1 所示。

表 3-1 View 类的主要子类

类 名	功能描述	事件监听器
TextView	文本视图	OnKeyListener
EditText	编辑文本框	OnEditorActionListener
Button	按钮	OnClickListener
Checkbox	复选框	OnCheckedChangeListener
RadioGroup	单选按钮组	OnCheckedChangeListener
Spinner	下拉列表	OnItemSelectedListener
AutoCompleteTextView	自动完成文本框	OnKeyListener
DataPicker	日期选择器	OnDateChangedListener
TimePicker	时间选择器	OnTimeChangedListener
DigitalClock	数字时钟	OnKeyListener
AnalogClock	模拟时钟	OnKeyListener
ProgessBar	进度条	OnProgressBarChangeListener
RatingBar	评分条	OnRatingBarChangeListener
SeekBar	搜索条	OnSeekBarChangeListener
GridView	网格视图	OnKeyDown,OnKeyUp
LsitView	列表视图	OnKeyDown,OnKeyUp
ScrollView	滚动视图	OnKeyDown,OnKeyUp

3.1.2 视图容器(ViewGroup)

ViewGroup 是 View 的容器，可将 View 添加到 ViewGroup 中，一个 ViewGroup 也可

以加入到另外一个 ViewGroup 里。

ViewGroup 类提供的主要方法如表 3-2 所示。

表 3-2　ViewGroup 类的主要方法

方法	功 能 描 述
ViewGroup()	构造方法
void addView(View child)	用于添加子视图
void bringChildToFront(View child)	将参数指定的视图移动到所有视图的前面显示
boolean clearChildFocus(View child)	清除参数指定的视图的焦点
boolean dispatchKeyEvent(KeyEvent event)	将参数指定的键盘事件分发给当前焦点路径的视图。分发判断事件时，按照焦点路径查找合适的视图。若本视图为焦点，则将键盘事件发送给自己；否则发送给焦点视图
boolean dispatchPopulateAccessibilityEvent (AccessibilityEvent event)	将参数指定的事件分发给当前焦点路径的视图
boolean dispatchSetSelected(boolean selected)	为所有的子视图调用 setSelected()方法

3.1.3　布局管理(Layout)

Layout 用于管理组件的布局格式，以规定界面中组件的呈现方式。Android 提供了多种布局，常用的有以下几种：

(1) LinearLayout：线性布局。该布局中子元素之间成线性排列，即在某一方向上的顺序排列，常见的有水平顺序排列、垂直顺序排列。

(2) RelativeLayout：相对布局。该布局是一种根据相对位置排列元素的布局方式，这种方式允许子元素指定他们相对于其他元素或父元素的位置(通过 ID 指定)。相对于线性布局，使用 RelativeLayout 布局可任意放置控件，没有规律性。需要注意线性布局不需要指定其参照物，而相对布局使用之前必须指定其参照物，只有指定参照物之后，才能定义其相对位置。

(3) TableLayout：表格布局。该布局将子元素的位置分配到表格的行或列中，即按照表格的顺序排列。一个表格布局有多个“表格行”，而每个表格行又包含表格单元。需要注意，表格布局并不是真正意义上的表格，只是按照表格的方式组织元素的布局，元素之间并没有实际表格中的分界线。

(4) AbsoluteLayout：绝对布局。该布局按照绝对坐标对元素进行布局。与相对布局相反，绝对布局不需要指定其参照物，而是使用整个手机界面作为坐标系，通过坐标系的两个偏移量(水平偏移量和垂直偏移量)来唯一指定其位置。

(5) FrameLayout：框架布局。该布局将所有子元素以层叠的方式显示，后加的元素会被放在最顶层，覆盖之前的元素。

(6) GridLayout：网格布局。该布局是 Android 4.0 新增的布局方式，能够同时对 X、Y 轴上的控件进行对齐，大大地简化了复杂布局的处理，并且在性能上也有大幅提升。

3.2 事件处理机制

Android 系统引用了 Java 的事件处理机制，包括事件、事件源和事件监听器三个事件模型：

(1) 事件(Event)：是一个描述事件源状态改变的对象，事件不是通过 new 运算符创建的，而是由用户操作触发的。事件可以是键盘事件、触摸事件等。事件一般作为事件处理方法的参数，以便从中获取事件的相关信息。

(2) 事件源(Event Source)：产生事件的对象，事件源通常是 UI 组件，例如单击按钮，则按钮就是事件源。

(3) 事件监听器(Event Listener)：当事件产生时，事件监听器用于对该事件进行响应和处理。监听器需要实现监听接口中定义的事件处理方法。

事件处理机制如图 3-1 所示。

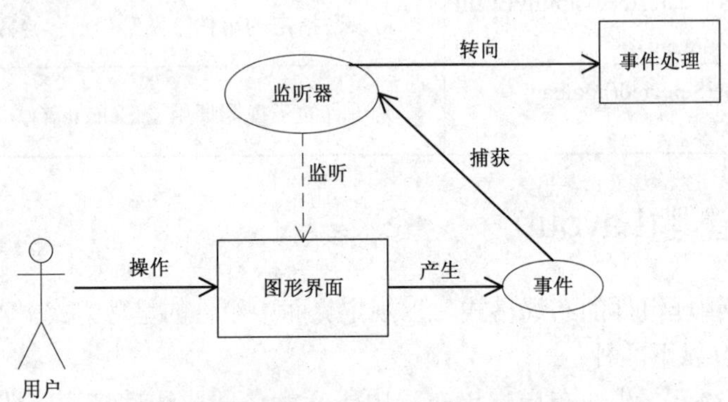

图 3-1　Android 事件处理机制示意图

Android 中常用的事件监听器如表 3-3 所示，这些事件监听器都定义在 android.view.View 中。

表 3-3　Android 中常用的事件监听器

事件监听器接口	事件	说　明
OnClickListener	单击事件	用户单击某个组件或者方向键时产生
OnFocusChangeListener	焦点事件	组件获得或者失去焦点时产生
OnKeyListener	按键事件	用户按下或释放设备上的某个按键时产生
OnTouchListener	触碰事件	设备具有触摸屏功能时，触碰屏幕时产生
OnCreateContextMenuListener	创建上下文菜单事件	创建上下文菜单时产生
OnCheckedChangeListener	选项事件	选择改变时触发

事件处理的基本步骤如下：

(1) 创建事件监听器。

(2) 在事件处理方法中编写事件处理代码。

(3) 在相应的组件上注册监听器。

【示例 3.1】 在 Android Studio 中新建项目 Ch03_3D1_Event，实现单击按钮改变界面背景颜色的功能。

首先编写界面布局文件 event_layout.xml，在其中添加两个 Button，代码如下：

```
<?xml version="1.0" encoding="utf-8"?>
<LinearLayout xmlns:android="http://schemas.android.com/apk/res/android"
        android:layout_width="match_parent"
        android:layout_height="match_parent"
        android:orientation="vertical" >
        <Button
                android:id="@+id/btn_Yellow"
                android:layout_width="wrap_content"
                android:layout_height="wrap_content"
                android:text="黄色"
                android:textColor="#fff"/>
        <Button
                android:id="@+id/btn_Blue"
                android:layout_width="wrap_content"
                android:layout_height="wrap_content"
                android:text="蓝色"
                android:textColor="#fff"/>
</LinearLayout>
```

然后在 res/values 目录下的 colors.xml 文件中添加 yellow 和 blue 两种颜色，代码如下：

```
<?xml version="1.0" encoding="UTF-8"?>
<resources>
<color name="colorPrimary">#3F51B5</color>
        <color name="colorPrimaryDark">#303F9F</color>
        <color name="colorAccent">#FF4081</color>
        <color name="yellow">#ffee55</color>
        <color name="blue">#0000ff</color>
</resources>
```

上述代码定义了两种颜色：黄色和蓝色。此时静态的资源引用文件 R.java 中会自动增加 color 资源，代码如下：

```
public final class R {
......
    public static final class color {
        public static final int blue=0x7f0a000b;
            public static final int yellow=0x7f0a003f;
        }
}
```

关于颜色资源文件的内容介绍参见实践篇第 2 章的知识拓展。

编写 EventActivity，代码如下：

```
public class EventActivity extends AppCompatActivity implements View.OnClickListener{

    private Button mBtnYellow;
    private Button mBtnBlue;

    @Override
    protected void onCreate(Bundle savedInstanceState) {
        super.onCreate(savedInstanceState);
        setContentView(R.layout.event_layout);
        initView();
    }

    private void initView() {
        //根据 ID 初始化界面中两个控件按钮
        mBtnYellow = (Button) findViewById(R.id.btn_Yellow);
        mBtnBlue = (Button)findViewById(R.id.btn_blue);
        //分别给两个按钮设置监听事件
        mBtnYellow.setOnClickListener(this);
        mBtnBlue.setOnClickListener(this);
    }

    @Override
    public void onClick(View view) {
        switch (view.getId()){
            case R.id.btn_Yellow:
                //设置背景颜色为黄色
                getWindow().setBackgroundDrawableResource(R.color.yellow);
                break;
            case R.id.btn_blue:
                //设置背景颜色为蓝色
                getWindow().setBackgroundDrawableResource(R.color.blue);
                break;
            default:
                break;
        }
    }
}
```

编写上述代码时，需注意以下几点：

(1) 让 EventActivity 实现 View.OnClickListener 监听方法，同时重写 onClick()方法，且不要忘记给控件设置监听事件(例如：mBtnYellow.setOnClickListener(this))，否则，onClick()方法设置的事件不会被触发。

(2) 可以通过 getWindow()方法获取屏幕窗口对象(Window)，再调用该对象的 setBackgroundDrawable Resource()方法设置窗口背景颜色，其参数 R.color.yellow 和 R.color.blue 是对颜色资源的引用，分别代表黄色和蓝色。

(3) 在 OnCreate()方法中，可以通过调用 initView()方法中的 findViewById()方法来初始化两个 Button 控件。

(4) 让 Activity 类实现 OnClickListener 接口的监听方法常用于在同一界面中有多个控件都要设置监听事件时，可以在 onClick()方法中使用 switch()条件语句处理相应的操作。

运行项目，效果如图 3-2 所示：当单击【黄色】按钮时，界面背景颜色变为黄色；当单击【蓝色】按钮时，界面背景颜色变为蓝色。

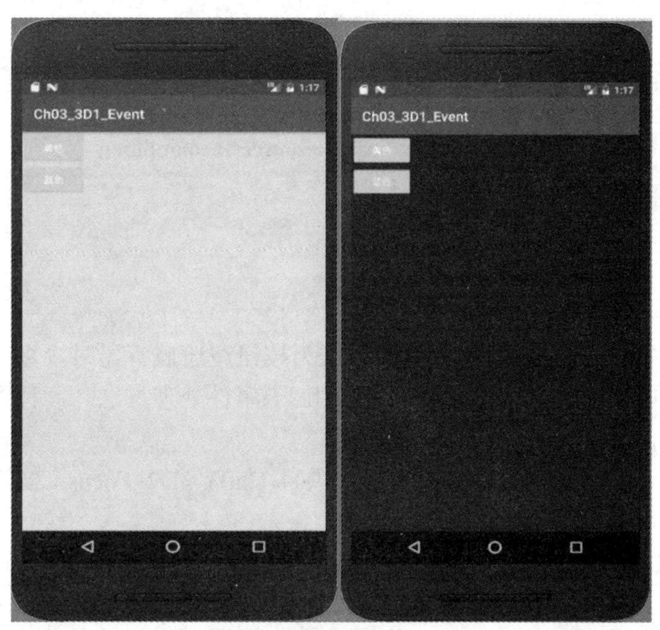

图 3-2　改变界面背景颜色

在事件处理的方式上，除采用上面这种事件监听器方式以外，还可以采用匿名方式，即在注册的同时实现监听器接口及其方法。采用匿名方式的事件处理代码如下：

```
public class EventActivity2 extends Activity {
    //声明两个按钮
    private Button btnYellow;
    private Button btnBlue;
    @Override
    public void onCreate(Bundle savedInstanceState) {
        super.onCreate(savedInstanceState);
```

```
        setContentView(R.layout.event_layout);
        //根据 ID 找到界面中的两个按钮组件
        btnYellow= (Button) this.findViewById(R.id.btn_Yellow);
        btnBlue = (Button) this.findViewById(R.id.btn_Blue);
        //注册监听器
        btnYellow.setOnClickListener(new OnClickListener() {
        @Override
        public void onClick(View v) {
            //设置背景颜色为黄色
        getWindow().setBackgroundDrawableResource(R.color.yellow);
        }
        });
        btnBlue.setOnClickListener(new OnClickListener() {
        @Override
        public void onClick(View v) {
            //设置背景颜色为蓝色
        getWindow().setBackgroundDrawableResource(R.color.blue);
        }
        });
    }
}
```

上述代码在调用 setOnClickListener()方法为按钮注册监听器时，直接通过匿名的方法实现了 OnClickListener 接口及其内部的 onClick()事件处理方法。此种方式比较简单，也较为常用，其运行结果与第一种方式相同。

【示例 3.2】 在 Android Studio 中新建项目 Ch03_3D2_Focus，实现单击输入框获取焦点自动填充内容的功能。

在项目的 activity_main.xml 文件中编写代码如下：

```
<?xml version="1.0" encoding="utf-8"?>
<LinearLayout xmlns:android="http://schemas.android.com/apk/res/android"
    xmlns:tools="http://schemas.android.com/tools"
    android:layout_width="match_parent"
    android:layout_height="match_parent"
    android:orientation="vertical"
    tools:context="com.ugrow.ch03_3d2_focus.MainActivity">

    <EditText
        android:id="@+id/et_name"
        android:layout_width="match_parent"
        android:layout_height="wrap_content"
```

```
        android:hint="请输入姓名..."/>
    <EditText
        android:id="@+id/et_age"
        android:layout_width="match_parent"
        android:layout_height="wrap_content"
        android:hint="请输入年龄..."/>
</LinearLayout>
```

在 MainActivity 中编写代码如下：

```
public class MainActivity extends AppCompatActivity {

    private EditText mEtName;
    private EditText mEtAge;

    @Override
    protected void onCreate(Bundle savedInstanceState) {
        super.onCreate(savedInstanceState);
        setContentView(R.layout.activity_main);
        mEtName = (EditText) findViewById(R.id.et_name);
        mEtAge = (EditText)findViewById(R.id.et_age);
        mEtName.setOnFocusChangeListener(new View.OnFocusChangeListener() {
            @Override
            public void onFocusChange(View view, boolean hasFocus) {
                if (hasFocus){
                    mEtName.setText("张三");
                    mEtName.setTextColor(Color.BLUE);
                }else {
                    mEtName.setText("");
                }
            }
        });
        mEtAge.setOnFocusChangeListener(new View.OnFocusChangeListener() {
            @Override
            public void onFocusChange(View view, boolean hasFocus) {
                if (hasFocus){
                    mEtAge.setText("18");
                    mEtAge.setTextColor(Color.BLUE);
                }else {
                    mEtAge.setText("");
                }
            }
```

```
        });
    }
}
```

上述代码首先在 activity_main.xml 中添加了两个输入框(EditText)控件，分别用来输入名字和年龄；然后在 MainActivity 中，通过 findViewById()方法初始化这两个控件，并对两个控件分别设置焦点改变监听事件 mEtName.setOnFocusChangeListener()，然后重写 onFocusChange()方法，该方法中的第二个参数 hasFocus 是 boolean 类型，可以通过 if 条件判断语句判断该参数的值，根据不同的值让输入框在获取焦点和失去焦点时实现对应的逻辑功能。

运行项目，效果如图 3-3 所示：打开应用程序时，上方输入框自动获取焦点，并自动填充字符串“张三”，下方输入框没有获取焦点；单击第二个输入框时，上方输入框失去焦点，下方输入框获取焦点，并自动填充字符串“18”。

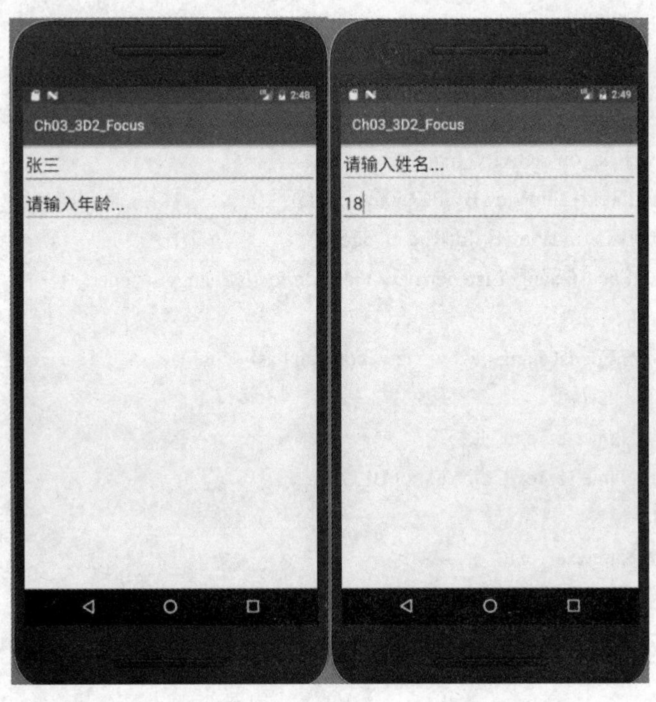

图 3-3　改变输入焦点并自动填充内容

3.3　布局管理(Layout)

Android 中提供了两种创建布局的方式：

(1) 在 XML 布局文件中声明。这种方式将需要显示的组件先在布局文件中进行声明，然后在程序中通过 setContentView(R.layout.XXX)方法将布局呈现在 Activity 中。这种方式是推荐使用的方式。

(2) 在程序中通过代码直接实例化布局及其组件。这种方式并不提倡使用，除非需要

动态改变界面中的组件及布局。

Android 的布局包括 LinearLayout、RelativeLayout、TableLayout 和 AbsoluteLayout 等多种类型，下面对其创建及应用方法逐一进行介绍。

3.3.1　线性布局(LinearLayout)

LinearLayout 是一种线性排列的布局，该布局中的子组件按照垂直或者水平方向排列，其方向由 android:orientation 属性控制，属性值有垂直(vertical)和水平(horizontal)两种，默认方向为水平方向。

线性布局的其他常用属性如下：

(1) android:layout_weight：该属性用于设置控件的权重，即各控件在水平或者垂直方向上平均分配，它表示比重，可实现百分比布局。在使用时，如果是在水平方向上设置权重，要将 android:layout_width 设置为 0dp；如果是在垂直方向上设置权重，要将 android:layout_height 设置为 0dp，否则权重容易受到高度或宽度的干扰而出现偏差。

(2) android:layout_gravity：该属性用于设置控件相对于父布局的对齐方式，可选值有 bottom、center_vertical、center 等。

【示例 3.3】 在 Android Studio 中新建项目 Ch03_3D3_Layout，演示线性布局 LinearLayout 的使用。

在项目的 res/layout 目录下创建线性布局文件 linear_layout.xml，编写代码如下：

```
<?xml version="1.0" encoding="utf-8"?>
<LinearLayout xmlns:android="http://schemas.android.com/apk/res/android"
    xmlns:tools="http://schemas.android.com/tools"
    android:layout_width="match_parent"
    android:layout_height="match_parent"
    android:background="#ededed"
    android:orientation="vertical">
    <LinearLayout
        android:layout_width="match_parent"
        android:layout_height="0dp"
        android:layout_weight="1"
        android:orientation="horizontal">
        <TextView
            android:layout_width="0dp"
            android:layout_weight="1"
            android:layout_height="match_parent"
            android:background="#aa0000"
            android:gravity="center_horizontal"
            android:text="red"
            android:textColor="#fff"/>
        <TextView
```

```
            android:layout_width="0dp"
            android:layout_weight="1"
            android:layout_height="match_parent"
            android:background="#00aa00"
            android:gravity="center_horizontal"
            android:text="green"
            android:textColor="#fff"/>
        <TextView
            android:layout_width="0dp"
            android:layout_weight="1"
            android:layout_height="match_parent"
            android:background="#0000aa"
            android:gravity="center_horizontal"
            android:text="blue"
            android:textColor="#fff"/>
        <TextView
            android:layout_width="0dp"
            android:layout_weight="1"
            android:layout_height="match_parent"
            android:background="#aaaa00"
            android:gravity="center_horizontal"
            android:text="yellow"
            android:textColor="#fff"/>
    </LinearLayout>
    <LinearLayout
        android:layout_width="match_parent"
        android:layout_height="0dp"
        android:layout_weight="1"
        android:orientation="vertical">
        <TextView
            android:layout_width="match_parent"
            android:layout_height="0dp"
            android:layout_weight="1"
            android:background="#aa0000"
            android:text="row one"
            android:textColor="#fff"
            android:textSize="15sp"/>
        <TextView
            android:layout_width="match_parent"
            android:layout_height="0dp"
```

```
            android:layout_weight="1"
            android:background="#00aa00"
            android:text="row two"
            android:textColor="#fff"
            android:textSize="15sp"/>
        <TextView
            android:layout_width="match_parent"
            android:layout_height="0dp"
            android:layout_weight="1"
            android:background="#0000aa"
            android:text="row three"
            android:textColor="#fff"
            android:textSize="15sp"/>
        <TextView
            android:layout_width="match_parent"
            android:layout_height="0dp"
            android:layout_weight="1"
            android:background="#aaaa00"
            android:text="row four"
            android:textColor="#fff"
            android:textSize="15sp"/>
    </LinearLayout>
</LinearLayout>
```

上述代码中使用了三个 LinearLayout：

(1) 第一个 LinearLayout 按照垂直方向布局，并包含其他两个 LinearLayout，是整个布局的主布局。

(2) 第二个 LinearLayout 按照水平方向布局，包含 4 个 TextView。

(3) 第三个 LinearLayout 按照垂直方向布局，也包含 4 个 TextView。

在 LayoutActivity 中设置应用 linear_layout.xml 布局文件，代码如下：

```
public class LayoutActivity extends AppCompatActivity {
    /** Called when the activity is first created. */
    @Override
    public void onCreate(Bundle savedInstanceState) {
        super.onCreate(savedInstanceState);
        setContentView(R.layout.linear_layout);
    }
}
```

上述代码通过调用 setContentView() 方法，将 linear_layout.xml 布局加载到 LayoutActivity 中，运行结果如图 3-4 所示。

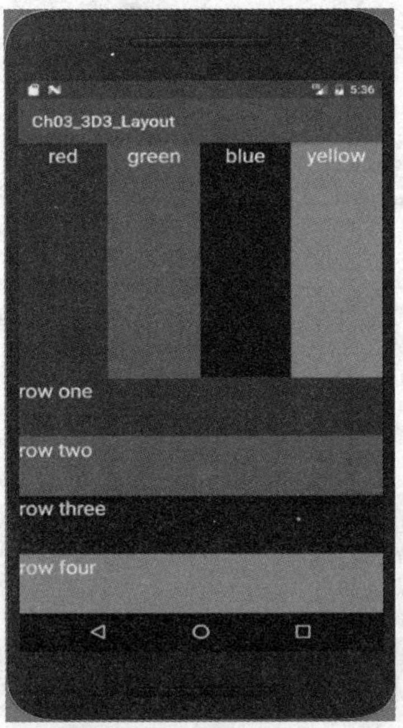

图 3-4　线性布局

3.3.2　相对布局(RelativeLayout)

RelativeLayout 是按照组件之间的相对位置(如在某个组件的左边、右边、上面和下面等)来进行的布局。

相对布局的常用属性如下：

(1) 用于表示兄弟控件之间的位置，该组属性的值是另一个控件的 ID：

- ✧ android:layout_toRightOf：表示该控件在某控件的右侧。
- ✧ android:layout_toLeftOf：表示该控件在某控件的左侧。
- ✧ android:layout_below：表示该控件在某控件的下方。
- ✧ android:layout_above：表示该控件在某控件的上方。

(2) 用于表示控件与父布局之间的对齐和位置关系，该组属性的值是 true 或者 false：

- ✧ android:layout_alignParentRight：表示该控件在父布局中右对齐。
- ✧ android:layout_alignParentTop：表示该控件在父布局中顶部对齐。
- ✧ android:layout_alignParentBottom：表示该控件在父布局中底部对齐。
- ✧ android:layout_alignParentLeft：表示该控件在父布局中左对齐。
- ✧ android:layout_centerHorizontal：表示该控件位于父布局的水平居中位置。
- ✧ android:layout_centerVertical：表示该控件位于父布局的垂直居中位置。
- ✧ android:layout_centerInParent：表示该控件位于父布局中心位置。

【示例 3.4】 在项目 Ch03_3D3_Layout 中，编写代码演示相对布局 RelativeLayout 的使用。

在项目的 res/layout 目录下创建线性布局文件 relative_layout.xml，编写代码如下：

```
<?xml version="1.0" encoding="utf-8"?>
<LinearLayout xmlns:android="http://schemas.android.com/apk/res/android"
    xmlns:tools="http://schemas.android.com/tools"
    android:layout_width="match_parent"
    android:layout_height="match_parent"
    android:background="#ededed"
    android:orientation="vertical">

    <RelativeLayout
        android:id="@+id/rl_01"
        android:layout_width="wrap_content"
        android:layout_height="wrap_content">

        <Button
            android:id="@+id/btn_a"
            android:layout_width="wrap_content"
            android:layout_height="wrap_content"
            android:text="A" />

        <Button
            android:id="@+id/btn_b"
            android:layout_width="wrap_content"
            android:layout_height="wrap_content"
            android:layout_toRightOf="@+id/btn_a"
            android:text="B" />

        <Button
            android:id="@+id/btn_c"
            android:layout_width="wrap_content"
            android:layout_height="wrap_content"
            android:layout_below="@+id/btn_a"
            android:text="C" />

        <Button
            android:id="@+id/btn_d"
            android:layout_width="wrap_content"
            android:layout_height="wrap_content"
            android:layout_below="@+id/btn_b"
            android:layout_toRightOf="@+id/btn_c"
```

```
        android:text="D" />
    </RelativeLayout>
</LinearLayout>
```

上述代码中共有四个按钮【A】【B】【C】【D】，其中：按钮【B】在按钮【A】的右边，需通过 android:layout_toRighOf 属性进行设置，属性值为“@+id/a”，表明参照物是按钮【A】；按钮【C】在按钮【A】的下面，需通过 android:layout_below 属性进行设置；按钮【D】在按钮【C】的右边，与按钮【B】类似，也需要通过 android:layout_toRighOf 属性进行设置，但此时属性值为“@+id/c”，表明参照物是按钮【C】。

在 LayoutActivity 中加载 relative_layout.xml 布局文件，代码如下：

```
setContentView(R.layout.relative_layout);
```

运行项目，结果如图 3-5 所示。

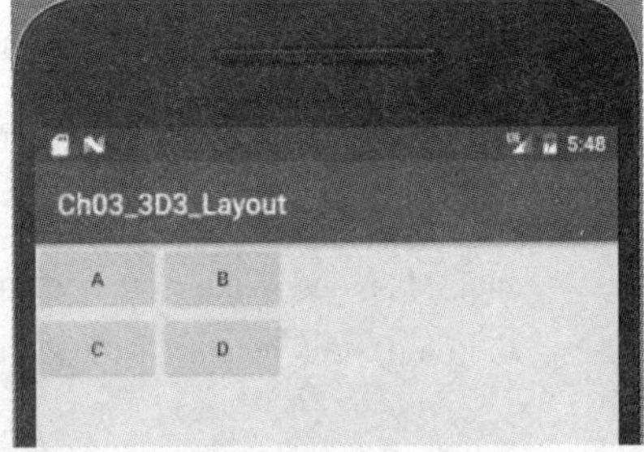

图 3-5　相对布局

3.3.3　表格布局(TableLayout)

TableLayout 以行、列表格的方式布局子组件。可以在 TableLayout 中使用 TableRow 对象来定义行。

【示例 3.5】 在项目 Ch03_3D3_Layout 中，编写代码演示表格布局 TableLayout 的使用。

在项目的 res/layout 目录下创建表格布局文件 table_layout.xml，编写代码如下：

```
<?xml version="1.0" encoding="utf-8"?>
<LinearLayout xmlns:android="http://schemas.android.com/apk/res/android"
    android:layout_width="match_parent"
    android:layout_height="match_parent"
    android:background="#ededed"
    android:orientation="vertical" >
<TableLayout
        android:id="@+id/TableLayout01"
        android:layout_width="match_parent"
        android:layout_height="wrap_content"
        android:collapseColumns="3"
        android:stretchColumns="1" >
<TableRow
            android:layout_width="wrap_content"
            android:layout_height="wrap_content" >
<Button
                android:id="@+id/a"
                android:layout_width="wrap_content"
```

```
            android:layout_height="wrap_content"
            android:text="A" >
</Button>
<Button
            android:id="@+id/b"
            android:layout_width="wrap_content"
            android:layout_height="wrap_content"
            android:text="B" >
</Button>
<Button
            android:id="@+id/c"
            android:layout_width="wrap_content"
            android:layout_height="wrap_content"
            android:text="C" >
</Button>
</TableRow>
<TableRow
        android:layout_width="wrap_content"
        android:layout_height="wrap_content" >
<Button
            android:id="@+id/d"
            android:layout_width="wrap_content"
            android:layout_height="wrap_content"
            android:text="D" >
</Button>
<Button
            android:id="@+id/e"
            android:layout_width="wrap_content"
            android:layout_height="wrap_content"
            android:text="E" >
</Button>
<Button
            android:id="@+id/f"
            android:layout_width="wrap_content"
            android:layout_height="wrap_content"
            android:text="F" >
</Button>
</TableRow>
<TableRow
        android:layout_width="wrap_content"
```

```
            android:layout_height="wrap_content" >
<Button
                android:id="@+id/g"
                android:layout_width="wrap_content"
                android:layout_height="wrap_content"
                android:text="G" >
</Button>
<Button
                android:id="@+id/h"
                android:layout_width="wrap_content"
                android:layout_height="wrap_content"
                android:text="H" >
</Button>
<Button
                android:id="@+id/i"
                android:layout_width="wrap_content"
                android:layout_height="wrap_content"
                android:text="I" >
</Button>
</TableRow>
</TableLayout>
</LinearLayout>
```

编写上述代码时，需注意以下两点：

(1) <TableLayout>元素定义了表格布局，该元素的 android:collapseColumns 属性用于指明表格的列数，此处设置表格的列数为 3；android:stretchColumns 属性用于指明表格的伸展列，指定的伸展列将进行拉伸以填满剩余的空间。注意列号从 0 开始，此处值为 1，代表第二列是伸展列。

(2) <TableRow>元素定义了表格中的行，所有的其他组件都放在该元素内。

在 LayoutActivity 中设置应用 table_layout.xml 布局文件，代码如下：

```
setContentView(R.layout.table_layout);
```

运行项目，结果如图 3-6 所示。若将<TableLayout>元素中的代码 android:stretchColumns="1"删除，即不指定伸展列，则项目运行结果如图 3-7 所示。

图 3-6 指定第二列为伸展列的布局

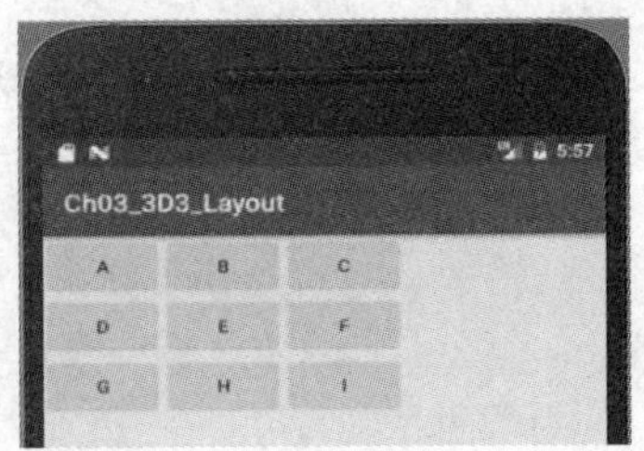

图 3-7 不指定伸展列的布局

3.3.4　绝对布局(AbsoluteLayout)

AbsoluteLayout 通过指定组件的绝对坐标来确定组件的位置。

【示例 3.6】 在项目 Ch03_3D3_Layout 中，编写代码演示绝对布局 AbsoluteLayout 的使用。

在项目的 res/layout 目录下创建绝对布局文件 absolute_layout.xml，编写代码如下：

```
<?xml version="1.0" encoding="utf-8"?>
<LinearLayout xmlns:android="http://schemas.android.com/apk/res/android"
    android:layout_width="match_parent"
    android:layout_height="match_parent"
    android:background="#ededed"
     android:orientation="vertical" >
<AbsoluteLayout
        android:id="@+id/AbsoluteLayout01"
        android:layout_width="wrap_content"
        android:layout_height="wrap_content" >
<Button
            android:id="@+id/Button01"
            android:layout_width="wrap_content"
            android:layout_height="wrap_content"
            android:layout_x="20px"
            android:layout_y="20px"
            android:text="A" />
<Button
            android:id="@+id/Button02"
            android:layout_width="wrap_content"
            android:layout_height="wrap_content"
            android:layout_x="150px"
            android:layout_y="20px"
            android:text="B" />
<Button
            android:id="@+id/Button03"
            android:layout_width="wrap_content"
            android:layout_height="wrap_content"
            android:layout_x="20px"
            android:layout_y="150px"
            android:text="C" />
<Button
```

```
        android:id="@+id/Button04"
        android:layout_width="wrap_content"
        android:layout_height="wrap_content"
        android:layout_x="150px"
        android:layout_y="150px"
        android:text="D" />
    </AbsoluteLayout>
</LinearLayout>
```

上述代码使用<AbsoluteLayout>元素定义绝对布局，该布局中有四个按钮，每个按钮的位置都通过 X、Y 轴坐标进行指定，其中 android:layout_x 属性用于指定 X 轴的坐标，android:layout_y 属性用于指定 Y 轴的坐标。

在 LayoutActivity 中设置应用 absolute_layout.xml 布局文件，代码如下：

```
setContentView(R.layout.absolute_layout);
```

运行项目，效果如图 3-8 所示。

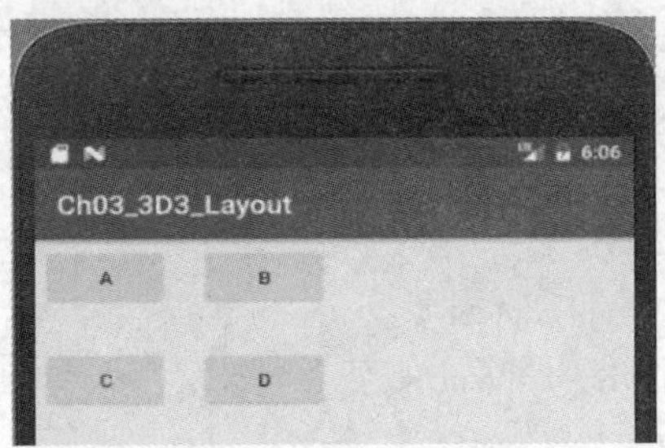

图 3-8　绝对布局

3.3.5　框架布局(FrameLayout)

FrameLayout 以层叠的方式显示子组件，后面加入的组件会覆盖前面加入的组件。

【示例 3.7】 在项目 Ch03_3D3_Layout 中，编写代码演示框架布局 FrameLayout 的使用。

在项目的 res/layout 目录下创建框架布局文件 frame_layout.xml，编写代码如下：

```
<FrameLayout xmlns:android="http://schemas.android.com/apk/res/android"
    android:layout_width="match_parent"
    android:layout_height="match_parent" >
<ImageView
        android:layout_width="wrap_content"
        android:layout_height="wrap_content"
        android:background="@mipmap/flower" />
<ImageView
        android:layout_width="wrap_content"
```

```
        android:layout_height="wrap_content"
        android:layout_gravity="center_vertical"
        android:background="@mipmap/ic_launcher" />
</FrameLayout>
```

上述代码使用<FrameLayout>元素定义框架布局，该布局中有两个图片控件，首先添加的是名称为“flower”的图片，后添加的是系统图标图片且垂直居中。

在 LayoutActivity 中加载 frame_layout.xml 布局文件，代码如下：

```
setContentView(R.layout.frame_layout);
```

运行项目，结果如图 3-9 所示：可以看到系统图标图片覆盖到了图片“flower”上面。

图 3-9　框架布局

3.3.6　网格布局(GridLayout)

GridLayout 以网格方式布局子组件，使子组件的 X、Y 轴自动对齐。

【示例 3.8】　在项目 Ch03_3D3_Layout 中，编写代码演示网格布局 GridLayout 的使用。

在项目的 res/layout 目录下创建布局文件 grid_layout.xml，编写代码如下：

```
<GridLayout xmlns:android="http://schemas.android.com/apk/res/android"
    android:layout_width="wrap_content"
    android:layout_height="wrap_content"
    android:columnCount="4"
    android:orientation="horizontal"
```

```
    android:rowCount="3" >
<Button
        android:layout_columnSpan="2"
        android:layout_gravity="fill"
        android:text="1.1" />
<Button android:text="1.2" />
<Button
        android:layout_gravity="fill"
        android:layout_rowSpan="2"
        android:text="1.3" />
<Button
        android:layout_columnSpan="3"
        android:layout_gravity="fill"
        android:text="2.1" />
<Button android:text="3.1" />
<Button android:text="3.2" />
<Button android:text="3.3" />
<Button android:text="3.4" />
</GridLayout>
```

上述代码使用<GridLayout>元素定义网格布局，该布局设置为 4 列 3 行，其中第一行第一列跨了 2 列，之后第四列跨了 2 行，第二行第一列跨了 3 列。

在 LayoutActivity 中加载 grid_layout.xml 布局文件，代码如下：

```
setContentView(R.layout.grid_layout);
```

运行项目，效果如图 3-10 所示。

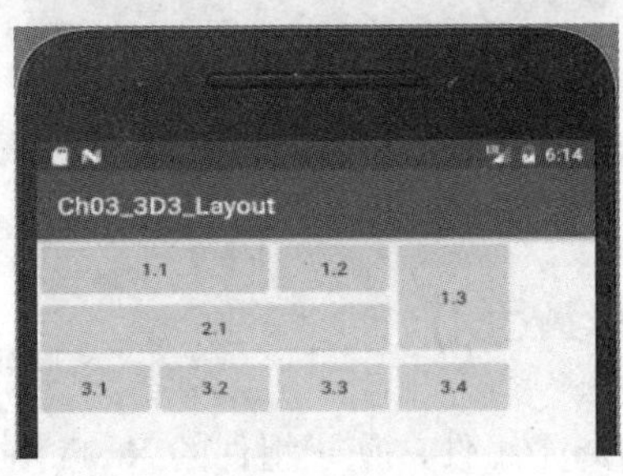

图 3-10　网格布局

3.4　提示信息(Toast)和对话框

在 Android 应用实际开发中，当用户完成某一操作时，为了提升用户体验，需要适当地使用一些话术提示用户，例如“再按一次退出程序”，实现此类功能常用到 Toast 和对话框，下面将分别讲解。

3.4.1　提示信息(Toast)

提示信息(Toast)是 Android 中用来显示提示信息的一种机制，与对话框不同，Toast 是没有焦点的，而且 Toast 显示时间有限，过一定的时间会自动消失。

Toast 类定义在 android.widget 包中，其常用的方法如表 3-4 所示。

表 3-4　Toast 常用方法

方　法	功 能 说 明
Toast(Context context)	构造函数
setDuration(int duration)	设置提示信息显示的时长，可以设置两种值：Toast.LENGTH_LONG 和 Toast.LENGTH_SHORT
setText(CharSequence s)	设置显示的文本
cancel()	关闭提示信息，即不显示
makeText(Context context, CharSequence text, int duration)	静态方法，用于直接创建一个带文本的提示信息，并指明时长
show()	显示提示信息

Toast 的创建步骤如下：

(1) 调用 Toast 的静态方法 makeText()创建一个指定文本和时长的提示信息。

(2) 调用 Toast 的 show()方法显示提示信息。

【示例 3.9】 在 Android Studio 中新建项目 Ch03_3D4_Toast，演示 Toast 的创建及显示方法。

在 ToastActivity 中编写如下代码：

```
public class ToastActivity extends AppCompatActivity {

    private Button mBtnLong;
    private Button mBtnShort;

    @Override
    protected void onCreate(Bundle savedInstanceState) {
        super.onCreate(savedInstanceState);
        setContentView(R.layout.activity_toast);
        initView();
    }

    private void initView() {
        mBtnLong = (Button) findViewById(R.id.btn_long);
        mBtnShort =(Button) findViewById(R.id.btn_short);
        //在按钮【长】上设置监听
        mBtnLong.setOnClickListener(new View.OnClickListener() {
            @Override
```

```
        public void onClick(View view) {
            //Toast.LENGTH_LONG 表示显示的时间较长
            Toast.makeText(ToastActivity.this,"我多显示一会",Toast.LENGTH_LONG).show();
        }
    });
    //在按钮【短】上设置监听
    mBtnShort.setOnClickListener(new View.OnClickListener() {
        @Override
        public void onClick(View view) {
            //Toast.LENGTH_SHORT 表示显示的时间较短
            Toast.makeText(ToastActivity.this,"我少显示一会",Toast.LENGTH_SHORT).show();
        }
    });
}
```

上述代码设置了【长】、【短】两个按钮。运行项目，效果如图 3-11 所示：当单击按钮【长】时，会显示一个时间较长的 Toast；而单击按钮【短】时，会显示一个时间较短的 Toast。

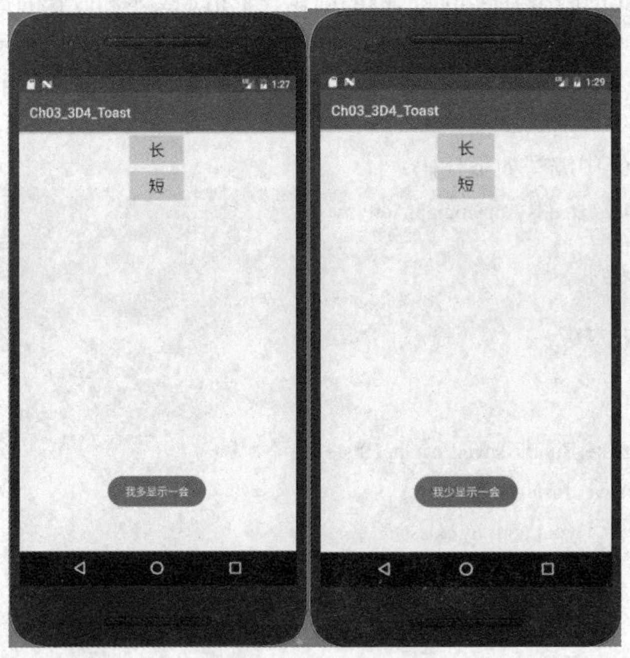

图 3-11　演示 Toast 功能

3.4.2　对话框

对话框是程序运行时的弹出窗口。例如，用户要删除一个联系方式时，就会弹出一个确认对话框。Android 系统中提供了四种对话框，如表 3-5 所示。

表 3-5　Android 提供的四种对话框

对　话　框	说　　明
AlertDialog	提示对话框
ProgressDialog	进度对话框
DatePickerDialog	日期选择对话框
TimePickerDialog	时间选择对话框

除了上面的四种系统定义的对话框外，用户还可以继承 android.app.Dialog，实现自己的对话框。

1. AlertDialog

AlertDialog 是一个提示对话框，可以要求用户做出选择，使用步骤如下：

(1) 获得 AlertDialog 的静态内部类 Builder，由该类来创建对话框。

(2) 通过 Builder 对象设置对话框的标题、按钮以及按钮将要响应的事件。

(3) 调用 Builder 对象的 create()方法创建对话框。

(4) 调用 Builder 对象的 show()方法显示对话框。

【示例 3.10】 在 Android Studio 中新建项目 Ch03_3D5_Dialog，演示 AlertDialog 的使用方法。

在布局文件 activity_alert_dialog.xml 中编写如下代码：

```
<?xml version="1.0" encoding="utf-8"?>
<LinearLayout xmlns:android="http://schemas.android.com/apk/res/android"
    xmlns:tools="http://schemas.android.com/tools"
    android:layout_width="match_parent"
    android:layout_height="match_parent"
    tools:context="com.ugrow.ch03_3d5_dialog.AlertDialogActivity">
     <TextView
        android:id="@+id/tv"
        android:layout_width="wrap_content"
        android:layout_height="wrap_content"
        android:layout_marginLeft=8dp
        android:text="取消删除"/>

    <Button
        android:id="@+id/btn_delete"
        android:layout_width="wrap_content"
        android:layout_height="wrap_content"
        android:layout_marginLeft=16sp
        android:text="删除"/>
```

```
</ LinearLayout >
```

在 AlertDialogActivity 中编写如下代码：

```
public class AlertDialogActivity extends AppCompatActivity {

    private TextView mTv;
    private Button mBtnDelete;

    @Override
    protected void onCreate(Bundle savedInstanceState) {
        super.onCreate(savedInstanceState);
        setContentView(R.layout.activity_alert_dialog);
        initView();
    }

    private void initView() {
        mTv = (TextView) findViewById(R.id.tv);
        mBtnDelete = (Button) findViewById(R.id.btn_delete);
        mBtnDelete.setOnClickListener(new View.OnClickListener() {
            @Override
            public void onClick(View view) {
                AlertDialog.Builder   builder = new AlertDialog.Builder(AlertDialogActivity.this);
                builder.setMessage("真的要删除该记录么？");
                builder.setPositiveButton("是", new DialogInterface.OnClickListener() {
                    @Override
                    public void onClick(DialogInterface dialogInterface, int i) {
                        mTv.setText("删除成功！");
                    }
                });
                builder.setNegativeButton("否", new DialogInterface.OnClickListener() {
                    @Override
                    public void onClick(DialogInterface dialogInterface, int i) {
                        mTv.setText("取消删除！");
                    }
                });
                builder.create().show();
            }
        });
    }
}
```

上述代码实现了提示对话框功能。运行项目，单击【删除】按钮，效果如图 3-12 所示：当单击提示对话框中的【是】按钮，则文本视图中会显示字符“删除成功！”；如果单击【否】按钮，则文本视图中会显示字符“取消删除！”。

图 3-12　提示对话框

2．ProgressDialog

ProgressDialog 是一种在当前界面弹出的一个置于所有界面元素之上的对话框，用于提示用户当前操作正在运行，让用户等待(例如请求数据时的加载动画)。

进度对话框 ProgressDialog 的使用步骤如下：

(1) 创建一个 ProgressDialog 实例。

(2) 通过 setCancelable()方法设置该对话框是否可以按返回键取消。

(3) 通过 setCanceledOnTouchOutside()方法设置在单击该对话框之外的区域时是否可以取消对话框的进度条。

(4) 通过 setIcon()方法设置对话框的图标。

(5) 通过 setTitle()方法设置对话框的标题。

(6) 通过 setMessage()方法设置对话框的提示信息。

(7) 通过调用该实例的 show()方法显示对话框。

【示例 3.11】 在项目 Ch03_3D5_Dialog 中新建 ProgressDialogActivity 类，演示 ProgressDialog 的使用方法。

在布局文件 activity_progress_dialog.xml 中编写代码如下：

```
<?xml version="1.0" encoding="utf-8"?>
<RelativeLayout xmlns:android="http://schemas.android.com/apk/res/android"
```

```
    xmlns:tools="http://schemas.android.com/tools"
    android:layout_width="match_parent"
    android:layout_height="match_parent"
    tools:context="com.ugrow.ch03_3d5_dialog.ProgressDialogActivity">

    <Button
        android:id="@+id/btn_progressDialog"
        android:layout_width="wrap_content"
        android:layout_height="wrap_content"
        android:layout_centerHorizontal="true"
        android:text="进度对话框"/>

</RelativeLayout>
```

在 ProgressDialogActivity 中编写代码如下：

```
public class ProgressDialogActivity extends AppCompatActivity {

    private Button mBtnProgress;

    @Override
    protected void onCreate(Bundle savedInstanceState) {
        super.onCreate(savedInstanceState);
        setContentView(R.layout.activity_progress_dialog);
        mBtnProgress = (Button) findViewById(R.id.btn_progressDialog);
        mBtnProgress.setOnClickListener(new View.OnClickListener() {
            @Override
            public void onClick(View view) {
                ProgressDialog dialog = new ProgressDialog(ProgressDialogActivity.this);
                dialog.setCancelable(true);// 设置是否可以通过点击返回键取消
                dialog.setCanceledOnTouchOutside(false);// 设置在单击对话框之外的区域是否取消
                                                        Dialog 进度条
                dialog.setIcon(R.mipmap.ic_launcher);//设置图片
                dialog.setTitle("提示");//设置标题
                dialog.setMessage("正在加载中，请稍后……");//设置消息
                dialog.show();//显示
            }
        });
    }
}
```

运行项目，单击【进度对话框】按钮，效果如图 3-13 所示。

图 3-13　进度对话框

3．DatePickerDialog

DatePickerDialog 是 Android 提供的一种日期选择对话框，使用步骤如下：

(1)　创建一个 DatePickerDialog 实例。

(2)　通过 setOnDateSetListener 给该实例设置日期监听事件，重写 onDateSet()方法。

【示例 3.12】　在项目 Ch03_3D5_Dialog 中，新建 DatePickerDialogActivity 类，演示 DatePickerDialog 的使用方法。

在布局文件 activity_date_picker_dialog.xml 中编写代码如下：

```
<?xml version="1.0" encoding="utf-8"?>
<RelativeLayout xmlns:android="http://schemas.android.com/apk/res/android"
    xmlns:tools="http://schemas.android.com/tools"
    android:layout_width="match_parent"
    android:layout_height="match_parent"
    tools:context="com.ugrow.ch03_3d5_dialog.DatePickerDialogActivity">

    <TextView
        android:id="@+id/tv_date"
        android:layout_width="wrap_content"
        android:layout_height="wrap_content"
        android:layout_centerHorizontal="true"
        android:text="选择日期"
        android:textSize="18sp"/>
    <Button
```

```
        android:id="@+id/btn_date"
        android:layout_width="wrap_content"
        android:layout_height="wrap_content"
        android:layout_centerHorizontal="true"
        android:layout_below="@+id/tv_date"
        android:layout_marginTop="24dp"
        android:text="日期对话框"/>
</RelativeLayout>
```

在 DatePickerDialogActivity 中编写代码如下：

```
public class DatePickerDialogActivity extends AppCompatActivity {

    private TextView mTvDate;
    private Button mBtnDate;
    private Calendar calendar;

    @Override
    protected void onCreate(Bundle savedInstanceState) {
        super.onCreate(savedInstanceState);
        setContentView(R.layout.activity_date_picker_dialog);
        mTvDate =(TextView) findViewById(R.id.tv_date);
        mBtnDate =(Button) findViewById(R.id.btn_date);
        calendar = Calendar.getInstance();
        mBtnDate.setOnClickListener(new View.OnClickListener() {
            @Override
            public void onClick(View view) {
                DatePickerDialog dialog = new DatePickerDialog(DatePickerDialogActivity.this);
                dialog.setOnDateSetListener(new DatePickerDialog.OnDateSetListener() {
                    @Override
                    public void onDateSet(DatePicker datePicker, int year, int monthOfYear, int dayOfMonth) {
                        mTvDate.setText(year+"年"+(monthOfYear+1)+"月"+dayOfMonth+"日");
                    }
                });
                dialog.show();
            }
        });
    }
}
```

上述代码中的 onDateSet()方法有四个参数，分别代表 DatePicker 的对象、年、月、日。注意：月参数(monthOfYear)的值比当前月少 1，需要在其基础上加 1 才能显示当前月份。

运行项目，单击【选择日期】按钮，会弹出日期选择框，单击其中的【确定】按钮，选择的日期就会在文本控件 TextView 上显示，如图 3-14 所示。

图 3-14　日期选择对话框

4. TimePickerDialog

TimePickerDialog 是 Android 提供的一种时间选择对话框，使用步骤如下：

(1) 通过 Calendar.getInstance()方法创建一个 Calendar 日历类对象。

(2) 通过该对象的 get(Calendar.HOUR_OF_DAY)方法和 get(Calendar.MINUTE)方法分别获取当天的小时和分钟。

(3) 创建一个 TimePickerDialog 实例。

(4) 调用该实例的 show()方法显示对话框。

【示例 3.13】 在项目 Ch03_3D5_Dialog 中，新建 TimePickerDialogActivity 类，演示 TimePickerDialog 的使用方法。

在布局文件 activity_time_picker.xml 中编写代码如下：

```
<?xml version="1.0" encoding="utf-8"?>
<RelativeLayout xmlns:android="http://schemas.android.com/apk/res/android"
    xmlns:tools="http://schemas.android.com/tools"
    android:layout_width="match_parent"
    android:layout_height="match_parent"
    tools:context="com.ugrow.ch03_3d5_dialog.TimePickerDialogActivity">
    <TextView
        android:id="@+id/tv_time"
        android:layout_width="wrap_content"
        android:layout_height="wrap_content"
        android:layout_centerHorizontal="true"
        android:text="选择时间"
```

```
        android:textSize="18sp"/>
    <Button
        android:id="@+id/btn_time"
        android:layout_width="wrap_content"
        android:layout_height="wrap_content"
        android:layout_centerHorizontal="true"
        android:layout_below="@+id/tv_time"
        android:layout_marginTop="24dp"
        android:text="时间对话框"/>
</RelativeLayout>
```

在 TimePickerDialogActivity 中编写代码如下：

```
public class TimePickerDialogActivity extends AppCompatActivity {

    private Button mBtnTime;
    private TextView mTvTime;
    private Calendar c;
    @Override
    protected void onCreate(Bundle savedInstanceState) {
        super.onCreate(savedInstanceState);
        setContentView(R.layout.activity_time_picker_dialog);
        mTvTime = (TextView) findViewById(R.id.tv_time);
        mBtnTime = (Button) findViewById(R.id.btn_time);
        //声明一个日历类的对象
        c = Calendar.getInstance();
        mBtnTime.setOnClickListener(new View.OnClickListener() {
            @Override
            public void onClick(View view) {
                /**
                 * TimePickerDialog 方法有五个参数：
                 * 第一个参数指定上下文对象；
                 * 第二个参数获取时间；
                 * 第三个参数(hour)和第四个参数(minute)为弹出的时间对话框初始显示的小时和分钟；
                 * 第五个参数为时间格式设置参数，值为 true 代表以 24 小时制显示时间。
                 */
                c.setTimeInMillis(System.currentTimeMillis());
                int hour = c.get(Calendar.HOUR_OF_DAY);
                int minute = c.get(Calendar.MINUTE);
                TimePickerDialog timeDialog = new TimePickerDialog(TimePickerDialogActivity.this,
                        new TimePickerDialog.OnTimeSetListener() {
```

```
                    @Override
                    public void onTimeSet(TimePicker view, int hourOfDay,
                                          int minute) {
                        c.setTimeInMillis(System.currentTimeMillis());
                        c.set(Calendar.HOUR_OF_DAY,hourOfDay);
                        c.set(Calendar.MINUTE, minute);
                        c.set(Calendar.SECOND, 0);
                        c.set(Calendar.MILLISECOND, 0);
                        SimpleDateFormat format = new SimpleDateFormat("yyyy 年 MM 月 dd 日
HH:mm");

                        mTvTime.setText(format.format(c.getTime()));
                    }
                }, hour, minute, true);
            //设置按钮的监听事件，注意第三个参数为 timeDialog 对象
            timeDialog.setButton(TimePickerDialog.BUTTON_POSITIVE,"确定",timeDialog);
            timeDialog.show();
        }
    });
}
}
```

上述代码通过 Calendar.getInstance()方法得到一个日历对象，并通过日历对象获取当前的时间。

运行项目，单击【选择时间】按钮，会弹出时间选择对话框，单击其中的【确定】按钮，选择的时间就会在文本控件 TextView 上显示，如图 3-15 所示。

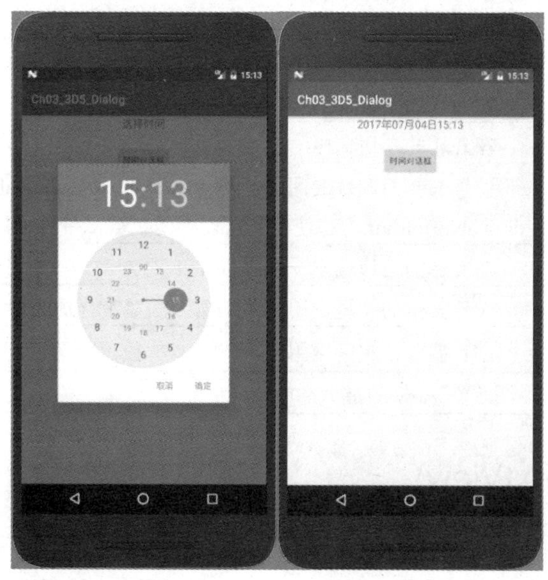

图 3-15　时间选择对话框

3.5 常用 Widget 组件

Widget 组件是窗体中使用的部件，都定义在 android.widget 包中，如 Button、TextView、EditText、CheckBox、RadioGroup、Spinner 等。

3.5.1 Widget 组件通用属性

对 Widget 组件进行 UI 设计时可以采用两种方式：xml 布局文件和 Java 代码。其中 xml 布局文件方式由于简单易用而被广泛使用。

所有的 Widget 组件几乎都属于 View 类，有些属性在这些组件之间是通用的，如表 3-6 所示。

表 3-6 Widget 组件通用属性

属性名称	描　述
android:id	设置控件的索引，Java 程序可通过 R.id.<索引>引用该控件
android:layout_height	设置布局高度，可以通过三种方式来指定高度：match_parent(和父元素相同)、wrap_content(随组件本身的内容调整)、通过指定 px 值来设置高度
android:layout_width	设置布局宽度，也可以采用三种方式：match_parent、wrap_content、指定 px 值
android:autoLink	设置当文本为 URL 链接时，文本是否显示为可单击的链接。可选值为 none/web/email/phone/map/all
android:autoText	如果设置，将自动执行输入值的拼写纠正
android:bufferType	设置 getText()方式取得的文本类别
android:capitalize	设置英文字母大写类型。需要弹出输入法才能看到
android:cursorVisible	设置光标为显示/隐藏，默认为显示
android:digits	设置允许输入哪些字符，如“1234567890.+-*/%\n()”
android:drawableBottom	在 text 的下方输出一个 drawable
android:drawableLeft	在 text 的左边输出一个 drawable
android:drawablePadding	设置 text 与图片的间隔，与 drawableLeft、drawableRight、drawableTop、drawableBottom 一起使用，可设置为负数，单独使用没有效果
android:drawableRight	在 text 的右边输出一个 drawable 对象
android:inputType	设置文本的类型，用于帮助输入法显示合适的键盘类型
android:cropToPadding	是否截取指定区域用空白代替；单独设置无效，需要与 scrollY 一起使用
android:maxHeight	设置 View 的最大高度

3.5.2 文本框(TextView)

TextView 即文本框控件，是屏幕中一块用于显示文本的区域。TextView 是一个不可编辑的文本框，常用来在屏幕中显示静态字符串，其功能类似于 Java 语言中 swing 包的 JLabel 组件。

TextView 类属于 android.widget 包且继承 android.view.View 类的方法和属性，同时又是 Button、CheckedTextView、Chronometer、DigitaClock 以及 EditText 的父类。TextView 类定义了文本框操作的基本方法，常用方法如表 3-7 所示。

表 3-7　TextView 类常用方法

方　　法	功 能 描 述
TextView()	TextView 的构造方法
getDefaultMovementMethod()	获取默认的箭头按键移动方式
getText()	取得文本内容
length()	获取 TextView 中文本长度
getEditableText()	取得文本的可编辑对象，通过该对象可对 TextView 的文本进行操作，如在光标之后插入字符
getLayout()	获取 TextView 的布局
getKeyListener()	获取键盘监听对象
setKeyListener()	设置键盘事件监听
setTransformationMethod()	设置文本是否显示为特殊字符
getCompoundPaddingBottom()	返回 TextView 的底部填充物
setCompoundDrawables()	设置 Drawable 图像显示的位置，在设置该 Drawable 资源之前需要调用 setBounds(Rect)
setCompoundDrawablesWithIntrinsicBounds()	设置 Drawable 图像显示的位置，但其边界不变
setPadding()	根据位置设置填充物
getAutoLinkMask()	返回自动链接的掩码
setTextColor()	设置文本显示的颜色
setHighlightColor()	设置选中时文本显示的颜色
setShadowLayer()	设置文本显示的阴影颜色
setHintTextColor()	设置提示文字的颜色
setLinkTextColor()	设置链接文本的颜色
setGravity()	设置当 TextView 超出了文本本身时，文本横向以及垂直对齐的方式

3.5.3　按钮(Button)

Button 即按钮控件，是最常用的控件之一。Button 类是 TextView 的子类，其常用方法如表 3-8 所示。

表 3-8　Button 常用方法

方　　法	功 能 描 述
getText()	获取按钮的文本内容
setText(CharSequence text)	设置按钮的文本内容
setOnClickListener()	对按钮的单击时间进行监听，为回调方法

3.5.4 编辑框(EditText)

EditText 即编辑框控件，其功能与 TextView 基本类似，主要区别是 EditText 可以编辑。EditText 是用户与系统之间的文本输入接口，可以把用户输入的数据传给系统，或从系统中获取用户需要的数据。

EditText 类是 TextView 类的子类，其提供了多种用于设置和控制文本框功能的方法，如表 3-9 所示。

表 3-9 EditText 常用方法

方 法	功 能 描 述
getText()	获取文本内容
selectAll()	获取输入的所有文本
setText(CharSequence text,TextView.BufferType type)	设置编辑框中的文本内容

3.5.5 复选框(CheckBox)

CheckBox 即复选框控件，在应用程序中为用户提供多项选择功能。

CheckBox 类提供了多种用于设置和控制复选框功能的方法，如表 3-10 所示。

表 3-10 CheckBox 常用方法

方 法	功 能 描 述
dispatchPopulateAccessibilityEvent()	在子视图创建时，分派一个辅助事件
isChecked()	判断组件状态是否勾选
onRestoreInstanceState()	设置视图恢复以前的状态，该状态由 onSaveInstanceState()方法生成
performClick()	执行 Click 动作，该动作会触发事件监听器
setButtonDrawable()	根据 Drawable 对象设置组件的背景
setChecked()	设置组件的状态。若参数为真，则设置组件为选中状态；否则设置组件为未选中状态
setOnCheckedChangeListener()	CheckBox 常用的设置事件监听器的方法，状态改变时调用该监听器
toggle()	改变按钮的当前状态
drawableStateChanged()	视图状态的变化影响到所显示可绘制的状态时调用该方法
onCreateDrawableState()	获取文本框为空时，文本框默认显示的字符串
onCreateDrawableState()	为当前视图生成新的 Drawable 状态

复选框是一种双状态按钮的特殊类型，复选框的状态只有两种：选中或者未选中状态，因此复选框的状态变化包含两种情况：

(1) 复选框由选中状态变成未选中状态。

(2) 复选框由未选中状态变成选中状态。

单击复选框，可触发复选框状态的改变。通过 setOnCheckedChangeListener()方法，可

以注册复选框组件状态改变监听器 OnCheckedChangeListener。

注 意　复选框状态彼此独立，可同时选择任意多个 CheckBox。

3.5.6　单选按钮组(RadioGroup)

RadioGroup 即单选按钮组控件，用于实现一组 RadioButton(单选按钮)，即有且仅有一个单选按钮被选中，在同一个单选按钮组中勾选一个单选按钮，则会取消该组中其他已经勾选的按钮的选中状态。

RadioGroup 类是 LinearLayout 类的子类，其提供了多种设置和控制单选按钮组功能的方法，如表 3-11 所示。

表 3-11　RadioGroup 常用方法

方 法	功 能 描 述
addView()	根据布局指定的属性添加一个子视图
check()	当传递-1 作为指定的选择标识符时，此方法同 clearCheck()方法作用等效
generateLayoutParams()	返回一个新的布局实例，这个实例是根据指定的属性集合生成的
setOnCheckedChangeListener()	注册单选按钮状态改变监听器
getCheckedRadioButtonId()	返回该单选按钮组中所选择的单选按钮的标识 ID

3.5.7　下拉列表(Spinner)

Spinner 即下拉列表控件，其功能类似于 RadioGroup，一个 Spinner 由多个 item 子元素组合而成，这些子元素之间相互影响，同时最多有一个子元素被选中。

Spinner 类是 LinearLayout 类的子类，其常用的方法如表 3-12 所示。

表 3-12　Spinner 常用方法

方 法	功 能 描 述
getBaseline()	获取组件文本基线
getPrompt()	获取被聚焦时的提示消息
performClick()	效果同单击一样，会触发监听事件 OnClickListener
setAdapter(SpinnerAdapter adapter)	设置选项，适配器(Adapter)用于给下拉列表提供选项数据
setPromptId()	设置对话框弹出的时候显示的文本
setOnItemSelectedListener()	设置下拉列表 item 被选中时触发的监听事件

Spinner 可以通过数组适配器读取 XML 中定义的子元素，这种设计方式被称为“适配器模式”。适配器模式建议定义一个包装类，包装有不兼容接口的对象，该包装类就是适配器(Adapter)，其包装的对象就是适配者(Adaptee)。适配器提供客户类需要的接口，适配器接口的实现就是把客户类的请求转化为对适配者的相应接口的调用，因此，适配器可以

使由于接口不兼容而不能交互的类一起工作。

例如，使用以下代码，可以创建一个下拉列表：

```
//获取下拉列表组件
Spinner position = (Spinner) findViewById(R.id.position);
// 创建一个下拉列表选项数组
String[] strs={"总裁","经理","秘书"};
//创建一个数组适配器
ArrayAdapter aa = new ArrayAdapter(this,
        android.R.layout.simple_spinner_dropdown_item, strs);
//设置下拉列表的适配器
position.setAdapter(aa);
```

【示例 3.14】 在 Android Studio 中新建项目 Ch03_3D6_Regist，通过注册界面演示 TextView、EditText、CheckBox、RadioGroup、Spinner 控件的使用方法。

首先编写布局文件 activity_main.xml，代码如下：

```
<?xml version="1.0" encoding="utf-8"?>
<LinearLayout xmlns:android="http://schemas.android.com/apk/res/android"
    xmlns:tools="http://schemas.android.com/tools"
    android:layout_width="match_parent"
    android:layout_height="match_parent"
    android:orientation="vertical"
    android:layout_margin="8dp"
    tools:context="com.ugrow.ch03_3d6_regist.MainActivity">
    <LinearLayout
        android:layout_width="match_parent"
        android:layout_height="40dp">
        <TextView
            android:layout_width="wrap_content"
            android:layout_height="match_parent"
            android:gravity="center_vertical"
            android:text="用户名称"
            android:textSize="14sp"
            android:textColor="#000"/>
        <EditText
            android:id="@+id/et_name"
            android:layout_width="match_parent"
            android:layout_height="match_parent"
            android:layout_marginLeft="8dp"
            android:hint="请输入您的用户名"
            android:textSize="14sp"
            android:textColor="#000"/>
```

```
</LinearLayout>
<LinearLayout
    android:layout_width="match_parent"
    android:layout_height="40dp">
    <TextView
        android:layout_width="wrap_content"
        android:layout_height="match_parent"
        android:gravity="center_vertical"
        android:text="用户密码"
        android:textSize="14sp"
        android:textColor="#000"/>
    <EditText
        android:id="@+id/et_code"
        android:layout_width="match_parent"
        android:layout_height="match_parent"
        android:layout_marginLeft="8dp"
        android:hint="请输入您的密码"
        android:inputType="textPassword"
        android:textSize="14sp"
        android:textColor="#000"/>
</LinearLayout>
<LinearLayout
    android:layout_width="match_parent"
    android:layout_height="wrap_content"
    android:layout_marginTop="8dp">
    <TextView
        android:layout_width="wrap_content"
        android:layout_height="wrap_content"
        android:text="性别        "
        android:textSize="14sp"
        android:textColor="#000"/>
    <RadioGroup
        android:layout_width="wrap_content"
        android:layout_height="wrap_content"
        android:layout_marginLeft="8dp">
        <RadioButton
            android:id="@+id/rb_man"
            android:layout_width="wrap_content"
            android:layout_height="wrap_content"
            android:text="男"/>
```

```
            <RadioButton
                android:id="@+id/rb_woman"
                android:layout_width="wrap_content"
                android:layout_height="wrap_content"
                android:text="女"/>
        </RadioGroup>
    </LinearLayout>
    <LinearLayout
        android:layout_width="match_parent"
        android:layout_height="40dp"
        android:layout_marginTop="8dp">
        <TextView
            android:layout_width="wrap_content"
            android:layout_height="match_parent"
            android:gravity="center_vertical"
            android:text="婚否          "
            android:textSize="14sp"
            android:textColor="#000"/>
        <ToggleButton
            android:id="@+id/tb_marry"
            android:layout_width="match_parent"
            android:layout_height="match_parent"
            android:layout_marginLeft="8dp"
            android:textColor="#000"/>
    </LinearLayout>
    <LinearLayout
        android:layout_width="match_parent"
        android:layout_height="40dp"
        android:layout_marginTop="8dp">
        <TextView
            android:layout_width="wrap_content"
            android:layout_height="match_parent"
            android:gravity="center_vertical"
            android:text="爱好          "
            android:textSize="14sp"
            android:textColor="#000"/>
        <CheckBox
            android:id="@+id/cb_read"
            android:layout_width="wrap_content"
            android:layout_height="wrap_content"
```

```
            android:layout_marginLeft="8dp"
            android:text="阅读"/>
        <CheckBox
            android:id="@+id/cb_swim"
            android:layout_width="wrap_content"
            android:layout_height="wrap_content"
            android:layout_marginLeft="8dp"
            android:text="游泳"/>
    </LinearLayout>
    <LinearLayout
        android:layout_width="match_parent"
        android:layout_height="40dp"
        android:layout_marginTop="8dp">
        <TextView
            android:layout_width="wrap_content"
            android:layout_height="match_parent"
            android:gravity="center_vertical"
            android:text="职务            "
            android:textSize="14sp"
            android:textColor="#000"/>
        <Spinner
            android:id="@+id/sp"
            android:layout_width="match_parent"
            android:layout_height="match_parent"
            android:layout_marginLeft="8dp"
            android:textColor="#000"/>
    </LinearLayout>
    <LinearLayout
        android:layout_width="match_parent"
        android:layout_height="40dp"
        android:layout_marginTop="8dp">
        <Button
            android:id="@+id/btn_cancel"
            android:layout_width="wrap_content"
            android:layout_height="match_parent"
            android:text="取消"
            android:textSize="14sp"
            android:textColor="#000"/>
        <Button
            android:id="@+id/btn_register"
```

```
            android:layout_width="match_parent"
            android:layout_height="match_parent"
            android:layout_marginLeft="8dp"
            android:text="注册"
            android:textColor="#000"/>
    </LinearLayout>
</LinearLayout>
```

上述代码中的<ToggleButton>是一个开关按钮。

然后编写 MainActivity.java 类，代码如下：

```
public class MainActivity extends AppCompatActivity {

    private EditText mEtName;
    private EditText mEtCode;
    private RadioButton mRbMan;
    private RadioButton mRbWoman;
    private ToggleButton mTbMarray;
    private CheckBox mCbRead;
    private CheckBox mCbSwim;
    private Spinner mSp;
    private Button mBtnCancel;
    private Button mBtnRegister;
    //声明一个下拉列表选项数组
    private String[] strs = {"CEO", "PM", "PL"};
    private ArrayAdapter arrayAdapter;

    @Override
    protected void onCreate(Bundle savedInstanceState) {
        super.onCreate(savedInstanceState);
        setContentView(R.layout.activity_main);
        initView();
    }

    private void initView() {
        //根据 ID 获取组件对象
        mEtName = (EditText) findViewById(R.id.et_name);
        mEtCode = (EditText) findViewById(R.id.et_code);
        mRbMan = (RadioButton) findViewById(R.id.rb_man);
        mRbWoman = (RadioButton) findViewById(R.id.rb_woman);
        mTbMarray = (ToggleButton) findViewById(R.id.tb_marry);
        mCbRead = (CheckBox) findViewById(R.id.cb_read);
```

```
        mCbSwim = (CheckBox) findViewById(R.id.cb_swim);
        mSp = (Spinner) findViewById(R.id.sp);
        mBtnCancel = (Button) findViewById(R.id.btn_cancel);
        mBtnRegister = (Button) findViewById(R.id.btn_register);
        arrayAdapter = new ArrayAdapter(this, android.R.layout.simple_spinner_dropdown_item, strs);
        mSp.setAdapter(arrayAdapter);
        //注册监听
        mBtnRegister.setOnClickListener(new View.OnClickListener() {
            @Override
            public void onClick(View view) {
                Log.i("TAG", "username：" + mEtName.getText().toString());
                Log.i("TAG", "password：" + mEtCode.getText().toString());
                if (mRbMan.isChecked()) {
                    Log.i("TAG", "sex：男");
                } else {
                    Log.i("TAG", "sex：女");
                }
                String temp = "like：";
                if (mCbRead.isChecked()) {
                    temp += "read";
                }
                if (mCbSwim.isChecked()){
                    temp += "swim";
                }
                Log.i("TAG",temp);
                if (mTbMarray.isChecked()){
                    Log.i("TAG","marriged：Yes");
                }else {
                    Log.i("TAG","marriged：No");
                }
                Log.i("TAG","position："+mSp.getSelectedItem().toString());
            }
        });
    }
}
```

上述代码实现了一个注册界面，当单击界面中的【注册】按钮时，会在 LogCat 窗口输出用户注册信息。

运行项目，效果如图 3-16 所示。

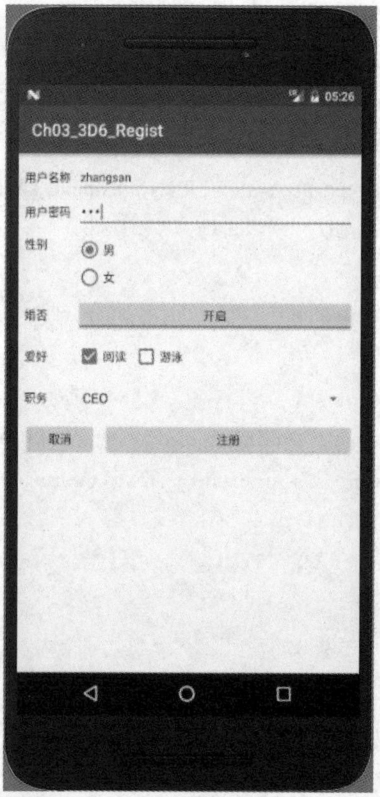

图 3-16　注册界面

单击界面中的【注册】按钮后，Logcat 输出信息如图 3-17 所示。

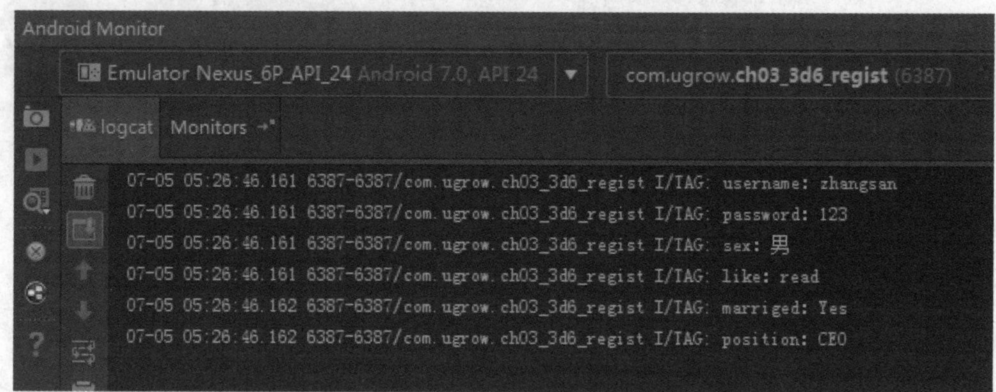

图 3-17　Log 输出信息

3.5.8　图片视图(ImageView)

ImageView 即图片视图控件，与 TextView 控件的功能基本类似，主要区别是可以显示的资源不同：ImageView 可显示图像资源，如图 3-18 所示，而 TextView 只能显示文本资源。

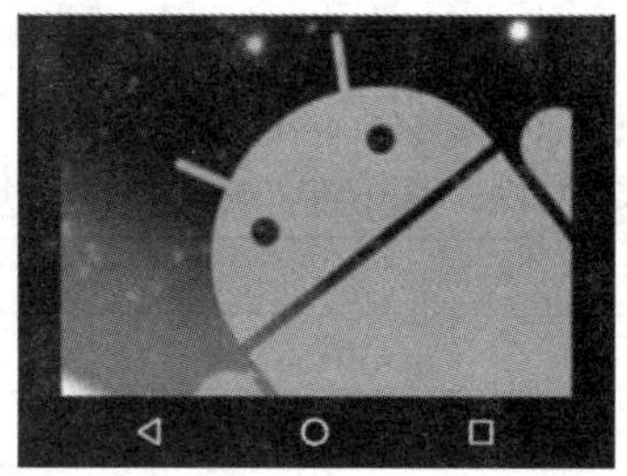

图 3-18　使用 ImageView 显示图片

ImageView 类的常用方法如表 3-13 所示。

表 3-13　ImageView 常用方法

方　法	功 能 描 述
ImageView()	ImageView 构造函数
setAdjustViewBounds(booleanab)	设置是否保持高宽比。需要结合 maxWidth 和 maxHeight 一起使用
getDrawable()	获取 Drawable 对象；若获取成功则返回 Drawable 对象，否则返回 null
getScaleType()	获取视图的填充方式
setImageBitmap(Bitmap bm)	设置位图
setAlpha(int alpha)	设置透明度，值范围为 0～255，其中 0 为完全透明，255 为完全不透明
setMaxHeight(int h)	设置控件的最大高度
setMaxWidth(int w)	设置控件的最大宽度
setImageURI(Uri uri)	设置图片地址，图片地址使用 URI 指定
setImageResource(int rid)	设置图片资源库
setColorFilter(int color)	设置颜色过滤，需要制定颜色过滤矩阵

ImageView 可通过两种方式显示图片资源：

(1) 通过 setImageBitmap()方法显示图片资源。

(2) 通过<ImageView> XML 元素的 android:src 属性，或 setImageResource(int)方法指定显示的图片资源。

使用 ImageView 显示图片的代码如下：

```
imageview = (ImageView)findViewById(R.id.imageview);
bitmap = BitmapFactory.decodeResource(this.getResources(), R.mipmap.motor);
imageview.setImageBitmap(bitmap);
```

3.5.9　滚动视图(ScrollView)

当手机界面上的元素超过屏幕最大的高度时，就需要设置滚动浏览的控件。而 ScrollView 控件提供了滚动浏览功能，可在屏幕上显示更多的内容，用户可通过滑动鼠标来实现 ScrollView 界面的滚动，类似于翻页功能。

ScrollView 的子元素可以包含复杂的布局，通常用到的子元素是垂直方向的 LinearLayout。注意 ScrollView 只支持垂直方向的滚动，不支持水平方向的滚动，且在

ScrollView 中只能包含一个直接的子元素。

ScrollView 的常用方法如表 3-14 所示。

表 3-14　ScrollView 常用方法

方　法	功 能 描 述
ScrollView()	ScrollView 构造函数
dispatchKeyEvent(KeyEvent event)	将参数指定的键盘事件分发给当前焦点的视图
addView (View child)	添加子视图
computeScroll()	更新子视图的值(mScrollX 和 mScrollY)
onTouchEvent (MotionEvent ev)	用于处理触摸屏幕时产生的运动事件
setOnTouchListener()	设置 ImageButton 单击事件监听
setColorFilter()	设置颜色过滤，需要制定颜色过滤矩阵
executeKeyEvent (KeyEvent event)	当接收到键盘事件时执行滚动操作
fullScroll (int direction)	将视图滚动到 direction 指定的方向
onInterceptTouchEvent (MotionEvent me)	用于拦截用户的触屏事件

【示例 3.15】 在 Android Studio 中新建项目 Ch03_3D07_ScrollView，演示 ScrollView 的使用。

在布局文件 activity_main.xml 中编写代码如下：

```
<?xml version="1.0" encoding="utf-8"?>
<LinearLayout xmlns:android="http://schemas.android.com/apk/res/android"
    xmlns:tools="http://schemas.android.com/tools"
    android:layout_width="match_parent"
    android:layout_height="match_parent"
    tools:context="com.ugrow.ch03_3d7_scrollview.MainActivity">

    <ScrollView
        android:layout_width="match_parent"
        android:layout_height="match_parent">

        <LinearLayout
            android:layout_width="match_parent"
            android:layout_height="wrap_content"
            android:orientation="vertical">

            <LinearLayout
                android:layout_width="match_parent"
                android:layout_height="wrap_content"
                android:orientation="vertical">

                <TextView
```

```
            android:layout_width="match_parent"
            android:layout_height="wrap_content"
            android:text="TextView" />

        <Button
            android:layout_width="match_parent"
            android:layout_height="wrap_content"
            android:text="Button" />
    </LinearLayout>

    <LinearLayout
        android:layout_width="match_parent"
        android:layout_height="wrap_content"
        android:orientation="vertical">

        <TextView
            android:layout_width="match_parent"
            android:layout_height="wrap_content"
            android:text="TextView" />

        <Button
            android:layout_width="match_parent"
            android:layout_height="wrap_content"
            android:text="Button" />
    </LinearLayout>

    <LinearLayout
        android:layout_width="match_parent"
        android:layout_height="wrap_content"
        android:orientation="vertical">

        <TextView
            android:layout_width="match_parent"
            android:layout_height="wrap_content"
            android:text="TextView" />

        <Button
            android:layout_width="match_parent"
            android:layout_height="wrap_content"
            android:text="Button" />
```

```
</LinearLayout>

<LinearLayout
    android:layout_width="match_parent"
    android:layout_height="wrap_content"
    android:orientation="vertical">

    <TextView
        android:layout_width="match_parent"
        android:layout_height="wrap_content"
        android:text="TextView" />

    <Button
        android:layout_width="match_parent"
        android:layout_height="wrap_content"
        android:text="Button" />
</LinearLayout>

<LinearLayout
    android:layout_width="match_parent"
    android:layout_height="wrap_content"
    android:orientation="vertical">

    <TextView
        android:layout_width="match_parent"
        android:layout_height="wrap_content"
        android:text="TextView" />

    <Button
        android:layout_width="match_parent"
        android:layout_height="wrap_content"
        android:text="Button" />
</LinearLayout>

<LinearLayout
    android:layout_width="match_parent"
    android:layout_height="wrap_content"
    android:orientation="vertical">

    <TextView
```

```
            android:layout_width="match_parent"
            android:layout_height="wrap_content"
            android:text="TextView" />

        <Button
            android:layout_width="match_parent"
            android:layout_height="wrap_content"
            android:text="Button" />
    </LinearLayout>

    <LinearLayout
        android:layout_width="match_parent"
        android:layout_height="wrap_content"
        android:orientation="vertical">

        <TextView
            android:layout_width="match_parent"
            android:layout_height="wrap_content"
            android:text="TextView" />

        <Button
            android:layout_width="match_parent"
            android:layout_height="wrap_content"
            android:text="Button" />
    </LinearLayout>

    <LinearLayout
        android:layout_width="match_parent"
        android:layout_height="wrap_content"
        android:orientation="vertical">

        <TextView
            android:layout_width="match_parent"
            android:layout_height="wrap_content"
            android:text="TextView" />

        <Button
            android:layout_width="match_parent"
            android:layout_height="wrap_content"
            android:text="Button" />
```

```
            </LinearLayout>
        </LinearLayout>
    </ScrollView>
</LinearLayout>
```

运行项目，效果如图 3-19 所示。

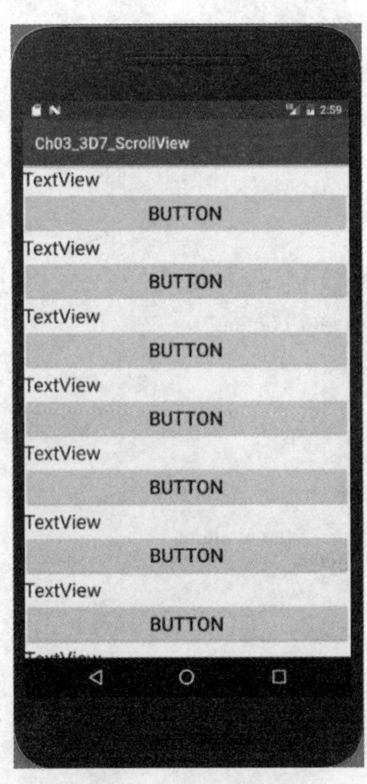

图 3-19 滚动视图

3.5.10 网格视图(GridView)

GridView 即网格视图控件，可以将其子元素组织成类似网格状的视图，视图排列方式与矩阵类似，当屏幕上有很多元素(文字、图片或其他元素)需要显示时，可以使用 GirdView。一个 GirdView 通常需要一个适配器(Adapter)，该适配器包含 GirdView 的子元素组件。

GridView 的常用属性如下：

(1) android:numColums：设置列数。

(2) android:horizontalSpacing：设置两列之间的间距。

(3) android:verticalSpacing：设置两行之间的间距。

(4) android:columWidth：设置列的宽度。

(5) android:StretchMode：设置缩放模式。

网格视图能够以数据网格形式显示子元素，并能够对这些子元素进行分页及自定义样式等操作，其常用的方法如表 3-15 所示。

表 3-15　GridView 常用方法

方　法	功 能 描 述
GridView()	GridView 构造函数
setGravity (int gravity)	设置此组件中的内容在组件中的位置
setColumnWidth(int)	设置网格视图的宽度
getAdapter ()	获取网格视图的适配器 Adapter
setAdapter (ListAdapter adapter)	设置网格视图对应的适配器
setStretchMode(int)	用于设置缩放模式(也可以通过属性 android:stretchMode 进行设置)可以设置为 NO_STRETCH 、STRETCH_SPACING 、STRETCH_SPACING_UNIFOR 或 STRETCH_COLUMN_WIDTH
onKeyMultiple(int keyCode, int repeatCount, KeyEvent event)	设置按键多次时的处理方法。当连续发生多次按键时，该方法会被调用，其中 keyCode 为按键对应的整型值，repeatCount 是按键的次数，event 是按键事件
setSelection(int p)	设置当前被选中的网格视图的子元素
onKeyUp(int keyCode, KeyEvent event)	设置释放按键时的处理方法。释放按键时，该方法会被调用，其中 keyCode 为按键对应的整型值，event 是按键事件
onKeyDown(int keyCode, KeyEvent event)	设置按键时的处理方法。按键时，该方法会被调用，其中 keyCode 为按键对应的整型值，event 是按键事件。注意用户按键的过程中，onKeyDown 先被调用，用户释放按键后调用 onKeyUp
setHorizontalSpacing(int c)	设置网格视图同一行子元素之间的水平间距
setNumColumns(int)	设置网格视图包含的子元素的列数
getHorizontalSpacing()	获取网格视图同一行子元素之间的水平间距
getNumColumns()	获取网格视图包含的子元素的列数
getSelection ()	获取当前被选中的网格视图的子元素

【示例 3.16】 在 Android Studio 中新建项目 Ch03_3D8_GridView，使用 GirdView 实现加载多项图片资源的功能。

在布局文件 activity_main.xml 中编写代码如下：

```
<?xml version="1.0" encoding="utf-8"?>
<LinearLayout xmlns:android="http://schemas.android.com/apk/res/android"
    xmlns:tools="http://schemas.android.com/tools"
    android:layout_width="match_parent"
    android:layout_height="match_parent"
    android:layout_margin="8dp"
    tools:context="com.ugrow.ch03_3d8_gridview.MainActivity">

    <GridView
        android:id="@+id/gv"
        android:layout_width="match_parent"
```

```
        android:layout_height="wrap_content"
        android:horizontalSpacing="24dp"
        android:numColumns="3"
        android:verticalSpacing="24dp" />

</LinearLayout>
```

在 MainActivity 中编写代码如下：

```
public class MainActivity extends AppCompatActivity {

    private GridView mGv;
    //图片资源
    private int[] images = {R.mipmap.a, R.mipmap.b, R.mipmap.c, R.mipmap.d, R.mipmap.e, R.mipmap.f,
R.mipmap.g, R.mipmap.h, R.mipmap.i,};

    @Override
    protected void onCreate(Bundle savedInstanceState) {
        super.onCreate(savedInstanceState);
        setContentView(R.layout.activity_main);
        mGv = (GridView) findViewById(R.id.gv);
        mGv.setAdapter(new MyAdapter());
        //设置监听事件
        mGv.setOnItemClickListener(new AdapterView.OnItemClickListener() {
            @Override
            public void onItemClick(AdapterView<?> adapterView, View view, int position, long l) {
                Toast.makeText(MainActivity.this,"position--->"+position,Toast.LENGTH_SHORT).show();
            }
        });
    }

    class MyAdapter extends BaseAdapter{

        @Override
        public int getCount() {
            return images.length;
        }

        @Override
        public Object getItem(int position) {
            return images[position];
        }
```

```
        @Override
        public long getItemId(int position) {
            return position;
        }

        @Override
        public View getView(int position, View view, ViewGroup viewGroup) {
            ViewHolder holder;
            if (view == null){
                holder = new ViewHolder();
                view = LayoutInflater.from(MainActivity.this).inflate(R.layout.item_gridview,null);
                holder.iv = (ImageView)view.findViewById(R.id.iv);
                view.setTag(holder);
            }else {
                holder= (ViewHolder) view.getTag();
            }
            //加载数据
            holder.iv.setImageResource(images[position]);
            return view;
        }
        class ViewHolder{
            ImageView iv;
        }
    }
}
```

因为 GridView 是适配器控件，所以加载数据时需要使用适配器。使用步骤分为三步：第一步，通过 findViewById()方法初始化控件；第二步，创建数据源和适配器；第三步，给 GridView 设置适配器，适配器需要加载 item 布局，从而设置每一个 item 显示的样式，代码如下：

```
<?xml version="1.0" encoding="utf-8"?>
<LinearLayout xmlns:android="http://schemas.android.com/apk/res/android"
    android:layout_width="match_parent"
    android:layout_height="match_parent"
    android:orientation="vertical">

    <ImageView
        android:id="@+id/iv"
        android:layout_width="wrap_content"
        android:layout_height="wrap_content"
```

```
        android:layout_gravity="center"
        android:scaleType="centerCrop"
        android:src="@mipmap/a" />

</LinearLayout>
```

运行项目，效果如图 3-20 所示：通过 mGv.setOnItemClickListener()方法，给 GridView 设置 item 的单击监听事件来实现相应的逻辑处理，即在单击某一个 item 时，以 Toast 的方式显示 item 的位置。

图 3-20　网格视图

3.5.11　列表视图(ListView)

ListView 即列表视图控件，用于将元素按照条目的方式自上而下列出来。它是 Android 最重要的控件之一，几乎每个 Android 应用都会用到 ListView。

ListView 有两大功能：① 将数据填充到布局中；② 处理用户的单击事件。实现一个列表视图必须具备 ListView、适配器以及子元素三个条件，其中适配器用于存储列表视图的子元素，列表视图则将子元素以列表的方式组织，使用户可通过滑动滚动条来显示界面之外的元素。

ListView 的常用方法如表 3-16 所示。

表 3-16　ListView 常用方法

方　法	功 能 描 述
ListView()	ListView 构造函数
getCheckedItemPosition()	返回当前被选中的子元素的位置
addFooterView (View view)	给列表视图添加脚注，通常脚注位于列表视图的底部，其中参数 View 为要添加脚注的视图
getMaxScrollAmount()	返回列表视图的最大滚动数量
getDividerHeight()	获取子元素之间分隔符的宽度(元素与元素之间的那条线)
setStretchMode(int)	设置缩放模式(也可以通过属性 android:stretchMode 进行设置)，可以设置为 NO_STRETCHSTRETCH_SPACING ， STRETCH_SPACING_UNIFOR 或 STRETCH_COLUMN_WIDTH
onKeyMultiple(int keyCode, int repeatCount, KeyEvent event)	设置按键多次时的处理方法。当连续发生多次按键时，该方法会被调用，其中 keyCode 为按键对应的整型值，repeatCount 是按键的次数，event 是按键事件
setSelection (int p)	设置当前被选中的列表视图的子元素
onKeyUp (int keyCode, KeyEvent event)	设置释放按键时的处理方法。释放按键时，该方法会被调用，其中 keyCode 为按键对应的整型值，event 是按键事件
onKeyDown (int keyCode, KeyEvent event)	设置按键时的处理方法。按键时，该方法会被调用，其中 keyCode 为按键对应的整型值，event 是按键事件。注意用户按键的过程中，onKeyDown 先被调用，然后用户释放按键后调用 onKeyUp
isItemChecked (int position)	判断指定位置的 position 元素是否被选中
addHeaderView (View view)	给列表视图添加头注，通常头注位于列表视图的顶部，其中的参数 View 为要添加头注的列表视图名称
getChoiceMode ()	返回当前的选择模式

【示例 3.17】 在 Android Studio 中新建项目 Ch03_3D9_ListView，使用 ListView 加载数据及设置 Item 的监听事件。

在布局文件 activity_main.xml 中编写代码如下：

```
<?xml version="1.0" encoding="utf-8"?>
<LinearLayout xmlns:android="http://schemas.android.com/apk/res/android"
    xmlns:tools="http://schemas.android.com/tools"
    android:layout_width="match_parent"
    android:layout_height="match_parent"
    android:layout_margin="8dp"
    tools:context="com.ugrow.ch03_3d9_listview.MainActivity">

    <ListView
        android:id="@+id/lv"
        android:layout_width="match_parent"
        android:layout_height="match_parent" />
```

```
</LinearLayout>
```

在 MainActivity 中编写代码如下：

```
public class MainActivity extends AppCompatActivity {

    private ListView mLv;
    private List<String> mList = new ArrayList<>();

    @Override
    protected void onCreate(Bundle savedInstanceState) {
        super.onCreate(savedInstanceState);
        setContentView(R.layout.activity_main);
        mLv = (ListView) findViewById(R.id.lv);
        for (int i = 0; i < 20; i++) {
            mList.add("标题：" + i);
        }
        mLv.setAdapter(new MyAdapter());
        //设置监听事件
        mLv.setOnItemClickListener(new AdapterView.OnItemClickListener() {
            @Override
            public void onItemClick(AdapterView<?> adapterView, View view, int position, long l) {
                Toast.makeText(MainActivity.this, "position--->" + position, Toast.LENGTH_SHORT).show();
            }
        });
    }

    class MyAdapter extends BaseAdapter {

        @Override
        public int getCount() {
            return mList.size();
        }

        @Override
        public Object getItem(int position) {
            return mList.get(position);
        }

        @Override
```

```
        public long getItemId(int position) {
            return position;
        }

        @Override
        public View getView(int position, View containView, ViewGroup viewGroup) {
            ViewHolder holder;
            if (containView == null) {
                holder = new ViewHolder();
                containView = LayoutInflater.from(MainActivity.this).inflate(R.layout.item_listview, null);
                holder.tv = (TextView) containView.findViewById(R.id.tv);
                containView.setTag(holder);
            } else {
                holder = (ViewHolder) containView.getTag();
            }
            //加载数据
            holder.tv.setText(mList.get(position));
            return containView;
        }

        class ViewHolder {
            TextView tv;
        }
    }
}
```

新建 ListView 的 item 布局文件 item_listview.xml，在其中编写代码如下：

```
<?xml version="1.0" encoding="utf-8"?>
<LinearLayout xmlns:android="http://schemas.android.com/apk/res/android"
    android:layout_width="match_parent"
    android:layout_height="match_parent"
    android:layout_margin="8dp">

    <LinearLayout
        android:layout_width="match_parent"
        android:layout_height="64dp">

        <ImageView
            android:layout_width="56dp"
            android:layout_height="match_parent"
            android:scaleType="centerCrop"
```

```
            android:src="@mipmap/e" />

        <TextView
            android:id="@+id/tv"
            android:layout_width="wrap_content"
            android:layout_height="match_parent"
            android:layout_marginLeft="8dp"
            android:gravity="center_vertical"
            android:text="标题"
            android:textColor="#000" />
    </LinearLayout>

</LinearLayout>
```

运行项目，并单击其中一个 item，效果如图 3-21 所示。

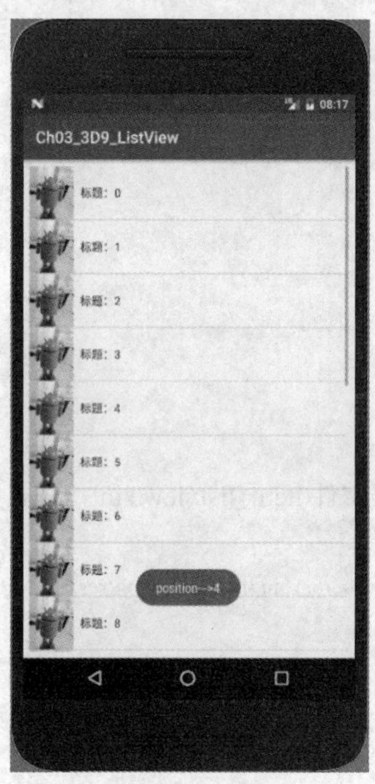

图 3-21　列表视图

3.5.12　滑动视图(RecyclerView)

RecyclerView 是 support.v7 包中的控件，可以说是 ListView 和 GridView 的增强升级版，可以在有限的窗口中展示大量数据集。

RecyclerView 架构提供了一种插拔式的体验，具备高度解耦及异常灵活的特性，其主要特点如下：

(1) 使用布局管理器 LayoutManager，可以控制 RecyclerView 显示的方式。LayoutManager 是一个抽象类，该类有 3 种实现类：LinearLayoutManager 类，用于实现水平方向和垂直方向的布局；GridLayoutManager 类，用于实现网格布局，类似于 GridView；StaggeredGridLayoutManager 类，用于实现瀑布流，该效果可以在 xml 文件中，也可以在 Java 代码中设置。

(2) 使用 ItemDecoration 类可以设置 item 的间距。

(3) 使用 ItemAnimator 可以设置 Item 的增加、删除、移动动画。

(4) 在实现 item 的单击和长按监听事件的方式上，RecyclerView 与 ListView 有所不同，ListView 可以直接设置监听事件，而 RecyclerView 要用接口回调的方式去实现。

【示例 3.18】 在 Android Studio 中新建项目 Ch03_3D10_RecyclerView，使用 RecyclerView 加载数据及设置 Item 的监听事件。

在 activity_main.xml 文件中编写代码如下：

```
<?xml version="1.0" encoding="utf-8"?>
<LinearLayout xmlns:android="http://schemas.android.com/apk/res/android"
    xmlns:tools="http://schemas.android.com/tools"
    android:layout_width="match_parent"
    android:layout_height="match_parent"
    tools:context="com.ugrow.ch03_3d10_recylerview.MainActivity">

    <android.support.v7.widget.RecyclerView
        android:id="@+id/recylerView"
        android:layout_width="match_parent"
        android:layout_height="match_parent"/>
</LinearLayout>
```

上述代码中只有一个 RecyclerView 控件，需要注意的是，RecyclerView 是 v7 包中的控件，在使用前需要在 build.gradle 中添加 RecyclerView 的依赖包。

在 MainActivity 中编写代码如下：

```
public class MainActivity extends AppCompatActivity implements MyRecyclerAdapter.OnChildClickListener {

    private RecyclerView mRecyclerView;
    private MyRecyclerAdapter myRecyclerAdapter;

    @Override
    protected void onCreate(Bundle savedInstanceState) {
        super.onCreate(savedInstanceState);
        setContentView(R.layout.activity_main);
        mRecyclerView = (RecyclerView) findViewById(R.id.recylerView);
        List<String> mList = new ArrayList<>();
        for (int i = 0; i < 20; i++) {
            mList.add("第" + i + "条数据");
```

```
        }
        //布局管理器
        LinearLayoutManager linearLayoutManager = new LinearLayoutManager(this,
LinearLayoutManager.VERTICAL, false);
        mRecyclerView.setLayoutManager(linearLayoutManager);
        //适配器
        myRecyclerAdapter = new MyRecyclerAdapter(this, mList);
        mRecyclerView.setAdapter(myRecyclerAdapter);
        //设置监听事件
        myRecyclerAdapter.setOnChildClickListener(this);

    }


    //监听事件
    @Override
    public void onChildClick(RecyclerView parent, View view, int position, String data) {
        Toast.makeText(MainActivity.this, data, Toast.LENGTH_SHORT).show();
    }
}
```

上述代码通过 findViewById() 方法初始化 RecyclerView；创建布局管理器 LinearLayoutManager，其第二个参数 LinearLayoutManager.VERTICAL 为垂直方向显示；创建一个自定义适配器 MyRecyclerAdapter，给 RecyclerView 设置适配器。让 MainActivity 实现在适配器中的回调方法，实现点击监听事件。

在 MyRecyclerAdapter 中编写代码如下：

```
public class MyRecyclerAdapter extends RecyclerView.Adapter<MyRecyclerAdapter.MyViewHolder>
implements View.OnClickListener {


    private Context context;
    private List<String> list;
    private OnChildClickListener listener;
    private RecyclerView recyclerView;

    public void setOnChildClickListener(OnChildClickListener listener) {
        this.listener = listener;
    }

    public MyRecyclerAdapter(Context context, List<String> list) {
        this.context = context;
        this.list = list;
    }
```

```
@Override
public MyViewHolder onCreateViewHolder(ViewGroup parent, int viewType) {
    View view = LayoutInflater.from(context).inflate(R.layout.item_recyclerview, parent, false);
    view.setOnClickListener(this);
    return new MyViewHolder(view);
}

@Override
public void onAttachedToRecyclerView(RecyclerView recyclerView) {
    super.onAttachedToRecyclerView(recyclerView);
    this.recyclerView = recyclerView;
}

@Override
public void onDetachedFromRecyclerView(RecyclerView recyclerView) {
    super.onDetachedFromRecyclerView(recyclerView);
    this.recyclerView = null;
}

@Override
public void onBindViewHolder(MyViewHolder holder, int position) {
    //数据绑定
    holder.text.setText(list.get(position));
}


@Override
public int getItemCount() {
    return list.size();
}

@Override
public void onClick(View v) {
    if (recyclerView != null && listener != null) {
        int position = recyclerView.getChildAdapterPosition(v);
        listener.onChildClick(recyclerView, v, position, list.get(position));
    }
}
```

```
    public static class MyViewHolder extends RecyclerView.ViewHolder {

        private TextView text;

        public MyViewHolder(View itemView) {
            super(itemView);
            text = (TextView) itemView.findViewById(R.id.tv_item);
        }
    }

    //接口回调
    public interface OnChildClickListener {
        void onChildClick(RecyclerView parent, View view, int position, String data);
    }
}
```

在 item_recyclerview.xml 文件中编写代码如下：

```
<?xml version="1.0" encoding="utf-8"?>
<LinearLayout xmlns:android="http://schemas.android.com/apk/res/android"
    android:layout_width="wrap_content"
    android:layout_height="wrap_content"
    android:layout_margin="8dp"
    android:orientation="vertical">

    <TextView
        android:id="@+id/tv_item"
        android:layout_width="match_parent"
        android:layout_height="match_parent"
        android:textSize="16sp"/>

</LinearLayout>
```

运行项目，单击某一条目 item，会以 Toast 的形式显示条目相关信息，如图 3-22 所示。

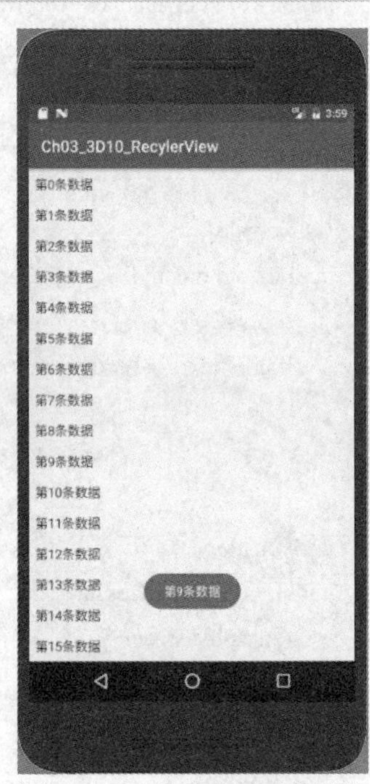

图 3-22　垂直方向显示数据的滑动视图

此时，如果将 MainActivity 布局管理器中的第二个参数修改如下：

```
        LinearLayoutManager linearLayoutManager = new LinearLayoutManager(this,
LinearLayoutManager.HORIZONTAL, false);//水平方向
        mRecyclerView.setLayoutManager(linearLayoutManager);
```

然后运行项目，则效果如图 3-23 所示。

图 3-23　水平方向显示数据的滑动视图

3.6　菜单

菜单是 UI 设计中经常使用的组件，可以展示不同的功能分组，实现了人机交互操作的人性化。Android 中的菜单分为三种类型：选项菜单(OptionMenu)、上下文菜单(ContextMenu)、弹出式菜单(PopupMenu)。

3.6.1　选项菜单(OptionMenu)

Android 手机上通常有一个 Menu 按键，当按下 Menu 时，会在屏幕底部弹出一个菜单，此菜单就是选项菜单。选项菜单提供了一种特殊的菜单显示方式，它没有对应的视图，即用户无法通过单击屏幕上的视图组件来加载选项菜单。

Android 开放了选项菜单的应用接口并屏蔽了其实现的复杂性，开发人员只需调用几个关键的方法就可以创建选项菜单，步骤如下：

(1)　重写 Activity 的 onCreateOptionsMenu()方法，当第一次打开菜单时，该方法会被自动调用。

(2)　调用 Menu 的 add()方法添加菜单项(MenuItem)。

(3)　当菜单项被选择时，覆盖 Activity 的 onOptionsItemSelected()方法来响应事件。

【示例 3.19】 在 Android Studio 中新建项目 Ch03_3D11_OptionMenu，演示选项菜单的使用方法。

在 OptionMenuActivity 中编写代码如下：

```
publicclass OptionMenuActivity extends Activity {
```

```
    privatefinalstaticintITEM = Menu.FIRST;

    @Override
    protectedvoid onCreate(Bundle savedInstanceState) {
        super.onCreate(savedInstanceState);
        setContentView(R.layout.activity_option_menu);
    }

    /**
     * 重写 onCreateOptionsMenu()方法，添加选项菜单
     */
    @Override
    publicboolean onCreateOptionsMenu(Menu menu) {
        // 添加菜单项
        menu.add(0, ITEM, 0, "开始");
        menu.add(0, ITEM + 1, 0, "退出");
        returntrue;
    }

    /**
     * 重写 onOptionsItemSelected()方法，响应选项菜单被单击的事件
     */
    publicboolean onOptionsItemSelected(MenuItem item) {
        switch (item.getItemId()) {
        caseITEM:
            // 设置 Activity 标题
            setTitle("开始游戏！");
            break;
        caseITEM + 1:
            setTitle("退出！");
            break;
        }
        returntrue;
    }
}
```

上述代码实现了选项菜单，单击 Menu 按键即可显示此菜单。

值得注意的是，目前多数手机已经取消了实体的触摸按键，改为虚拟按键(例如 Nexus 4)，此类手机的菜单通常会显示到标题栏(ActionBar)中，▣按钮并不是菜单按键，而是显示当前运行的程序列表，因此只能通过单击标题栏中右侧的⋮按钮显示选项菜单。

使用虚拟按键显示选项菜单的效果如图 3-24 所示。

图 3-24　使用虚拟按键显示菜单

单击选项菜单中的【开始】和【退出】项，Activity 的标题栏会显示不同的提示信息，如图 3-25 所示。

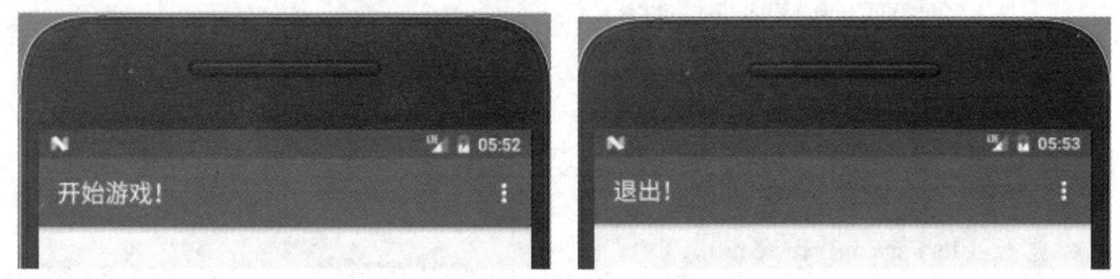

图 3-25　单击选项菜单后的标题栏效果

3.6.2　上下文菜单(ContextMenu)

ContextMenu 上下文菜单是 android.view.Menu 的子类，提供了用于创建和添加菜单的接口，其常用的方法如表 3-17 所示。

表 3-17　ContextMenu 常用方法

方　法	功 能 描 述
setHeaderIcon(int iconRes)	设置上下文菜单的图标
setHeaderIcon(Drawable icon)	设置上下文菜单的图标
setHeaderTitle(CharSequence title)	设置上下文菜单的标题
setHeaderTitle(int titleRes)	设置上下文菜单的标题
add(int groupId, int itemId, int order, CharSequence title)	添加子菜单

创建上下文菜单的步骤如下：

(1) 重写 Activity 的 onCreateContextMenu()方法，调用 Menu 的 add()方法添加菜单项(MenuItem)。

(2) 重写 onContextItemSelected()方法，响应菜单单击事件。

(3) 在 Activity 的 onCreate()方法中，调用 registerForContextMenu()方法，为视图注册上下文菜单。

【示例 3.20】 在 Android Studio 中新建项目 Ch03_3D12_ContextMenu，演示上下文菜单的使用方法。

在 ContextMenuActivity 中编写代码如下：

```
public class ContextMenuActivityextends Activity {
        //菜单 ID 常量
        private static final int ITEM1 = Menu.FIRST;
        private static final int ITEM2 = Menu.FIRST + 1;
        private static final int ITME3 = Menu.FIRST + 3;
        private static final int ITME4 = Menu.FIRST + 4;
        private static final int ITME5 = Menu.FIRST + 5;
        //声明文本视图
        private TextView myTV;

        public void onCreate(Bundle savedInstanceState) {
                super.onCreate(savedInstanceState);
                setContentView(R.layout.activity_context_menu);
                //根据 ID 获取文本视图
                myTV = (TextView) findViewById(R.id.tv);
                //在文本视图上注册上下文菜单
                registerForContextMenu(myTV);
        }

        /**
         * 重写 onCreateOptionsMenu()方法，添加选项菜单
         */
        public boolean onCreateOptionsMenu(Menu menu) {
                //添加菜单项
                menu.add(0, ITEM1, 0, "开始");
                menu.add(0, ITEM2, 0, "退出");
                return true;
        }

        /**
         * 重写 onOptionsItemSelected()方法，响应选项菜单被单击事件
         */
        public boolean onOptionsItemSelected(MenuItem item) {
```

```
        switch (item.getItemId()) {
        case ITEM1:
                //设置 Activity 标题
                setTitle("开始游戏！");
                break;
        case ITEM2:
                setTitle("退出！");
                break;
        }
        return true;
}

/**
 * 重写 onCreateContextMenu()方法，添加上下文菜单
 */
@Override
public void onCreateContextMenu(ContextMenu menu, View v,
                ContextMenuInfo menuInfo) {
        //添加菜单项
        menu.add(0, ITME3, 0, "红色背景");
        menu.add(0, ITME4, 0, "绿色背景");
        menu.add(0, ITME5, 0, "白色背景");
}
/**
 * 重写 onContextItemSelected()方法，响应上下文菜单被单击事件
 */
@Override
public boolean onContextItemSelected(MenuItem item) {
        switch (item.getItemId()) {
        case ITME3:
                //设置文本视图的背景颜色
                myTV.setBackgroundColor(Color.RED);
                break;
        case ITME4:
                myTV.setBackgroundColor(Color.GREEN);
                break;
        case ITME5:
                myTV.setBackgroundColor(Color.WHITE);
                break;
```

```
            }
            return true;
        }
}
```

上述代码创建了一个上下文菜单。运行项目，在 TextView 组件上长时间按住不动，就会显示该上下文菜单，如图 3-26 所示。

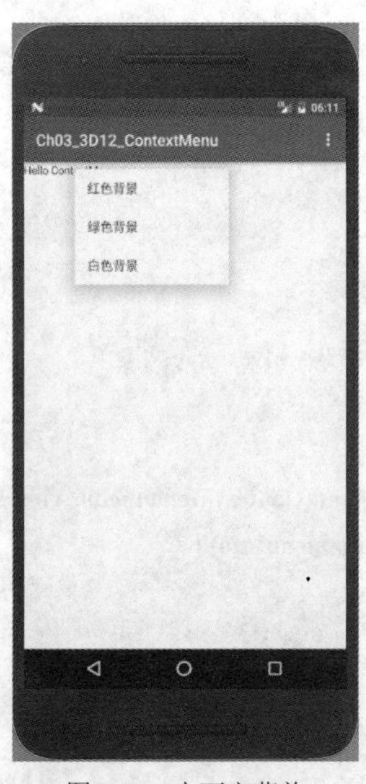

图 3-26　上下文菜单

单击不同的菜单项，TextView 的背景颜色将被改变，如图 3-27 所示。

图 3-27　单击上下文菜单项后的 TextView 效果

3.6.3　弹出式菜单(PopupMenu)

PopupMenu 是弹出式菜单，可以在指定控件上弹出。默认情况下，PopupMenu 会显示在组件的上方或下方。PopupMenu 可显示多个菜单项，并为菜单项增加子菜单。

创建弹出式菜单的步骤如下：

(1) 调用 new PopupMenu(Context context,View view)方法创建弹出式菜单，其中的参数 view 为单击弹出菜单的组件名称。

(2) 调用 MenuInflater 的 inflater()方法，将菜单资源添加到 PopupMenu 中。

(3) 调用 PopupMenu 的 show()方法，显示弹出式菜单。

【示例 3.21】 在 Android Studio 中新建项目 Ch03_3D13_PopupMenu，演示弹出式菜单的使用方法。

在布局文件 activity_popup_menu.xml 中编写代码如下：

```
<?xml version="1.0" encoding="utf-8"?>
<RelativeLayout xmlns:android="http://schemas.android.com/apk/res/android"
    xmlns:tools="http://schemas.android.com/tools"
    android:layout_width="match_parent"
    android:layout_height="match_parent"
    tools:context="com.ugrow.ch03_3d13_popupmenu.PopupMenuActivity">

    <Button
        android:id="@+id/btn_popup"
        android:layout_width="wrap_content"
        android:layout_height="wrap_content"
        android:layout_centerHorizontal="true"
        android:text="弹出式菜单"
        android:textColor="#000"/>

</RelativeLayout>
```

在 res 文件夹下新建 menu 文件夹，在其中创建布局文件 popup_menu.xml，编写以下代码：

```
<?xml version="1.0" encoding="utf-8"?>
<menu xmlns:android="http://schemas.android.com/apk/res/android">
    <item
        android:id="@+id/look"
        android:title="查找"/>

    <item
        android:id="@+id/add"
        android:title="添加"/>
    <item
        android:id="@+id/change"
        android:title="编辑"/>

    <item
        android:id="@+id/exit"
        android:title="退出"/>
</menu>
```

在 PopupMenuActivity 中编写以下代码：

```
public class PopupMenuActivity extends AppCompatActivity {

    private Button mBtnPopup;
    PopupMenu popup = null;

    @Override
    protected void onCreate(Bundle savedInstanceState) {
        super.onCreate(savedInstanceState);
        setContentView(R.layout.activity_popup_menu);
        mBtnPopup = (Button) findViewById(R.id.btn_popup);
        mBtnPopup.setOnClickListener(new View.OnClickListener() {
            @Override
            public void onClick(View view) {
                //创建 PopupMenu 对象
                popup = new PopupMenu(PopupMenuActivity.this, mBtnPopup);
                //将 R.menu.popup_menu 菜单资源加载到 popup 菜单中
                getMenuInflater().inflate(R.menu.popup_menu, popup.getMenu());
                //为 popup 菜单的菜单项单击事件绑定事件监听器
                popup.setOnMenuItemClickListener(new PopupMenu.OnMenuItemClickListener() {

                    @Override
                    public boolean onMenuItemClick(MenuItem item) {
                        switch (item.getItemId()) {
                            case R.id.look:
                                //使用 Toast 显示用户单击的菜单项
                                Toast.makeText(PopupMenuActivity.this, "您单击了【" + item.getTitle() + "】
菜单项", Toast.LENGTH_SHORT).show();
                                break;
                            case R.id.add:
                                Toast.makeText(PopupMenuActivity.this, "您单击了【" + item.getTitle() + "】
菜单项", Toast.LENGTH_SHORT).show();
                                break;
                            case R.id.change:
                                Toast.makeText(PopupMenuActivity.this, "您单击了【" + item.getTitle() + "】
菜单项", Toast.LENGTH_SHORT).show();
                                break;
                            case R.id.exit:
                                //隐藏该对话框
                                popup.dismiss();
                                break;
```

```
                    default:
                        break;

                }
                return false;
            }

        });
        popup.show();
    }
  });
}
@Override
public boolean onCreateOptionsMenu(Menu menu) {
    getMenuInflater().inflate(R.menu.popup_menu, menu);
    return true;
}
}
```

运行项目，单击界面中的【弹出式菜单】按钮，会弹出菜单选项，单击某项时，以 Toast 的形式显示单击的内容，如图 3-28 所示。

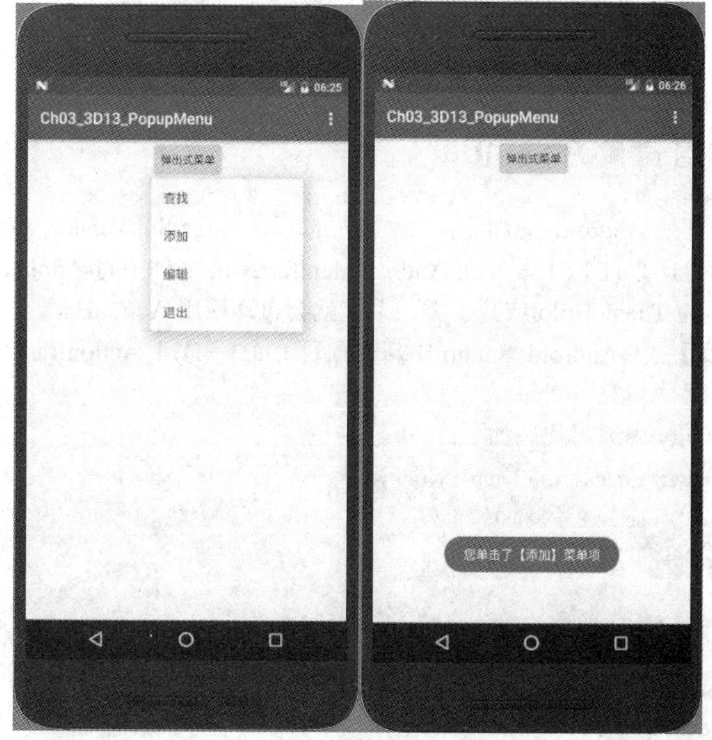

图 3-28　弹出式菜单

3.7 ActionBar

ActionBar 是 Android 3.0 中新增的导航栏功能控件，它的主要功能是标识用户当前操作页面的位置，并提供额外的操作按钮，以便于用户操作和界面导航。ActionBar 的优点是能提供全局统一的 UI 界面，且会自动适应各种不同大小的屏幕，从而使用户能快速习惯使用任意一款使用了 ActionBar 的软件。

ActionBar 的基本样式如图 3-29 所示，其中：

(1) 标签 1 所示为 ActionBar 的图标，用于标识当前页面的位置。

(2) 标签 2 所示为 Action Button，一般将常用的功能放到这里。

(3) 标签 3 所示为 OverFlow Button，应用的选项菜单，如果 ActionBar 没有足够的空间，Action Button 也将自动添加到这里。

(4) 标签 4 所示为 Tabs ActionBar，为应用提供了统一的 Tabs，类似于选项卡样式，便于页面切换。

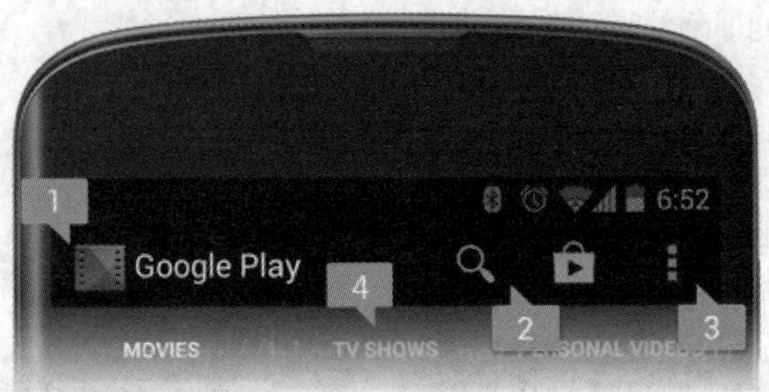

图 3-29 ActionBar 的基本样式

3.7.1 显示与隐藏 ActionBar

ActionBar 是在 Android 3.0 版本中才加入的。将 targetSdkVersion 或 minSdkVersion 的版本号设置为 11 或 11 以上，并在 AndroidManifest.xml 文件中将 Application 或 Activity 的主题样式指定为 ThemeHolo 或其子类，系统就会自动应用 ActionBar。

【示例 3.22】 在 Android Studio 中新建项目 Ch03_3D14_ActionBar，演示 ActionBar 的使用方法。

在 MainActivity 中编写代码如下：

```
public class MainActivity extends AppCompatActivity {
    private ActionBar actionBar = null;
    private Button btn = null;

    @Override
    protected void onCreate(Bundle savedInstanceState) {
        super.onCreate(savedInstanceState);
        setContentView(R.layout.activity_main);
```

```
            //获取应用的 ActionBar
            actionBar = getSupportActionBar();

            btn = (Button) findViewById(R.id.btn);
            btn.setOnClickListener(new OnClickListener() {

                  @Override
                  public void onClick(View arg0) {
                        if (actionBar.isShowing()) {
                              //隐藏 ActionBar
                              actionBar.hide();
                              btn.setText("Show");
                        } else {
                              //显示 ActionBar
                              actionBar.show();
                              btn.setText("Hide");
                        }
                  }
            });
      }
}
```

上述代码首先通过 getSupportActionBar()方法获取应用的 ActionBar 对象，然后通过调用 ActionBar 的 show()方法显示 ActionBar，通过调用 hide()方法将其隐藏。

此外还有另一种隐藏 ActionBar 的方式，即将指定的主题样式改为 Theme.APPCompat.Light.NoActionBar 即可。

3.7.2　修改图标和标题

ActionBar 的标题默认会显示应用名称，可以通过修改 AndroidManifest.xml 文件中每个 Activity 的属性自定义 Activity 的标题，代码如下：

```
<activity
      android:name=".MainActivity"
      android:label="示例"
      >
...
</activity>
```

项目运行之前与运行之后的效果对比如图 3-30 所示。

图 3-30　修改 ActionBar 标题

3.7.3　添加 Action 按钮

ActionBar 可以根据应用程序当前的功能提供与其相关的 Action 按钮，这些按钮会以图标或文字的形式直接显示在 ActionBar 上，Action 按钮既可以通过配置文件添加，也可以通过代码动态添加，下面介绍通过配置文件添加 Action 按钮的方法。

在 res/menu 文件夹中创建配置文件 main.xml，编写代码如下：

```
<menu xmlns:android="http://schemas.android.com/apk/res/android"
    xmlns:app="http://schemas.android.com/apk/res-auto">
    <item
        android:id="@+id/action_search"
        android:icon="@android:drawable/ic_menu_search"
        app:showAsAction="ifRoom|withText"
        android:title="查询"/>
    <item
        android:id="@+id/action_add"
        android:icon="@android:drawable/ic_menu_add"
        app:showAsAction="ifRoom|withText"
        android:title="添加"/>
</menu>
```

上述代码中的每一个<item>都代表了一个 Action 按钮，每个<item>均有自己的属性，常用属性如下：

(1) 属性 id 为 Action 按钮的唯一标识。

(2) 属性 icon 指定 Action 按钮显示的图片。

(3) 属性 title 指定 Action 按钮显示的文本。

(4) 属性 showAsAction 指定 Action 按钮 n 的显示位置，有以下参数可选：always 表示永远显示在 ActionBar 中，若 ActionBar 没有足够空间，则该 Action 按钮无法显示；ifRoom 表示如果 ActionBar 有足够的空间，则显示在 ActionBar 中，否则显示在 OverFlow 中；never 表示将永远显示在 OverFlow 中。

如果需要同时显示图片和文本，那么可以设置 android:showAsAction="ifRoom|withText"，但是，如果同时设置了图片和文本，实际显示时通常不会显示文本，因为 ActionBar 空间有限，所以会优先显示图片。

菜单配置文件创建后，下一步需要将其绑定到 Activity 中，重写 onCreateOptionsMenu (Menu menu)方法，在 MainActivity 中添加代码如下：

```
@Override
```

```
public Boolean onCreateOptionsMenu (Menu menu) {
        getMenuInflater().inflate(R.menu.main, menu);
        return true;
}
```

然后为 Action 按钮添加事件，重写 onOptionsItemSelected(MenuItem item)方法，在 MainActivity 中添加代码如下：

```
@Override
public boolean onOptionsItemSelected(MenuItem item) {
        int id = item.getItemId();
        switch (id) {
        case R.id.action_add:
                Toast.makeText(this, "add", Toast.LENGTH_SHORT).show();
                break;
        case R.id.action_search:
                Toast.makeText(this, "Search", Toast.LENGTH_SHORT).show();
                break;
        }
        returnsuper.onOptionsItemSelected(item);
}
```

运行项目，效果如图 3-31 所示。

图 3-31　在 ActionBar 中添加 Action 按钮

注意：当添加的 Action 按钮太多导致 ActionBar 放不下，或者将 showAsAction 设置为 never，此时这些 Action 按钮就会被自动添加到 OverFlow 中，但运行程序后可能会发现 OverFlow 按钮没有显示出来，而原本添加到 OverFlow 中的 Action 按钮被显示到了屏幕底部的菜单中(需要单击手机的 menu 按键显示)。这是因为 Android 会判断当前手机设备的按键，如果是物理按键，Action 按钮就会显示到底部的菜单中；如果是虚拟按键，则会显示在 OverFlow 中。很显然，这样的设计会造成一个问题，就是用户体验不统一，此时可以使用反射技术来修改此情况，只需在 onCreate()方法中调用即可，方法如下：

在 MainActivity 中编写代码如下：

```
/**
 * 强制显示 overflow menu
 */
private void getOverflowMenu() {
        try {
```

```
        ViewConfiguration config = ViewConfiguration.get(this);
        Field menuKeyField = ViewConfiguration.class
                .getDeclaredField("sHasPermanentMenuKey");
        if (menuKeyField != null) {
            menuKeyField.setAccessible(true);
            menuKeyField.setBoolean(config, false);
        }
    } catch (Exception e) {
        e.printStackTrace();
    }
}
```

3.7.4 添加导航按钮

如果要从子页面回到主页面，可通过单击返回键的操作实现。ActionBar 也提供了此项功能，只需调用 ActionBar 的 setDisplayHomeAsUpEnabled(true)方法即可，代码如下：

```
actionBar.setDisplayHomeAsUpEnabled(true);
```

同时，需要在 onOptionsItemSelected(MenuItem item)中对这个按钮进行监听，此按钮的 ID 为 android.R.id.home，代码如下：

```
...
case android.R.id.home:
    Toast.makeText(this, "Home", Toast.LENGTH_SHORT).show();
    break;
...
```

运行项目，效果如图 3-32 所示。

图 3-32 在 ActionBar 中添加导航按钮

3.7.5 添加 ActionView

前面添加了一个查询的 Action 按钮，但在实际使用中会发现一个问题：如果想要实现查询功能，必须重新创建一个 View 来接收输入的查询信息，操作比较繁琐。而 ActionView 解决了这一问题，它能够在不切换或不增加界面的情况下来完成比较丰富的操作。下面我们就使用 ActionView 来完善这个搜索功能。

在 onCreateOptionsMenu(Menu menu)中加入以下 MainActivity 代码：

```
@Override
public boolean onCreateOptionsMenu(Menu menu) {
```

```
        getMenuInflater().inflate(R.menu.main,menu);
        MenuItem item = menu.findItem(R.id.action_search);
        //动态添加 View
        SearchView searchView = new SearchView(this);
        //设置提示文字
        SpannableString spanText = new SpannableString("请输入名字");
        //设置文字颜色
        spanText.setSpan(new ForegroundColorSpan(Color.WHITE), 0, spanText.length(),
Spannable.SPAN_INCLUSIVE_EXCLUSIVE);
        searchView.setQueryHint(spanText);
        item.setActionView(searchView);
        // 添加监听事件
        searchView.setOnQueryTextListener(new SearchView.OnQueryTextListener() {

            @Override
            public boolean onQueryTextSubmit(String arg0) {
                // 当提交查询时，输出查询内容到屏幕
                Toast.makeText(MainActivity.this, arg0, Toast.LENGTH_SHORT).show();
                return false;
            }

            @Override
            public boolean onQueryTextChange(String arg0) {
                return false;
            }
        });
        return true;
}
```

上述代码实现了查询功能，其中，onQueryTextSubmit(String arg0)是输入完毕并提交时执行的回调方法，可以提取输入内容并加以处理；onQueryTextChange(String arg0)是输入内容发生改变时执行的回调方法。

图 3-33　在 ActionBar 中添加搜索框

运行项目，效果如图 3-33 所示：当单击搜索按钮时，ActionBar 中的其他 ActionButton 和标题全部隐藏，取而代之的是一个搜索框。

3.8　适配器(Adapter)

适配器(Adapter)是连接数据(Data)和适配器控件(AdapterView)的桥梁，通过适配器能够将数据源的数据添加到适配器控件中，有效地实现数据与适配器控件的分离，使适配器控件与数据的绑定及数据的修改更加简便。常用的适配器有 ArrayAdapter、SimpleAdapter、SimpleCursorAdapter 和 BaseAdapter。

下面在 Android Studio 中新建项目 Ch03_3D15_Adapter，分别使用四种适配器实现 ListView 数据的加载功能。

3.8.1 数组适配器(ArrayAdapter)

ArrayAdapter 又称数组适配器，是比较简单且经常使用的一种数组适配器，只能展示一行文字，无需自定义布局，使用系统提供的布局即可，它将数据放入一个数组以便显示。

ArrayAdapter 提供了多种构造函数来生成数组适配器，其常用的函数如下所示：

```
ArrayAdapter(Context context,int resource,int textViewResId)
ArrayAdapter(Context context,int textViewResId,T[] objects)
ArrayAdapter(Context context,int textViewResId,List<T> objects)
```

其中的参数说明如下：

(1) context：上下文环境，在 Activity 中一般使用 this。

(2) resource：资源 ID。

(3) textViewResId：文本视图资源 ID，如下拉列表组件的 ID。

(4) objects：泛型集合/数组。

ArrayAdapter 的使用步骤如下：

(1) 定义一个数组或集合，用于存放 ListView 中 item 的内容。

(2) 使用 ArrayAdapter 的构造方法创建一个 ArrayAdapter 对象。ArrayAdapter 有多个构造方法，最常用的是带有三个参数的方法：第一个参数是上下文对象；第二个参数是 ListView 每一行(item)的布局资源 ID；第三个参数是数据源。

(3) 使用 ListView 的 setAdapter()方法绑定 ArrayAdapter。

在项目 Ch03_3D15_Adapter 中，新建布局文件 activity_array_adapter.xml，编写代码如下：

```
<?xml version="1.0" encoding="utf-8"?>
<LinearLayout xmlns:android="http://schemas.android.com/apk/res/android"
    xmlns:tools="http://schemas.android.com/tools"
    android:layout_width="match_parent"
    android:layout_height="match_parent"
    tools:context="com.ugrow.ch03_3d15_adapter.ArrayAdapterActivity">
    <ListView
        android:id="@+id/lv"
        android:layout_width="match_parent"
        android:layout_height="match_parent" />
</LinearLayout>
```

在 ArrayAdapterActivity 中编写代码如下：

```
public class ArrayAdapterActivity extends AppCompatActivity {

    private ListView mLv;
    private List<String> mList = new ArrayList<>();

    @Override
    protected void onCreate(Bundle savedInstanceState) {
        super.onCreate(savedInstanceState);
        setContentView(R.layout.activity_array_adapter);
        //根据 ID 找到控件
        mLv = (ListView) findViewById(R.id.lv);
        //数据源
        for (int i = 0; i < 20; i++) {
            mList.add("第" + i + "条数据");
        }
        //设置适配器
        ArrayAdapter<String> adapter = new
ArrayAdapter<String>(this,android.R.layout.simple_list_item_1,mList);
        mLv.setAdapter(adapter);
    }
}
```

运行项目，效果如图 3-34 所示。

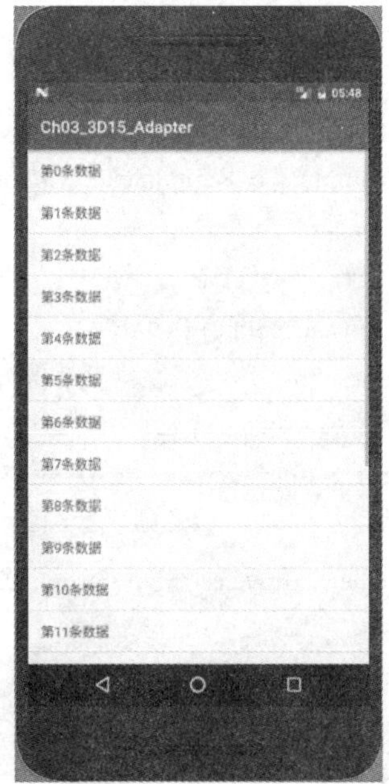

图 3-34　使用 ArrayAdapter 加载数据

3.8.2 简单适配器(SimpleAdapter)

SimpleAdapter 比 ArrayAdapter 具备更好的扩展性，可以自定义各种各样的布局，除了放置文本，还可以放置 ImageView(图片)与 Button(按钮)等。

SimpleAdapter 提供了多种构造函数来生成适配器，其常用的函数如下所示：

```
SimpleAdapter(Context context,List<? Extends Map<String,?>> data,int resource,String[] from,int[] to);
```

其中的参数说明如下：

(1) context：表示上下文对象或者环境对象。

(2) data：表示数据源。往往采用 List<Map<String, Object>>集合对象。

(3) resource：自定义 ListView 中每个 item 的布局文件。用 R.layout.文件名的形式来调用。

(4) from：其实是数据源中 Map 的 key 组成的一个 String 数组。

(5) to：表示数据源中 Map 的 value 要放置在 item 中的哪个控件位置上。其实就是自定义 item 布局文件中每个控件的 ID，通过 R.id.id 名字的形式来调用。

SimpleAdapter 的使用步骤如下：

(1) 定义一个集合，用于存放 ListView 中 item 的内容。

(2) 定义一个 item 的布局。

(3) 创建一个 SimpleAdapter 对象。

(4) 使用 ListView 的 setAdapter()方法绑定 SimpleAdapter。

在项目 Ch03_3D15_Adapter 中，新建 SimpleAdapterActivity，编写代码如下：

```
public class SimpleAdapterActivity extends AppCompatActivity {

    private ListView mLv;
    private int[] imgs = new int[]{R.mipmap.a, R.mipmap.b, R.mipmap.c, R.mipmap.d, R.mipmap.e, R.mipmap.f,
R.mipmap.g,R.mipmap.a, R.mipmap.b, R.mipmap.c, R.mipmap.d, R.mipmap.e, R.mipmap.f, R.mipmap.g};

    @Override
    protected void onCreate(Bundle savedInstanceState) {
        super.onCreate(savedInstanceState);
        setContentView(R.layout.activity_array_adapter);
        mLv = (ListView) findViewById(R.id.lv);
        // 创建数据源
        List<Map<String, Object>> list = new ArrayList<Map<String, Object>>();
        for (int i = 0; i < imgs.length; i++) {
            Map<String, Object> map = new HashMap<String, Object>();
            map.put("username", "张三" + i);
            map.put("pwd", "123456" + i);
            map.put("imgId", imgs[i]);
            list.add(map);
```

```
        }
        SimpleAdapter adapter = new SimpleAdapter(this, list, R.layout.item_simple_adapter, new String[]
                {"username", "pwd", "imgId"}, new int[]{R.id.tv_name, R.id.tv_pwd, R.id.iv});
        // 给 ListView 设置适配器
        mLv.setAdapter(adapter);
    }
}
```

运行项目，效果如图 3-35 所示。

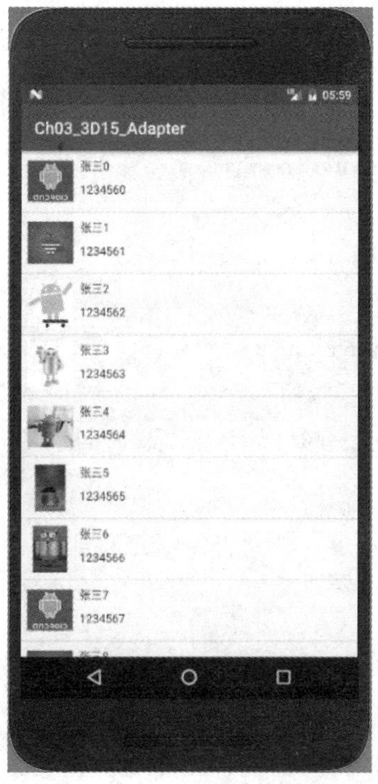

图 3-35　使用 SimpleAdapter 加载数据

3.8.3　简单游标适配器(SimpleCursorAdapter)

SimpleCursorAdapter 可以看作 SimpleAdapter 与数据库的简单结合，能方便地把数据库中的内容以列表的形式展现出来。

SimpleCursorAdapter 提供了多种构造函数来生成适配器，其常用的函数如下所示：

```
SimpleCursorAdapter (Context context,int layout,Cursor c,String[] from,int[] to,int flags);
```

其中的参数说明如下：

(1) context：表示上下文对象或者环境对象。

(2) layout：item 的布局，其中至少要有参数 to 代表的数组中的所有视图。

(3) c：数据库游标，如果游标不可用，则可以为空。

(4) from：由数据源中的 key 组成的一个 String 数组。

(5) to：表示数据源中 Map 的 value 要放置在 item 中的哪个控件位置上，是由自定义

item 布局文件中每个控件的 ID 组成的数组，可以通过“R.id.ID 名”的形式来调用。

(6) flags：用来确定适配器行为的标志。

SimpleCursorAdapter 的使用步骤如下：

(1) 创建数据库，在其中插入数据。

(2) 读取数据库数据，得到数据对象 Cursor。

(3) 使用 ListView 的 setAdapter()方法绑定 SimpleCursorAdapter。

在项目 Ch03_3D15_Adapter 中，新建 SimpleCursorAdapterActivity 类，编写代码如下：

```
public class SimpleCursorActivity extends AppCompatActivity {
    DatabaseManager dbManager;

    @Override
    protected void onCreate(Bundle savedInstanceState) {
        super.onCreate(savedInstanceState);
        setContentView(R.layout.activity_array_adapter);
        //得到一个数据层操作对象
        dbManager = new DatabaseManager(this);
        //因为在数据库建立的时候已经创建了一个表，所以这里可以直接插入数据
        String insertSql = "insert into test_table (name,age) values('kale',20)";
        //用循环的方式来插入 20 条数据
        for (int i = 0; i < 20; i++) {
            dbManager.executeSql(insertSql);
        }

        //这个游标查询到的数据中必须有一个列名为_id，否则会报错
        String sql = "select _id,name,age from test_table";

        //得到一个Cursor，用于放入适配器中
        Cursor cursor = dbManager.executeSql(sql, null);
        SimpleCursorAdapter adapter = new SimpleCursorAdapter(this,
                R.layout.item_simple_cursor, cursor, new String[] { "name", "age" },
                new int[] { R.id.list_name, R.id.list_phone }, 0);

        ListView listView = (ListView) findViewById(R.id.lv);
        listView.setAdapter(adapter);
    }

}
```

运行项目，效果如图 3-36 所示。

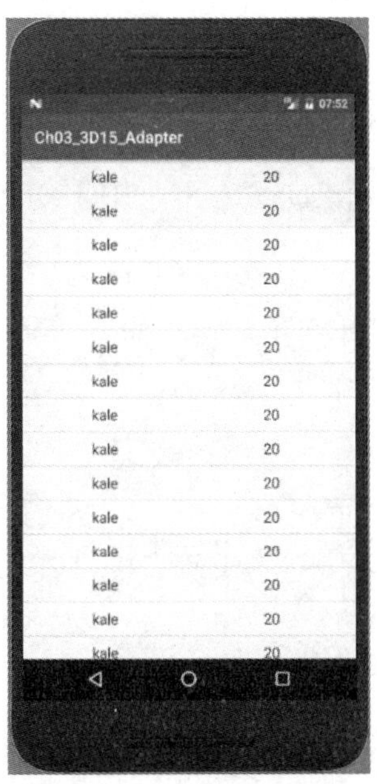

图 3-36　使用 SimpleCursorAdapter 加载数据

3.8.4　自定义适配器(BaseAdapter)

BaseAdapter 是自定义适配器，在实际开发中使用最多，具备实用性更好和灵活性更强的优点。

BaseAdapter 的使用步骤如下：

(1) 定义一个集合，用于存放 ListView 中 item 的内容。

(2) 定义一个 item 布局文件。

(3) 创建一个内部类或者外部类 MyAdapter，继承 BaseAdapter，重写未实现的方法。

(4) 使用 ListView 的 setAdapter()方法绑定自定义的适配器 MyAdapter 对象。

在项目 Ch03_3D15_Adapter 中，新建 BaseAdapterActivity，编写代码如下：

```
public class BaseAdapterActivity extends AppCompatActivity {

    private ListView mLv;
    private List<String> mlist = new ArrayList<>();

    @Override
    protected void onCreate(Bundle savedInstanceState) {
        super.onCreate(savedInstanceState);
        setContentView(R.layout.activity_array_adapter);
        mLv = (ListView) findViewById(R.id.lv);
        for (int i = 0; i < 20; i++) {
            mlist.add("自定义第" + i + "条数据");
```

```
        }
        mLv.setAdapter(new MyAdapter());
    }
    class MyAdapter extends BaseAdapter{

        @Override
        public int getCount() {
            return mlist.size();
        }

        @Override
        public Object getItem(int position) {
            return mlist.get(position);
        }

        @Override
        public long getItemId(int position) {
            return position;
        }

        @Override
        public View getView(int position, View view, ViewGroup viewGroup) {
            ViewHolder holder;
            if (view == null){
                holder = new ViewHolder();
                view = LayoutInflater.from(BaseAdapterActivity.this).inflate(R.layout.item_base_layout,null);
                holder.tv=(TextView)view.findViewById(R.id.tv_base);
                view.setTag(holder);
            }else {
                holder=(ViewHolder) view.getTag();
            }
            holder.tv.setText(mlist.get(position));
            return view;
        }
        class ViewHolder{
            TextView tv;
        }
    }
}
```

其中，自定义适配器时需要注意以下几点：

(1) 继承 BaseAdapter，需要实现 getCount()、getItem()、getItemId()、getView()四个方法。

(2) 定义内部类 ViewHolder，将 item 布局文件中的控件都定义成该类的属性。

(3) 需要构建一个布局填充器对象 LayoutInflater，调用该对象的 infalter()方法填充 item 布局文件，并将返回一个 view 对象。

(4) 调用 view 对象的 findViewById()方法初始化 item 布局中的控件，将控件对象赋值给 ViewHolder 类中的属性。

(5) 给 view 对象设置标签，也就是使用 setTag()方法，将 ViewHolder 对象作为标签贴在 view 对象上；也可从 view 对象上取回作为标签的 ViewHolder 对象。

运行项目，效果如图 3-37 所示。

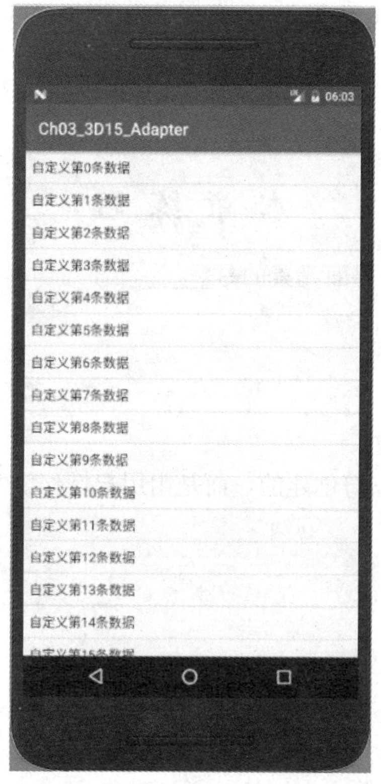

图 3-37　使用 BaseAdapter 加载数据

本 章 小 结

通过本章的学习，读者应当了解：

- ✧ Android 的用户界面主要由 View、ViewGroup 和 Layout 几个部分构成。
- ✧ Android 系统引用了 Java 的事件处理机制，包括事件、事件源和事件监听器三个事件模型。
- ✧ Android 中提供了两种创建布局的方式：XML 布局文件和代码直接实现。
- ✧ Android 的布局包含 LinearLayout、RelativeLayout、TableLayout、Absolute

Layout、FrameLayout 和 GridView 等多种类型。

- ✧ 提示信息(Toast)是 Android 中用来显示提示信息的一种机制，与对话框不同，Toast 是没有焦点的，而且 Toast 显示时间有限，过一定的时间会自动消失。
- ✧ Android 系统中提供了四种对话框：AlertDialog、ProgressDialog、DatePickerDialog 和 TimePickerDialog。
- ✧ 常用的 Widget 组件有：按钮(Button)、文本框(TextView)、编辑框(EditText)、复选框(CheckBox)、单选按钮组(RadioGroup)、下拉列表(Spinner)。
- ✧ Android 的菜单有三种：选项菜单(OptionsMenu)、上下文菜单(ContextMenu)和弹出式菜单(PopupMenu)。
- ✧ ActionBar 的主要元素包括：图标和标题、ActionButton、OverFlow、ActionView、Tabs。
- ✧ 常用的适配器主要有 ArrayAdapter、SimpleAdapter、SimpleCursorAdapter 和 BaseAdapter 四种。

本 章 练 习

1. 以下不属于 Android 用户界面元素的是______。

   A. 视图组件

   B. 视图容器组件

   C. 布局管理

   D. 资源引用

2. ______不是通过 new 运算符创建的，而是由用户操作触发的。

   A. 事件

   B. 事件源

   C. 监听器

   D. 事件处理方法

3. Spinner 是______组件。

   A. 文本框

   B. 滚动视图

   C. 下拉列表

   D. 列表视图

4. 简述选项菜单和上下文菜单的创建步骤。

5. 修改示例 3.16 中注册窗口的代码，实现在用户单击注册时使用 Toast 显示其注册信息的效果。

第4章　意图(Intent)

本章目标

- 了解 Intent 的功能及作用
- 掌握 Intent 常用的属性及方法
- 熟悉 Activity 之间的消息传递机制
- 掌握 Activity 权限的设置方法

4.1 Intent 概述

大多数传统类型的手机应用程序之间相互独立、互相隔离，应用程序与硬件和原生组件之间没有交互的行为。然而交互机制是扩展手机应用功能所必须的，通过应用之间的交互，手机才能支持复杂的应用。鉴于此种需求，Android 系统提供了用于开发应用程序交互功能的组件，这些组件包括广播接收器(BroadcastReceiver)、意图(Intent)、适配器(Adapter)以及内容提供器(ContentProvider)。

Intent 是 Android 的核心组件，它利用消息实现应用程序间的交互，这种消息描述了应用中某一次操作的动作以及数据，系统通过该 Intent 的描述找到对应的组件，并将 Intent 传递给需调用的组件，从而完成组件的调用。

4.1.1 Intent 组成属性

Intent 由动作、数据、分类、类型、组件和扩展信息等 7 部分内容组成，每个组成部分都由相应的属性来表示。同时，Intent 还提供了设置和获取相应属性的方法，如表 4-1 所示。

表 4-1 Intent 属性及对应方法

组成	属性	设置属性方法	获取属性方法
动作	Action	setAction()	getAction()
数据	Data	setData()	getData()
分类	Category	addCategory()	
类型	Type	setType()	getType()
组件	Component Name	setComponent() setClass() setClassName()	getComponentName()
扩展信息	Extra	putExtra()	通过 getXXXExtra()方法获取不同数据类型的数据，如 int 类型数据使用 getIntExtra()，字符串数据则使用 getStringExtra()；通过 getExtras()方法获取 Bundle 包
标记	Flag	setFlags()	Intent 通过 setFlags()方法来添加控制标记，还可以设置 Activity 的启动模式

1. Action 属性

Action 属性用于描述 Intent 要完成的动作，并对要执行的动作进行简要地描述。Intent 类定义了一系列 Action 属性常量，用来表示一套标准动作，如 ACTION_CALL(打电话)、ACTION_EDIT(编辑)等。根据使用动作的组件不同，可以将这套动作分为 Activity 动作和 Broadcast 动作。表 4-2 列举了常用的 Action 属性常量。

表 4-2　常用的 Action 属性常量

Action 常量	行为描述	使用组件(分类)
ACTION_CALL	打电话，即直接呼叫 Data 中所带的电话号码	
ACTION_ANSWER	接听来电	Activity
ACTION_SEND	由用户指定发送方式进行数据发送操作	
ACTION_SENDTO	根据不同的 Data 类型，通过对应的软件发送数据	
ACTION_VIEW	根据不同的 Data 类型，通过对应的软件显示数据	
ACTION_EDIT	显示可编辑的数据	
ACTION_MAIN	指定应用程序的入口	
ACTION_SYNC	同步服务器与移动设备之间的数据	
ACTION_BATTERY_LOW	警告设备电量低	Broadcast
ACTION_HEADSET_PLUG	插入或者拔出耳机	
ACTION_SCREEN_ON	打开移动设备屏幕	
ACTION_TIMEZONE_CHANGED	移动设备时区发生变化	

2. Data 属性

Intent 的 Data 属性用于指定执行动作的 URI 和 MIME，常用的 Data 属性常量如表 4-3 所示。

表 4-3　常用的 Data 属性常量

Data 属性	说　明	示　例
tel://	号码数据格式，后跟电话号码	tel://123
mailto://	邮件数据格式，后跟邮件收件人地址	mailto://dh@163.com
smsto://	短信数据格式，后跟短信接收号码	smsto://123
content://	内容数据格式，后跟需要读取的内容	content://contacts/people/1
file://	文件数据格式，后跟文件路径	file://sdcard/mymusic.mp3
geo://latitude,longitude	经纬数据格式，在地图上显示经纬度所指定的位置	geo://180,65

Action 和 Data 通常匹配使用，不同的 Action 由不同的 Data 数据指定，表 4-4 列举了一些常见的应用。

表 4-4　Action 和 Data 属性匹配应用

Action 属性	Data 属性	描　述
ACTION_VIEW	content://contacts/people/1	显示_id 为 1 的联系人信息
ACTION_EDIT	content://contacts/people/1	编辑_id 为 1 的联系人信息
ACTION_VIEW	tel:123	显示电话为 123 的联系人信息
ACTION_VIEW	http://www.google.com	在浏览器中浏览该网页
ACTION_VIEW	file://sdcard/mymusic.mp3	播放 MP3

3. Category 属性

Category 属性用于指定一个执行 Action 动作的环境。Intent 中定义了一系列 Category 属性常量，常用的 Category 属性常量如表 4-5 所示。

表 4-5　常用的 Category 属性常量

Category 属性	说　明
CATEGORY_DEFAULT	默认的执行方式，按照普通 Activity 的执行方式执行
CATEGORY_HOME	该组件为 Home Activity
CATEGORY_LAUNCHER	优先级最高的 Activity，通常与入口 ACTION_MAIN 配合使用
CATEGORY_BROWSABLE	可以使用浏览器启动
CATEGORY_GADGET	可以内嵌到另外的 Activity 中

4. ComponentName 属性

ComponentName 属性用于指明 Intent 的目标组件的类名称。通常 Android 会根据 Intent 中包含的其他属性的信息(比如 Action、Data/Type、Category)进行查找，最终找到一个与之匹配的目标组件。但是，如果指定了 ComponentName 这个属性，Intent 就会根据组件名直接查找到相应的组件，而不再执行上述查找过程。指定 ComponentName 属性后，Intent 的其他属性都是可选的。根据 Intent 寻找目标组件时所采用的方式不同，可以将 Intent 分为两类：

(1) 显式 Intent：这种方式通过直接指定 ComponentName 来实现。

(2) 隐式 Intent：这种方式通过 Intent Filter 过滤实现，过滤时通常根据 Action、Data 和 Category 属性进行匹配查找。

显式 Intent 通过 setComponent()、setClassName()或 setClass()设置组件名，例如：

```
//创建一个 Intent 对象
Intent intent = new Intent();
//将 Intent对象的目标组件指定为 SecondActivity
intent.setClass(MainActivity.this, SecondActivity.class);
```

上面代码首先使用 Intent 的构造函数生成一个 Intent 对象，并利用 setClass()方法将其目标组件设置为 SecondActivity。

setClass()方法的原型如下：

```
setClass(Context packageContext, Class<?> cls);
```

该方法包含两个参数：

✧ Context packageContext 为当前环境，例如 MainActivity.this。

✧ Class<?> cls 为目标组件类型，例如 SecondActivity.class。

5. Extra 属性

Extra 属性用于添加一些附加信息，可以使用 Intent 对象的 putExtra()方法来添加附加信息。例如，将一个人的姓名附加到 Intent 对象中，代码如下：

```
Intent intent = new Intent();
intent.putExtra("name","zhangsan");
```

使用 Intent 对象的 getXXXExtra()方法可以获取该对象中的附加信息。例如，若要将

上面代码存入 Intent 对象中的人名获取出来，因为存入的是字符串，就可以使用getStringExtra()方法获取数据，代码如下：

```
String name=intent.getStringExtra("name");
```

6．Type 属性

Type 属性用于指定 Data 所指定的 URI 对应的 MIME 类型。MIME 类型由类型与子类型两个字符串组成，中间用“/”隔开(例如：abc/xyz)。

浏览器会自动使用指定的应用程序来打开某 MIME 类型的 Data，多用于指定一些客户端自定义的文件名或一些媒体文件的打开方式。

例如，使用 Type 播放视频，代码如下：

```
Intent intent = new Intent();
Uri uri = Uri.parse(“file:///sdcard/media.mp4”);
intent.setAction(Intent.ACTION_VIEW);
intent.setDataAndType(uri, “video/*”);
startActivity(intent);
```

7．Flag 属性

Flag 属性用于设定 Activity 的启动模式，与在清单文件中设置 launchMode 的属性值所实现的效果相同。

例如：

```
Intent intent = new Intent(this,SecondActivity.class);
Intent.setFlags(Intent.FLAG_ACTIVITY_CLEAR_TOP);
startActivity(intent);
```

上述代码中：

(1) Intent.FLAG_ACTIVITY_CLEAR_TOP：效果与在清单文件中将 Activity 启动模式的 launchMode 属性值设置为 singleTask 相同。

(2) Intent.FLAG_ACTIVITY_SINGLE_TOP：效果与在清单文件中将 Activity 启动模式的 launchMode 属性值设置为 singleTop 相同。

(3) Intent.FLAG_ACTIVITY_NEW_TASK：效果与在清单文件中将 Activity 启动模式的 launchMode 属性值设置为 singleInstance 相同。

【示例 4.1】 在 Android Studio 中新建项目 Ch04_4D1_Intent，使用 Intent 属性实现显示跳转、拨打电话、发送短信、打开网页的操作。

在 activity_main.xml 文件中编写以下代码：

```
<?xml version="1.0" encoding="utf-8"?>
<LinearLayout xmlns:android="http://schemas.android.com/apk/res/android"
    xmlns:tools="http://schemas.android.com/tools"
    android:layout_width="match_parent"
    android:layout_height="match_parent"
    android:orientation="vertical"
    android:padding="8dp"
    tools:context="com.ugrow.ch04_4d1_intent.MainActivity">
```

```
    <Button
        android:id="@+id/btn_start"
        android:layout_width="wrap_content"
        android:layout_height="wrap_content"
        android:text="显示跳转" />
    <Button
        android:id="@+id/btn_call"
        android:layout_width="wrap_content"
        android:layout_height="wrap_content"
        android:text="拨打电话" />
    <Button
        android:id="@+id/btn_send"
        android:layout_width="wrap_content"
        android:layout_height="wrap_content"
        android:text="发送短息" />
    <Button
        android:id="@+id/btn_web"
        android:layout_width="wrap_content"
        android:layout_height="wrap_content"
        android:text="打开网页" />

</LinearLayout>
```

在 MainActivity 中编写以下代码：

```
public class MainActivity extends AppCompatActivity implements View.OnClickListener {

    private Button mBtnStart;
    private Button mBtnCall;
    private Button mBtnSend;
    private Button mBtnWeb;
    private static final int MY_PERMISSIONS_REQUEST_CALL_PHONE = 1;

    @Override
    protected void onCreate(Bundle savedInstanceState) {
        super.onCreate(savedInstanceState);
        setContentView(R.layout.activity_main);
        initView();
    }

    private void initView() {
```

```
        mBtnStart = (Button) findViewById(R.id.btn_start);
        mBtnStart.setOnClickListener(this);
        mBtnCall = (Button) findViewById(R.id.btn_call);
        mBtnCall.setOnClickListener(this);
        mBtnSend = (Button) findViewById(R.id.btn_send);
        mBtnSend.setOnClickListener(this);
        mBtnWeb = (Button) findViewById(R.id.btn_web);
        mBtnWeb.setOnClickListener(this);
    }

    @Override
    public void onClick(View v) {
        switch (v.getId()) {
            case R.id.btn_start:
                Intent intent = new Intent();
                ComponentName componentName = new ComponentName(MainActivity.this,
SecondActivity.class);
                intent.setComponent(componentName);
                startActivity(intent);
                break;
            case R.id.btn_call:
                //动态添加权限， 判断Android版本是否大于23
                if (Build.VERSION.SDK_INT >= Build.VERSION_CODES.M){
                    if (ContextCompat.checkSelfPermission(this, Manifest.permission.CALL_PHONE) !=
PackageManager.PERMISSION_GRANTED) {
                        ActivityCompat.requestPermissions(this, new
String[]{Manifest.permission.CALL_PHONE}, MY_PERMISSIONS_REQUEST_CALL_PHONE);
                    } else {
                        callPhone();
                    }
                }else {
                    callPhone();
                }

                break;
            case R.id.btn_send:
                Intent intent3 = new Intent(Intent.ACTION_SENDTO, Uri.parse("sms://10086"));
                //intent3.putExtra("sms_body", "你好啊");// 发送内容
                startActivity(intent3);
                break;
```

```
            case R.id.btn_web:
                intent = new Intent(Intent.ACTION_VIEW);
                intent.setData(Uri.parse("http://www.baidu.com"));
                startActivity(intent);
                break;
            default:
                break;
        }
    }

    public void callPhone() {
        Intent intent = new Intent(Intent.ACTION_CALL);
        Uri data = Uri.parse("tel:" + "10086");
        intent.setData(data);
        startActivity(intent);
    }

    @Override
    public void onRequestPermissionsResult(int requestCode, String[] permissions, int[] grantResults) {

        if (requestCode == MY_PERMISSIONS_REQUEST_CALL_PHONE) {
            if (grantResults[0] == PackageManager.PERMISSION_GRANTED) {
                callPhone();
            } else {
                // Permission Denied
                Toast.makeText(MainActivity.this, "Permission Denied", Toast.LENGTH_SHORT).show();
            }
            return;
        }
        super.onRequestPermissionsResult(requestCode, permissions, grantResults);
    }
}
```

上述代码中的拨打电话、打开网页等操作都需要添加相应的权限。而随着 Android 版本的不断提高，相应的权限要求也随之提升，因此，在 Java 代码中动态添加权限之前，还需要在清单文件 AndroidManifest.xml 中添加相应的权限，示例如下：

```
<?xml version="1.0" encoding="utf-8"?>
<manifest xmlns:android="http://schemas.android.com/apk/res/android"
    package="com.ugrow.ch04_4d1_intent">
    <!--打电话的权限-->
    <uses-permission android:name="android.permission.CALL_PHONE"></uses-permission>
```

```
    <!--访问网络权限-->
    <uses-permission android:name="android.permission.INTERNET"></uses-permission>

    <application
        android:allowBackup="true"
        android:icon="@mipmap/ic_launcher"
        android:label="@string/app_name"
        android:roundIcon="@mipmap/ic_launcher_round"
        android:supportsRtl="true"
        android:theme="@style/AppTheme">
        <activity android:name=".MainActivity">
            <intent-filter>
                <action android:name="android.intent.action.MAIN" />

                <category android:name="android.intent.category.LAUNCHER" />
            </intent-filter>
        </activity>
        <activity android:name=".SecondActivity"></activity>
    </application>

</manifest>
```

4.1.2　使用 Intent 启动组件

Android 应用程序的三个核心组件——活动(Activity)、广播接收器(BroadcastReceiver)以及服务(Service)都可通过 Intent 来启动或激活。对于这三种不同的组件，Intent 提供了不同的启动方法，如表 4-6 所示。

表 4-6　Intent 启动不同组件的方法

核心组件	调用方法	作　用
Activity	Context.startActivity() Activity.startActivityForRestult()	启动一个 Activity 或使一个已存在的 Activity 去做新的工作
Service	Context.startService()	初始化一个 Service 或传递一个新的操作给当前正在运行的 Service
	Context.bindService()	绑定一个已存在的 Service
BroadcastReceiver	Context.sendBroadcast() Context.sendOrderedBroadcast() Context.sendStickyBroadcast()	对所有想接受消息的 Broadcast Receiver 传递消息

多 Activity 的 Android 应用程序可通过 startActivity()方法指定相应的 Intent 对象来启动另外一个 Activity。

【示例 4.2】 在 Android Studio 中新建项目 Ch04_4D2_Activity，通过 Intent 实现多 Activity 的 Android 应用的启动。

在 activity_main.xml 文件中编写以下代码：

```
<?xml version="1.0" encoding="utf-8"?>
<LinearLayout xmlns:android="http://schemas.android.com/apk/res/android"
    xmlns:tools="http://schemas.android.com/tools"
    android:layout_width="match_parent"
    android:layout_height="match_parent"
    android:orientation="vertical"
    android:padding="8dp"
    tools:context="com.ugrow.ch04_4d2_activity.MainActivity">

    <TextView
        android:layout_width="match_parent"
        android:layout_height="wrap_content"
        android:text="第一个 Activity"
        android:textColor="#000"/>

    <RadioGroup
        android:id="@+id/rg"
        android:layout_width="wrap_content"
        android:layout_height="wrap_content"
        android:orientation="vertical">


        <RadioButton
            android:id="@+id/rb_android"
            android:layout_width="wrap_content"
            android:layout_height="wrap_content"
            android:text="Android" />

        <RadioButton
            android:id="@+id/rb_symbian"
            android:layout_width="wrap_content"
            android:layout_height="wrap_content"
            android:text="Symbian" />

        <RadioButton
            android:id="@+id/rb_other"
            android:layout_width="wrap_content"
```

```
            android:layout_height="wrap_content"
            android:text="Other" />

        <Button
            android:id="@+id/btn_submit"
            android:layout_width="wrap_content"
            android:layout_height="wrap_content"
            android:text="提交" />
    </RadioGroup>
</LinearLayout>
```

在 MainActivity 中编写以下代码：

```
public class MainActivity extends AppCompatActivity {

    private RadioGroup mRg;
    private RadioButton mRbAndroid;
    private RadioButton mRbSymbian;
    private RadioButton mRbOther;
    private Button mBtnSubmit;

    @Override
    protected void onCreate(Bundle savedInstanceState) {
        super.onCreate(savedInstanceState);
        setContentView(R.layout.activity_main);
        initView();
    }

    private void initView() {
        mRg = (RadioGroup) findViewById(R.id.rg);
        mRbAndroid =(RadioButton) findViewById(R.id.rb_android);
        mRbSymbian = (RadioButton)findViewById(R.id.rb_symbian);
        mRbOther = (RadioButton)findViewById(R.id.rb_other);
        mBtnSubmit = (Button) findViewById(R.id.btn_submit);
        mBtnSubmit.setOnClickListener(new View.OnClickListener() {
            @Override
            public void onClick(View view) {
                //创建Intent对象
                Intent intent = new Intent();
                //指定Intent的目标组件名称
                intent.setClass(MainActivity.this,SecondActivity.class);
```

```
                //启动Activity
                startActivity(intent);
            }
        });
    }
}
```

上述代码中，MainActivity 使用 activity_main.xml 文件生成程序界面布局，该布局包含了一个 TextView 组件、一个 RadioGroup 组件和一个 Button 组件。其中，RadioGroup 中包含选择操作系统的三个单选按钮；Button 上注册了按钮单击事件的监听器，当单击该按钮时，程序会使用 Intent 对象调用另外一个 Activity：首先使用显示 Intent 的方式，通过调用 setClass()方法设置其目标组件为 SecondActivity；然后调用 startActivity()方法启动 Intent 指定的 Activity。

新建 SecondActivity，在其中编写代码如下：

```
public class SecondActivity extends AppCompatActivity {

    private Button mBtnBack;

    @Override
    protected void onCreate(Bundle savedInstanceState) {
        super.onCreate(savedInstanceState);
        setContentView(R.layout.activity_second);
        initView();
    }

    private void initView() {
        mBtnBack = (Button) findViewById(R.id.btn_back);
        mBtnBack.setOnClickListener(new View.OnClickListener() {
            @Override
            public void onClick(View view) {
                //关闭该 Activity
                finish();
            }
        });
    }
}
```

新建 activity_second.xml，编写代码如下：

```
<?xml version="1.0" encoding="utf-8"?>
<LinearLayout xmlns:android="http://schemas.android.com/apk/res/android"
    xmlns:tools="http://schemas.android.com/tools"
    android:layout_width="match_parent"
```

```
    android:layout_height="match_parent"
    android:orientation="vertical"
    tools:context="com.ugrow.ch04_4d2_activity.SecondActivity">

    <TextView
        android:layout_width="fill_parent"
        android:layout_height="wrap_content"
        android:text="第二个 Activity"
        android:textColor="#000" />
   <TextView
        android:id="@+id/tv_selected"
        android:layout_width="match_parent"
        android:layout_height="wrap_content"
        android:layout_marginTop="8dp"
        android:textColor="#000" />

    <Button
        android:id="@+id/btn_back"
        android:layout_width="wrap_content"
        android:layout_height="wrap_content"
        android:text="返回" />


</LinearLayout>
```

上述代码中，SecondActivity 使用 activity_second.xml 生成程序界面，该布局包含一个 TextView 和一个 Button 组件(【返回】按钮)，单击该按钮时，会关闭当前的 Activity。

另外需要注意，在使用 Android Studio 创建 Android 工程时，系统在 AndroidManifest.xml 中自动生成了主 Activity(MainActivity)的定义，但没有生成关于 SecondActivity 的定义，因此需要在 AndroidManifest.xml 中添加 SecondActivity 的相关配置，否则在系统运行时，会因找不到 SecondActivity 而出现异常终止的错误，如图 4-1 所示。

```
Error
me: FATAL EXCEPTION: main
    Process: com.ugrow.ch04_4d2_activity, PID: 20197
    android.content.ActivityNotFoundException: Unable to find explicit activity class {com.ugrow.ch04_4d2_activity/com.ugrow.c
        at android.app.Instrumentation.checkStartActivityResult(Instrumentation.java:1805)
        at android.app.Instrumentation.execStartActivity(Instrumentation.java:1523)
        at android.app.Activity.startActivityForResult(Activity.java:4224)
        at android.support.v4.app.BaseFragmentActivityJB.startActivityForResult(BaseFragmentActivityJB.java:50)
```

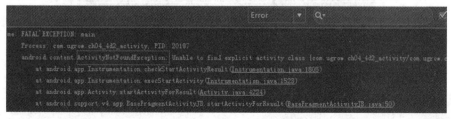

图 4-1　因找不到 SecondActivity 而出现异常终止错误

在 AndroidManifest.xml 中添加 SecondActivity 的配置，代码如下：

```
<?xml version="1.0" encoding="utf-8"?>
<manifest xmlns:android="http://schemas.android.com/apk/res/android"
```

```
        package="com.ugrow.ch04_4d2_activity" >
            <application
            android:allowBackup= "true"
            android:icon="@mipmap/ic_launcher"
            android:label="@string/app_name">
            <activity android:name=".MainActivity">
                <intent-filter>
                    <action android:name="android.intent.action.MAIN" />
                    <category android:name="android.intent.category.LAUNCHER" />
                </intent-filter>
            </activity>
            <activity android:name=".SecondActivity"/>
        </application>
</manifest>
```

启动该 Android 程序，单击 MainActivity 中的【提交】按钮，将切换到 SencondActivity 界面，如图 4-2 所示。

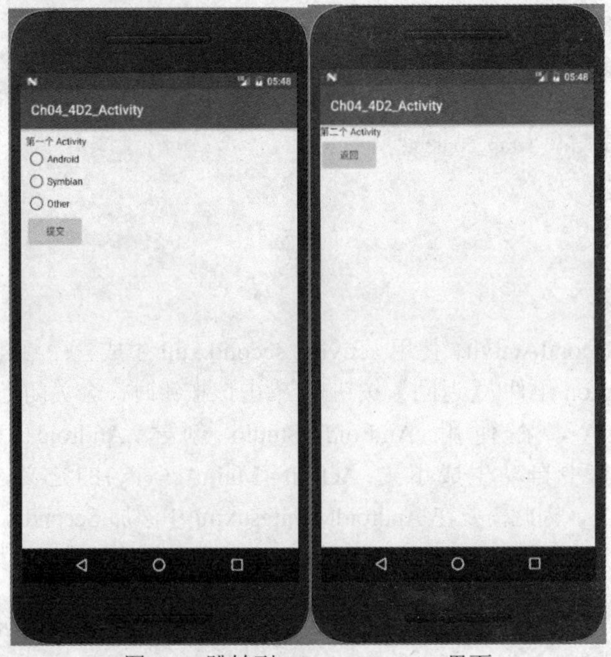

图 4-2　跳转到 SencondActivity 界面

4.2　Intent 数据传递

Intent 的 Extra 属性用于添加一些附加信息，利用该属性可以进行数据的传递。将传递的数据存放到 Extra 属性中有如下两种方式：

(1) 一种是直接调用 putExtra()方法。该方式是将数据添加到 Extra 属性中，然后可以通过调用 getXXXExtra()方法进行获取。这种方式比较简单、直接，主要在数据量比较少

的情况下使用。

(2) 另一种是先将数据封装到 Bundle 包中。Bundle 是一个“键/值”映射的哈希表，当数据量比较多时，可以使用 Bundle 存放数据，然后通过 putExtras()方法将 Bundle 对象添加到 Extra 属性中，再使用 getExtras()方法获取存放的 Bundle 对象，最后读取 Bundle 包中的数据。这种方式是间接通过 Bundle 包对数据先进行封装，再进行传递，实现起来比较繁琐，因此主要在数据量较多的情况下使用。

4.2.1　Intent 传值

下面通过一个示例演示使用 Intent 传递数据的方法。

【示例 4.3】 在项目 Ch04_4D2_Activity 中，修改代码使用上述第一种方式实现多个 Activity 间的数据传递。

在 MainActivity 中编写代码如下：

```
public class MainActivity extends AppCompatActivity {

    private RadioGroup mRg;
    private RadioButton mRbAndroid;
    private RadioButton mRbSymbian;
    private RadioButton mRbOther;
    private Button mBtnSubmit;

    @Override
    protected void onCreate(Bundle savedInstanceState) {
        super.onCreate(savedInstanceState);
        setContentView(R.layout.activity_main);
        initView();
    }

    private void initView() {
        mRg = (RadioGroup) findViewById(R.id.rg);
        mRbAndroid = (RadioButton) findViewById(R.id.rb_android);
        mRbSymbian = (RadioButton) findViewById(R.id.rb_symbian);
        mRbOther = (RadioButton) findViewById(R.id.rb_other);
        mBtnSubmit = (Button) findViewById(R.id.btn_submit);
        mBtnSubmit.setOnClickListener(new View.OnClickListener() {
            @Override
            public void onClick(View view) {
                //创建 Intent 对象
                Intent intent = new Intent();

                //指定 Intent 的目标组件名称
```

```
                intent.setClass(MainActivity.this, SecondActivity.class);
                //根据用户选择不同的单选按钮，向 Intent 对象的 Extra 属性中存放不同的值
                if (mRbAndroid.isChecked()) {
                    intent.putExtra("selected", mRbAndroid.getText().toString());
                } else if (mRbSymbian.isChecked()) {
                    intent.putExtra("selected", mRbSymbian.getText().toString());
                } else if (mRbOther.isChecked()) {
                    intent.putExtra("selected", mRbOther.getText().toString());
                } else {
                    intent.putExtra("selected", "");
                }
                //启动 Activity
                startActivity(intent);
            }
        });
    }
}
```

上述代码中，MainActivity 包含一个 RadioGroup 和一个 Button 组件。单击 Button 按钮后，会将用户选择的不同选项的值的数据保存到 Intent 对象的 Extra 属性中，并通过 Intent 对象的 Extra 属性将数据传递给相应的 Activity。

在 SecondActivity 中编写代码如下：

```
public class SecondActivity extends AppCompatActivity {

    private Button mBtnBack;
    private TextView mTvSelected;
    private Intent intent;
    private String selected;

    @Override
    protected void onCreate(Bundle savedInstanceState) {
        super.onCreate(savedInstanceState);
        setContentView(R.layout.activity_second);
        initView();
    }

    private void initView() {
        mBtnBack = (Button) findViewById(R.id.btn_back);
        mTvSelected = (TextView) findViewById(R.id.tv_selected);
        intent = getIntent();
        if (intent.hasExtra("selected")) {
```

```
            selected = intent.getStringExtra("selected");
            if (!selected.equals("")) {
                mTvSelected.setText(selected + "被选中");
            } else {
                mTvSelected.setText("没有选中任何系统");
            }
        }
        mBtnBack.setOnClickListener(new View.OnClickListener() {
            @Override
            public void onClick(View view) {
                //关闭该 Activity
                finish();
            }
        });

    }
}
```

上述代码中，SecondActivity 通过调用 getIntent()方法获取传递过来的 Intent 对象，再通过调用 Intent 对象的 getStringExtra()方法获取其 Extra 属性中名为“selected”的内容，即 MainActivity 传递过来的用户选择的系统信息，并将信息显示在文本控件中。

运行该 Android 程序，效果如图 4-3 所示：单击 MainActivity 界面的【提交】按钮，此时手机界面会切换到另外一个 Activity(SecondActivity)的界面，该界面会显示前一个页面中被选中的单选按钮的值。

图 4-3　使用 Intent 对象实现 Activity 间的传值

4.2.2 Bundle 传值

在介绍利用 Bundle 包实现 Activity 之间的数据传递的方法之前，首先要了解 Bundle 类中的常用方法，如表 4-7 所示。

表 4-7　Bundle 类常用方法

方　法	功 能 描 述
Object get(String key)	获取关键字 key 对应的数据
boolean getBoolean(String key)	获取关键字 key 对应的布尔值，若找不到关键字的记录，则返回 false
boolean getBoolean (String key, boolean defaultValue)	获取关键字 key 对应的布尔值，若找不到关键字的记录，则返回 defaultValue
Bundle getBundle(String key)	获取关键字 key 对应的 Bundle 对象，若找不到关键字的记录，则返回 null
char getChar (String key)	获取关键字 key 对应的 char 值，若找不到关键字的记录，则返回 0
char getChar (String key, char defaultValue)	获取关键字 key 对应的 char 值，若找不到关键字的记录，则返回 defaultValue
boolean hasFileDescriptors()	设置 Bundle 对象是否包含文件描述符，返回 true 则 Bundle 对象包含文件描述符，否则不包含
void putAll (Bundle map)	将 map 插入到该 Bundle 对象中
void putBoolean (String key, boolean value)	将布尔值 value 插入到该 Bundle 对象中，若关键字 key 已存在，则原有值被 value 替代
void putBundle (String key, Bundle value)	将 Bundle 对象 value 插入到该 Bundle 对象中
void putByte (String key, byte value)	将字节值 value 插入到该 Bundle 对象中
void remove (String key)	移除关键字为 key 的记录
int size ()	获取 Bundle 对象的关键字个数

由上述表格可知，Bundle 传值方式主要通过 putXXX()方法将不同数据类型封装到 Bundle 对象中，再通过 getXXX()方法获取相应数据类型的数据。

【示例 4.4】 修改项目 Ch04_4D2_Activity 中的代码，使用 Bundle 传值实现示例 4.3 所示功能。

首先将 MainActivity 的代码进行以下修改：

```
public class MainActivity extends AppCompatActivity {

    private RadioGroup mRg;
    private RadioButton mRbAndroid;
    private RadioButton mRbSymbian;
    private RadioButton mRbOther;
```

```
private Button mBtnSubmit;

@Override
protected void onCreate(Bundle savedInstanceState) {
    super.onCreate(savedInstanceState);
    setContentView(R.layout.activity_main);
    initView();
}

private void initView() {
    mRg = (RadioGroup) findViewById(R.id.rg);
    mRbAndroid = (RadioButton) findViewById(R.id.rb_android);
    mRbSymbian = (RadioButton) findViewById(R.id.rb_symbian);
    mRbOther = (RadioButton) findViewById(R.id.rb_other);
    mBtnSubmit = (Button) findViewById(R.id.btn_submit);
    mBtnSubmit.setOnClickListener(new View.OnClickListener() {
        @Override
        public void onClick(View view) {
            //创建 Intent 对象
            Intent intent = new Intent();
            //指定 Intent 的目标组件名称
            intent.setClass(MainActivity.this, SecondActivity.class);
            //创建 Bundle 对象，该对象用于记录被传送的数据
            Bundle bundle = new Bundle();
            //根据用户选择的不同单选按钮，向 Intent 对象的 Extra 属性中存入不同的值
            if (mRbAndroid.isChecked()) {
                bundle.putString("selected", mRbAndroid.getText().toString());
            } else if (mRbSymbian.isChecked()) {
                bundle.putString("selected", mRbSymbian.getText().toString());
            } else if (mRbOther.isChecked()) {
                bundle.putString("selected", mRbOther.getText().toString());
            } else {
                bundle.putString("selected", "");
            }
            //将 Bundle 对象数据封装到 Intent 对象中，通过该 Intent 对象将数据传递给相应的 Activity
            intent.putExtras(bundle);
            //启动 Activity
            startActivity(intent);
        }
    });
```

```
    }
}
```

上述代码通过 Bundle 的 putString()方法将被选中的单选按钮的文本值封装到该 Bundle 对象中，然后通过调用 Intent 对象的 putExtras()方法，将 Bundle 对象捆绑到 Intent 对象中。

然后将 SecondActivity 的代码进行以下修改：

```
public class SecondActivity extends AppCompatActivity {

    private Button mBtnBack;
    private TextView mTvSelected;
    private Intent intent;
    private String selected;
    private Bundle bundle;

    @Override
    protected void onCreate(Bundle savedInstanceState) {
        super.onCreate(savedInstanceState);
        setContentView(R.layout.activity_second);
        initView();
    }

    private void initView() {
        mBtnBack = (Button) findViewById(R.id.btn_back);
        mTvSelected = (TextView) findViewById(R.id.tv_selected);
        intent = getIntent();
        bundle = intent.getExtras();
        if (bundle != null) {
            selected = bundle.getString("selected");
            if (!selected.equals("")) {
                mTvSelected.setText(selected + "被选中");
            } else {
                mTvSelected.setText("没有选中任何系统");
            }
        }
        mBtnBack.setOnClickListener(new View.OnClickListener() {
            @Override
            public void onClick(View view) {
                //关闭该 Activity
                finish();
            }
```

```
        });

    }
}
```

上述代码通过 getExtras()方法获取 Intent 中的 Bundle 对象，然后调用 Bundle 对象的 getString()方法获取指定 key 的值，最后将值显示在文本控件中，其运行结果与图 4-3 所示相同。由此，可以看到 Bundle 的使用虽然相对复杂一些，但传递数据比较多时，可以通过循环遍历将数据提取出来，比较方便。

4.3　设置 Activity 权限

Android 系统开放了许多底层应用供用户调用，同其他系统不同，Android 系统有自己特殊的调用底层应用的方式，它会在运行时检查该用户程序是否有权限调用该底层应用，因此需要通过某种方式设置 Activity 权限才能运行相应的应用。这种方式提供了程序使用系统应用的安全性保证，底层应用只有用相应的权限才能被用户程序使用，否则程序运行会出现错误。

在 AndroidManifest.xml 中可以配置应用程序的权限。例如，打电话应用需要调用系统提供的电话底层处理 ACTION_CALL 行为，这时需要在 AndroidManifest.xml 中的<uses-permission>添加打电话的权限属性，代码如下：

```
<uses-permission android:name="android.permission.CALL_PHONE" />
/*这样用户就能使用 ACTION_CALL 来激活打电话的应用。如果不在清单文件(AndroidManifest.xml)中设置许可，则运行电话应用时会弹出提示用户缺少相应权限的异常错误。*/
```

当运行未设置 android.permission.CALL_PHONE 权限的 ACTION_CALL 应用时，系统会弹出安全异常错误(java.lang.SecurityException)的提示，如图 4-4 所示。

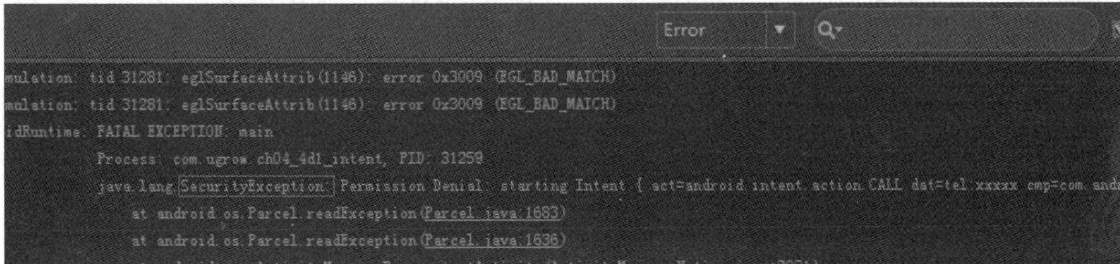

图 4-4　缺少相应权限的异常错误

可以看到，异常错误提示了发生异常的详细原因，包括触发异常的行为以及所需要的许可，从而可以判断出用户程序出现异常错误是因为使用了 android.Intent.action.CALL 行为，而这种行为需要 android.permission. CALL_PHONE 的权限。

需要注意，<uses-permission>标签包含在<manifest>中，并与<application>标签属于同一级别，代码如下：

```
<?xml version="1.0" encoding="utf-8"?>
<manifest>
```

```
        <uses-permission android:name="android.permission.CALL_PHONE" />
        <application>
                <activity>
                        <intent-filter>
                        </intent-filter>
                </activity>
        </application>
</manifest>
```

Android 系统提供了很多权限，用户使用相应底层服务时，需要在 AndroidManifest.xml 中添加相应的权限。Android 系统提供的主要权限如表 4-8 所示。

表 4-8　Android 主要权限列表

权 限 名 称	权 限 功 能
android.permission.ACCESS_CHECKIN_PROPERTIES	允许读写 checkin 数据库中的表 properties
android.permission.ACCESS_COARSE_LOCATION	允许程序通过访问 Cell ID 或 WIFI 热点来获取粗略的位置
android.permission.BLUETOOTH	允许程序同匹配的蓝牙设备建立连接
android.permission.CALL_PHONE	允许程序拨打电话，无需通过拨号器的用户界面确认
ndroid.permission.CLEAR_APP_CACHE	允许用户清除该设备上的所有安装程序的缓存
android.permission.CLEAR_APP_USER_DATA	允许程序清除用户数据
android.permission.CONTROL_LOCATION_UPDATES	允许启用/禁止无线模块的位置更新
android.permission.PROCESS_OUTGOING_CALLS	允许程序监视、修改或者删除已拨电话
android.permission.READ_INPUT_STATE	允许程序获取当前按键状态
android.permission.REBOOT	请求用户设备重启的操作
android.permission.RECEIVE_BOOT_COMPLETED	允许一个程序接收到系统启动后的广播 ACTION_BOOT_COMPLETED
android.permission.RECEIVE_MMS	允许程序处理收到 MMS 彩信
android.permission.RECEIVE_SMS	允许程序处理收到短信息
android.permission.SET_TIME_ZONE	允许程序设置系统时区
android.permission.SET_WALLPAPER	允许程序设置手机壁纸
android.permission.STATUS_BAR	允许程序打开、关闭或禁用状态栏及图标
android.permission.WRITE_CALENDAR	允许程序写入但不读取用户日历
android.permission.WRITE_CONTACTS	允许程序写入但不读取用户联系人数据
android.permission.WRITE_GSERVICES	允许程序修改 Google 服务地图
android.permission.WRITE_SETTINGS	允许程序读取或修改系统设置
android.permission.WRITE_SMS	允许程序修改短信
android.permission.DELETE_CACHE_FILES	允许程序删除缓存文件
android.permission.DELETE_PACKAGES	允许程序删除包

续表一

权 限 名 称	权 限 功 能
android.permission.DEVICE_POWER	允许访问底层电源管理
android.permission.DISABLE_KEYGUARD	允许程序禁用键盘锁
android.permission.DUMP	允许程序获取系统服务的状态 dump 信息
android.permission.GET_ACCOUNTS	允许访问 Accounts Service 中账户列表
android.permission.GET_PACKAGE_SIZE	允许程序获取任何 package 占用空间大小
android.permission.GET_TASKS	允许程序获取当前或最近运行的任务的概要信息
android.permission.HARDWARE_TEST	允许访问程序系统硬件
android.permission.INTERNET	允许程序打开网络套接字
android.permission.MODIFY_AUDIO_SETTINGS	允许程序修改系统音频设置
android.permission.MODIFY_PHONE_STATE	允许修改电话状态，如充电
android.permission.MOUNT_UNMOUNT_FILESYSTEMS	允许挂载和反挂载移动设备
android.permission.SET_ACTIVITY_WATCHER	允许程序监视和控制系统 Activities 的启动
android.permission.SET_ALWAYS_FINISH	允许程序控制 Activity 在处于后台时是否立即结束
android.permission.SET_DEBUG_APP	配置一个用于调试的程序
android.permission.SET_ORIENTATION	允许通过底层应用设置屏幕方向
android.permission.SET_PREFERRED_APPLICATIONS	允许程序修改默认程序列表
android.permission.SET_PROCESS_FOREGROUND	允许程序强制将当前运行程序转到前台运行
android.permission.SET_PROCESS_LIMIT	允许设置最大的系统当前运行进程数量
android.permission.ACCESS_LOCATION_EXTRA_COMMANDS	允许应用程序使用额外的位置提供命令
android.permission.ACCESS_MOCK_LOCATION	允许程序创建用于测试的模拟位置
android.permission.ACCESS_NETWORK_STATE	允许程序获取网络状态信息
android.permission.ACCESS_SURFACE_FLINGER	允许程序获取 SurfaceFlinger 底层特性
android.permission.ACCESS_WIFI_STATE	允许程序获取 WIFI 网络信息
android.permission.ADD_SYSTEM_SERVICE	允许程序发布系统级服务
android.permission.BATTERY_STATS	允许程序更新手机电池统计信息
android.permission.BLUETOOTH_ADMIN	允许程序发现和配对蓝牙设备
android.permission.BROADCAST_PACKAGE_REMOVED	允许程序广播一个已经被移除的包的消息
android.permission.BROADCAST_STICKY	允许一个程序广播带数据的 Intents
android.permission.CAMERA	请求使用照相设备
android.permission.CHANGE_COMPONENT_ENABLED_STATE	允许一个程序启用或禁用其他组件
android.permission.CHANGE_CONFIGURATION	允许一个程序修改当前设置

续表二

权 限 名 称	权 限 功 能
android.permission.CHANGE_NETWORK_STATE	允许程序改变网络连接状态
android.permission.CHANGE_WIFI_STATE	允许程序改变 WIFI 连接状态
android.permission.READ_SYNC_SETTINGS	允许程序读取同步设置
android.permission.READ_CONTACTS	允许程序读取用户联系人数据

本 章 小 结

通过本章的学习，读者应当了解：

✧ Intent 由动作、数据、分类、类型、组件、扩展信息和标记等内容组成。

✧ Action 属性用于描述 Intent 要完成的动作，对要执行的动作进行一个简要描述。

✧ Category 属性指明一个执行 Action 的分类。

✧ ComponentName 属性用于指明 Intent 的目标组件的类名称。

✧ 多 Activity 的 Android 应用程序可通过 startActivity()方法指定相应的 Intent 对象来启动另外一个 Activity。

✧ Intent 的 Extra 属性用于添加一些附加信息，利用该属性可以进行数据的传递。

✧ 将传递的信息存放到 Extra 属性中有以下两种方式：一种是直接将信息添加到 Extra 属性中，另一种是将数据封装到 Bundle 包中。

✧ 新建 Activity 实例后，需要在配置文件 AndroidManifest.xml 中设置 Activity 权限。

本 章 练 习

1. 下列 Intent 的 Action 属性中，用来标识应用程序入口的是______。

   A．ACTION_CALL　　　　B．ACTION_VIEW

   C．ACTION_MAIN　　　　D．ACTION_SCREEN_ON

2. Android 系统为终端用户提供了开发应用程序交互功能的组件，这些组件包括______。

   A．广播接收器　　B．意图　　C．适配器　　D．内容提供器

3. 下列关于启动 Intent 的说法正确的是______。

   A．Context.startActivity()用于启动 Activity

   B．Context.startService()用于启动 Service

   C．Context.sendBroadcast()用于发送广播

   D．Context.startBroadcast()用于开始广播

4. Intent 由______、数据、______、类型、组件和______等内容组成，每个组成部分都由相应的属性进行表示，并提供设置和获取相应属性的方法。

5. 编写程序，使用两种方式实现 Activity 之间的数据传递。

第5章　广播(Broadcast)

本章目标

- 了解 Android Broadcast 工作机制及特点
- 掌握广播的三大要素
- 掌握广播接收者的注册方式
- 了解广播的优先级

5.1 Broadcast 简介

广播(Broadcast)在 Android 中有着广泛的应用。有一些操作完成以后会发送广播，比如发送一条短信，或者打出一个电话，如果某个程序接收了这个广播，就会进行相应的处理。

Android 中的广播和传统意义上的电台广播有相似之处，因为发送方并不关心接收方是否接收数据，也不关心接收方如何处理数据。广播可以被一个或者多个应用程序所接收，也可能不被任何应用程序所接收。

Android 中的广播是非常灵活的，因为 Android 中的每个应用程序都可以对自己感兴趣的广播进行注册，这样该程序就只会接收到自己关心的广播内容，这些广播可能是来自于系统的，也可能是来自于其他应用程序的。Android 提供了一套完整的 API，允许应用程序自由地发送和接收广播。

5.1.1 Broadcast 三要素

在 Android 中，任何一条发送成功并实现了相应功能的广播都具备以下三大要素：

(1) 广播(Broadcast)：发送广播的操作本身，一种被广泛用于在应用程序间传输信息的机制，使用 sendBroadcast()方法发送广播。

(2) 广播接收者(BroadcastReceiver)：对发送出来的 Broadcast 进行过滤接收并响应的一类组件。

(3) 意图(Intent)：用于保存广播相关信息的媒介，已在本书第 4 章中详细讲解。

5.1.2 Broadcast 生命周期

广播(Broadcast)的生命周期是由广播接收者(BroadcastReceiver)决定的：BroadcastReceiver 仅在执行 onReceive()方法时处于活跃状态，onReceive()方法执行完毕后即变为失活状态；而广播(Broadcast)的生命周期亦从回调 onReceive()方法开始，到该方法返回结果后结束。

广播的生命周期只有十秒左右，如果在 onReceive()方法中执行超过十秒的操作，程序就会报 ANR(Application No Response 缩写，即程序无响应)的错误信息。因此，如果响应某个广播信息需要的时间较长，则通常会将其放入一个衍生的子线程中完成，而不在主线程内完成，以保证用户交互的流畅。

5.1.3 Broadcast 分类

Android 中的广播主要可分为两种类型：标准广播和有序广播。

1. 标准广播

一种完全异步执行的广播，发送方使用 sendBroadcast()方法发出广播。广播发出后，

发送的内容几乎同时到达多个广播接收者，广播接收者之间没有先后顺序，亦不相互影响，这种广播的效率较高，但广播的传播无法被截断。

一个标准广播的示例如下：

```
Intent intent = new Intent();
Intent.setAction("...");
sendBroadCast(intent);
```

2. 有序广播

一种同步执行的广播，发送方使用 sendOrderedBroadcast()方法发出广播。广播发出后，同一时刻只有一个广播接收者能够收到这条广播，当该接收者将这条广播处理完毕后，广播才会继续传递。所以此类广播的接收者是有先后顺序的，需要为各广播接收者提前设置优先级，优先级高的接收者会先接收到广播，且该接收者可以使用 abortBroadcast()方法终止广播的继续发送。

一个有序广播的示例如下：

```
Intent intent = new Intent();
Intent.setAction("...");
sendOrderedBroadCast(intent,null);
```

5.2 BroadcastReceiver

BroadcastReceiver 是 Android 应用程序四大组件之一，是为了实现系统广播功能而提供的一种组件，其对广播事件的处理机制是系统级别的。例如，手机开机后会发出一条广播，电池的电量发生变化时会发出一条广播，时间或时区发生改变时也会发出一条广播，而如果想要接收到这些广播，就需要定义一个 BroadcastReceiver 来接收该广播。

使用 BroadcastReceiver 来接收广播时，前提是要将该 BroadcastReceiver 进行注册。当有广播事件产生时，Android 操作系统会通知已经注册的 BroadcastReceiver 产生了一个什么事件，每个接收者先判断是不是自己需要的事件，如果是，再进行相应的处理。

5.2.1 BroadcastReceiver 注册

BroadcastReceiver 用于监听和接收被广播的事件(Intent)，为达到这个目的，必须先将 BroadcastReceiver 进行注册，注册的方式有两种：一种是在清单文件 AndroidManifest.xml 中注册，称为静态注册；另一种是在代码中注册，称为动态注册。

1. 静态注册

静态注册方式即是在清单文件 AndroidManifest.xml 的 application 标签中定义一个 receiver，并设置要接收的 action。

静态注册方式的特点是：不管 BroadcastReceiver 所在的应用程序是否处于活跃状态，该 BroadcastReceiver 都会进行监听。

一个静态注册的示例如下：

```
<receiver android:name="”MyReceiver”>
        <intent-filter>
        <action android:name="MyReceiver_Action"/>
        </intent-filter>
</receiver>
```

上述代码中，MyReceiver 是继承自 BroadcastReceiver 的类，其重写了 onReceive()方法，在该方法中对广播进行处理，并通过<intent-filter>标签为 receiver 设置过滤器，接收指定 action 的广播。

【示例 5.1】 在 Android Studio 中新建项目 Ch05_5D1_BroadCast，演示静态注册 BroadcastReceiver 的操作。

在 activity_main.xml 中编写以下代码：

```
<?xml version="1.0" encoding="utf-8"?>
<LinearLayout xmlns:android="http://schemas.android.com/apk/res/android"
        xmlns:tools="http://schemas.android.com/tools"
        android:layout_width="match_parent"
        android:layout_height="match_parent"
        tools:context="com.ugrow.ch05_5d1_broadcast.MainActivity">

        <Button
                android:id="@+id/btn"
                android:layout_width="wrap_content"
                android:layout_height="wrap_content"
                android:text="静态注册"
                android:textColor="#000"/>
</LinearLayout>
```

在 MainActivity 中编写以下代码：

```
public class MainActivity extends AppCompatActivity {

        private Button mBtn;

        @Override
        protected void onCreate(Bundle savedInstanceState) {
                super.onCreate(savedInstanceState);
                setContentView(R.layout.activity_main);
                mBtn = (Button) findViewById(R.id.btn);
                mBtn.setOnClickListener(new View.OnClickListener() {
                @Override
        public void onClick(View view) {
                        Intent intent = new Intent();
```

```
                    intent.putExtra("name","静态注册");
                    intent.setAction("broadcast");
                    sendBroadcast(intent);

            }
        });
    }
}
```

上述代码中，使用 Intent 意图对象传递数据，并通过 sendBroadcast()方法发送广播。

在 AndroidManifest.xml 中编写以下代码：

```
<?xml version="1.0" encoding="utf-8"?>
<manifest xmlns:android="http://schemas.android.com/apk/res/android"
    package="com.ugrow.ch05_5d1_broadcast">

    <application
        android:allowBackup="true"
        android:icon="@mipmap/ic_launcher"
        android:label="@string/app_name"
        android:roundIcon="@mipmap/ic_launcher_round"
        android:supportsRtl="true"
        android:theme="@style/AppTheme">
        <activity android:name=".MainActivity">
            <intent-filter>
                <action android:name="android.intent.action.MAIN" />

                <category android:name="android.intent.category.LAUNCHER" />
            </intent-filter>
    </activity>
    <receiver android:name=".MyReceiver">
        <intent-filter>
            <action android:name="broadcast"></action>
        </intent-filter>
    </receiver>
    </application>

</manifest>
```

在 com.ugrow.ch05_5d1_broadcast 文件夹下新建 MyReceiver 类，继承 BroadcastReceiver 类，并重写 onReceive()方法：

```
public class MyReceiver extends BroadcastReceiver{
```

```
    @Override
    public void onReceive(Context context, Intent intent) {
         String action = intent.getAction();
        if ("broadcast".equals(action)){
            String name = intent.getStringExtra("name");
            Toast.makeText(context,name,Toast.LENGTH_SHORT).show();
        }
    }
}
```

上述代码获取广播发送的 Intent 中的 action，并判断其是否与注册 BroadcastReceiver 时在清单文件中指定的 action 的 name 属性值一致，如果一致，则使用 BroadcastReceiver 中的 Intent 获取广播传递的消息内容。

编辑完毕，运行应用程序，单击【静态注册】按钮，即会在屏幕上显示字符“静态注册”，如图 5-1 所示。

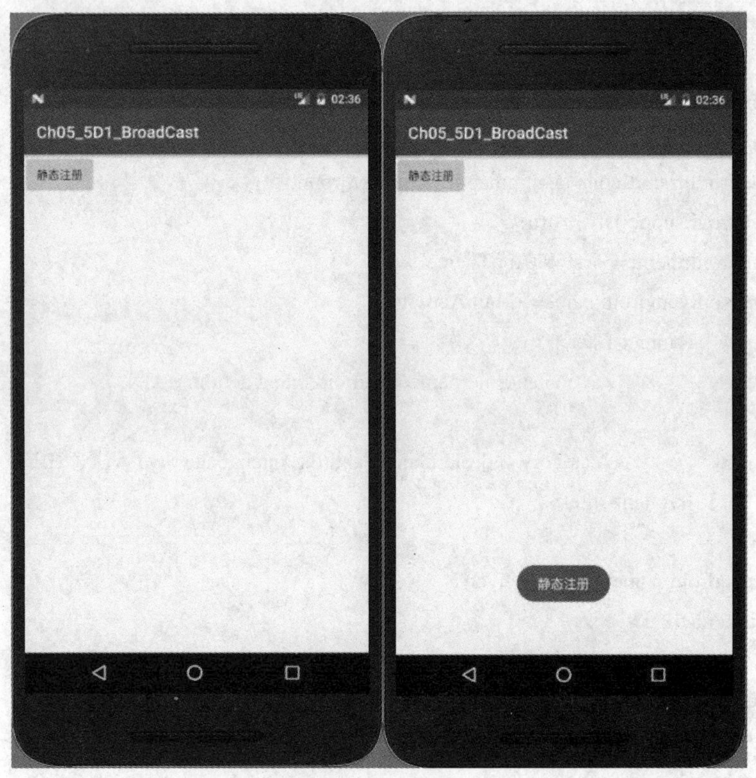

图 5-1　静态注册

2．动态注册

动态注册方式需要在 Activity 里面调用方法 registerReceiver()进行注册，该方法有两个参数，一个是 receiver，另一个是 intentFilter，与静态注册方式类似，也需要向该方法中传入要接收的 action。

动态注册方式的特点有两个，分别是：① 必须在程序启动之后才能接收到广播，因为注册的逻辑是写在 onCreate()方法中的；② 由于手机内存有限，为节约内存，防止内存泄漏，广播接收完毕后，需要将动态注册的 BroadcastReceiver 销毁，销毁操作通过在 onDestroy()方法中调用 unregisterReceiver()方法实现。

一个动态注册的示例如下：

```
MyReceiver receiver = new MyReceiver();
//创建过滤器，并指定 action，使之用于接收相同 action 的广播
IntentFilter filter = new IntentFilter("MyReceiver_Action");
//注册广播接收器
registerReceiver(receiver,filter);
```

【示例 5.2】 在 Android Studio 中新建项目 Ch05_5D2_BroadCast，演示动态注册 BroadcastReceiver 的操作。

在 activity_main.xml 中编写以下代码：

```
<?xml version="1.0" encoding="utf-8"?>
<LinearLayout xmlns:android="http://schemas.android.com/apk/res/android"
    xmlns:tools="http://schemas.android.com/tools"
    android:layout_width="match_parent"
    android:layout_height="match_parent"
    tools:context="com.ugrow.ch05_5d2_broadcast.MainActivity">

    <Button
        android:id="@+id/btn"
        android:layout_width="wrap_content"
        android:layout_height="wrap_content"
        android:text="动态注册"
        android:textColor="#000"/>
    </LinearLayout>
```

在 MainActivity 中编写以下代码：

```
public class MainActivity extends AppCompatActivity {

    private Button mBtn;
    private MyReceiver myReceiver;

    @Override
    protected void onCreate(Bundle savedInstanceState) {
        super.onCreate(savedInstanceState);
        setContentView(R.layout.activity_main);
        mBtn = (Button) findViewById(R.id.btn);
        myReceiver = new MyReceiver();
```

```
        //为 BroadcastReceiver 指定 action，即要监听的消息名字
        IntentFilter intentFilter = new IntentFilter("filter");
        //注册监听
        registerReceiver(myReceiver, intentFilter);

        mBtn.setOnClickListener(new View.OnClickListener() {
            @Override
            public void onClick(View view) {
                Intent intent = new Intent("filter");
                intent.putExtra("TAG", "动态注册");
                sendBroadcast(intent);
            }
        });
    }

class MyReceiver extends BroadcastReceiver {

        @Override
        public void onReceive(Context context, Intent intent) {
            if ("filter".equals(intent.getAction())) {
            String tag = intent.getStringExtra("TAG");
            Toast.makeText(context, tag, Toast.LENGTH_SHORT).show();
            }
        }
    }

    //注意：在 onDestroy 方法中，必须注销广播，否则有内存泄漏的风险！！！
    @Override
    protected void onDestroy() {
        super.onDestroy();
        if (myReceiver != null) {
            unregisterReceiver(myReceiver);
        }
    }
}
```

上述代码新建了一个内部类 MyReceiver，继承 BroadcastReceiver，重写了 onReceiver()方法，然后在 onCreate()方法中使用 registerReceiver()方法动态注册 BroadcastReceiver。

编辑完毕，运行应用程序，单击【动态注册】按钮，在屏幕上显示字符串“动态注册”，如图 5-2 所示。

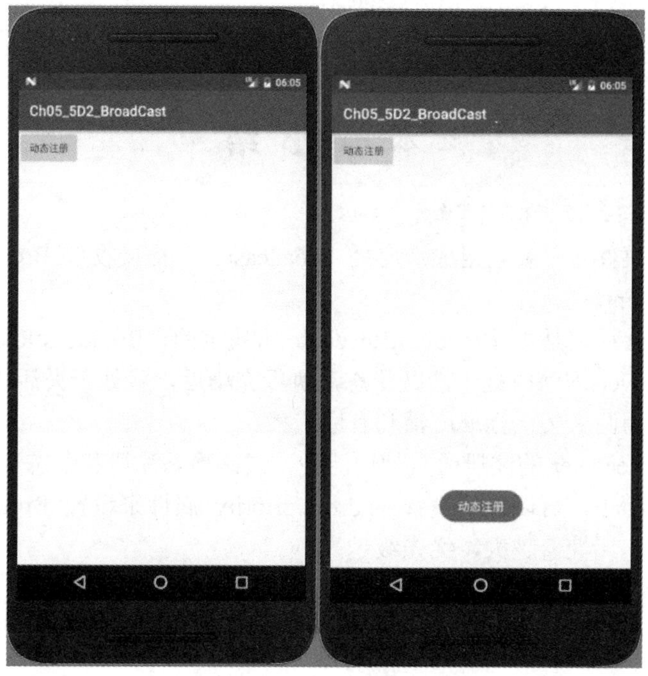

图 5-2　动态注册

5.2.2 BroadcastReceiver 优先级

当 BroadcastReceiver 注册完成后，可以通过设置 BroadcastReceiver 的优先级来规定广播内容的传递顺序。

在清单文件 AndroidManifest.xml 中可以设置 BroadcastReceiver 的优先级，示例如下：

```
<receiver android:name=".MyReceiver">
    <intent-filter android:priority="100">
    <action android:name="broadcast"></action>
    </intent-filter>
</receiver>
```

优先级的声明需要通过<intent-filter>标签中的 android:priority 属性进行设置，该属性的值越大，该 BroadcastReceiver 的优先级越高；除此之外也可以调用 IntentFilter 对象的 setPriority()方法进行设置。

有序广播的 BroadcastReceiver 可以终止广播的传播，广播的传播一旦终止，后面的 BroadcastReceiver 就无法接收到该广播；但是，优先级高的 BroadcastReceiver 可以向优先级低的 BroadcastReceiver 发送新的广播内容。

高优先级 BroadcastReceiver 发送新广播的示例代码如下：

```
Bundle bundle = new Bundle();
bundle.putString("NEW","我的优先级比你高...");
setResultExtras(bundle);
```

低优先级 BroadcastReceiver 接收新广播的示例代码如下：

```
Bundle bundle = getResultExtras(true);
String   newString = bundle.getString("NEW","");
```

本 章 小 结

通过本章的学习，读者应当了解：

✧ 一个广播事件由三要素组成：广播 Broadcast、广播接收者 BroadcastReceiver 和意图 Intent。

✧ 广播的生命周期是由 BroadcastReceiver 来决定的，BroadcastReceiver 仅在执行 onReceive()方法时处于活跃状态，执行完毕后，就处于失活状态。

✧ 广播可分为两大类：标准广播和有序广播。

✧ BroadcastReceiver 的注册通常使用两种方式：静态注册和动态注册。

✧ 在有序广播中，可以通过设置 android:priority 属性来设置 BroadcastReceiver 的优先级，属性值越大，优先级越高。

本 章 练 习

1. BroadcastReceiver 在执行下列哪个方法时处于活跃状态____。

   A. sendBroadcast()

   B. onRecive()

   C. registerReceiver()

   D. unregisterReceiver

2. 广播的三大要素包括：______、______和______。
3. BroadcastReceiver 的注册方式分为：______和______。
4. 简述广播的生命周期。
5. 简述如何设置 BroadcastReceiver 的优先级。
6. 编写代码，分别使用静态注册和动态注册实现广播数据的传递。

第 6 章　服务(Service)

本章目标

- 了解 Android Service 的工作机制及特点
- 了解 Service 和 Activity 的不同之处
- 掌握如何创建、启动和停止 Service
- 熟悉 Android 常用的系统服务
- 掌握 NotificationManager 和 Notification 的使用方法

6.1 Service 简介

按照工作的方式，Android 应用程序可分为前台应用程序和后台服务程序两种。Activity 对应的程序是前台程序，可使用 startActivity()方法将 Intent 指定的活动转到前台运行，即将活动控制权由当前活动转到 Intent 指定的活动；Service 对应的程序是后台服务程序(后台服务程序往往需要运行较长时间，甚至可能会从系统启动时开始运行到系统关闭时结束)，其功能类似于 Linux 系统中的守护进程，可使用 startService()方法将指定的应用转到后台运行，即不改变当前运行程序的控制权。

Service 作为 Android 四大组件之一，是一个可以在后台执行长时间运行操作而不提供用户界面的应用组件。Service 可由其他应用组件启动，而且，即使用户切换到其他应用，Service 仍将在后台继续运行。此外，Service 可以绑定组件，与之进行交互，甚至可以执行进程间的通信(IPC)。例如，Service 可以处理网络事务、播放音乐，执行文件 I/O 或与内容提供程序交互，而所有的这一切均可在后台进行。

Service 分为两种类型：

(1) 本地服务(Local Service)：这种服务主要在应用程序内部使用，用于实现应用程序本身的任务，比如自动下载程序。

(2) 远程服务(Remote Service)：这种服务主要在应用程序之间使用，一个应用程序可以使用远程服务调用其他的应用程序，例如天气预报。

Android 提供了一些特殊的 Service 类，如 AbstractInputMethodService、AccessibilityService、IntentService、RecognitionService 以及 WallpaperService。以 AccessibilityService 类为例，当 AccessibilityEvent 事件(比如焦点变化、按钮被单击等)发生后，AccessibilityService 会被自动调用。

6.2 Service 特点

Service 的特点如下：

(1) Service 在后台运行，不可以与用户直接交互。

(2) 一个 Service 不是一个单独的线程。和其他组件一样，默认情况下，Service 中的所有代码都是运行在主线程中。

(3) Service 的使用频率虽然不如 Activity 那么高。但一般都需要用 Service 执行耗时较长的操作，例如播放音乐、下载文件、上传文件等等。但是因为 Service 默认运行在主线程中，因此不能直接用它来执行耗时的请求或者动作，而是最好在 Service 中启动一个新线程来执行耗时的任务。

(4) 需要通过某一个 Activity 或其他 Context 对象来启动 Service。例如采用 context.startService()或 context.bindService()方式。

(5) Service 很大程度上充当了应用程序后台线程管理器的角色。如果在 Activity 中新开启一个线程，当该 Activity 关闭后，该线程依然在工作，但是与开启它的 Activity 失去

联系，也就是说此时的这个线程处于失去管理的状态。而如果使用 Service，则可以对后台运行的线程进行有效地管理。

Service 与 Activity 既有相同之处，也有所区别。

✧ 相同点：使用 Activity 时需要在配置文件中声明<activity>标签，使用 Service 同样需要在配置文件中声明<service>标签，且两者都具有一定的生命周期。

✧ 不同点：Activity 是与用户交互的组件，可以看到 UI 界面，而 Service 是在后台运行的，无需界面。

6.3 实现 Service

自定义一个 Service 类比较简单，只要继承 Service 类，实现其生命周期中的方法就可以。一个定义好的 Service 必须在 AndroidManifest.xml 配置文件中通过<service>标签声明才能使用。实现 Service 应用的步骤如下：

(1) 创建一个 Service 类并配置。

(2) 启动或绑定 Service。

(3) 停止 Service。

6.3.1 创建 Service 类

创建一个 Service 类时，需要继承 android.app.Service 类，并且重写其 onCreate()、onStart()以及 onDestroy()等方法。这些方法在 Service 生命周期中的不同阶段被调用：

(1) onCreate()方法用来初始化 Service，标志着 Service 生命周期的开始。

(2) onStart()方法用来启动一个 Service，代表 Service 进入了运行的状态。

(3) onDestroy()方法用来释放 Service 占用的资源，标志着 Service 生命周期的结束。

创建 Service 类的代码如下：

```
//继承 Service 类
public class MyService extends Service {
	@Override
	public IBinder onBind(Intent intent) {
		/*这个方法会在 Service 被绑定到其他程序上时被调用。onBind 将返回给客户端一个 IBind
接口实例，IBind 允许客户端回调服务，比如获取 Service 运行的状态或其他操作。*/
		return null;
	}
	@Override
	public void onCreate() {
		//创建服务
		super.onCreate();
	}
	@Override
```

```
    public void onStart() {
        //启动服务
        super.onStart();
    }
    @Override
    public void onDestroy() {
        //释放 Service 资源的代码，例如释放内存
        super.onDestroy();
    }
}
```

上述代码中的回调方法包括：

(1) onCreate()：首次创建服务时系统将调用此方法(在调用 onStartCommand()或 onBind()方法之前)。如果服务已在运行，则不会调用此方法。

(2) onStartCommand()：当另一个组件(如 Activity)通过调用 startService()方法请求启动服务时，系统将调用此方法，一旦执行此方法，服务即会启动并可在后台无限期运行。如果重写了此方法，则在服务完成后，需要通过调用 stopSelf()或 stopService()方法来停止服务。

(3) onBind()：当另一个组件想调用 bindService()方法与服务绑定时，系统调用此方法。在此方法的实现中，必须通过返回的 IBind 提供一个接口，供客户端与服务进行通信。

(4) onDestroy()：当服务不再使用且将销毁时，系统将调用此方法。应当在服务中实现此方法，来清理所有服务使用的资源，如线程、注册的监听器、接收器等。这是服务接收的最后一个调用方法。

要想使用上述代码中定义的 Service 类，必须在 AndroidManifest.xml 配置文件中使用<service>标签声明该 Service，并在<service>标签中添加<intent-filter>指定如何访问该 Service，配置内容如下：

```
<!—指定 Service 的类名-->
<service android:name=".MyService">
    <intent-filter>
    <!--定义 Service 的名字，根据该名字用于启动或停止服务-->
        <action android:name=" MY_SERVICE" />
    </intent-filter>
</service>
```

6.3.2 启动 Service

Service 类创建后，可以通过两种方式启动 Service。

(1) 启动方式：使用 Context.startService()方法启动 Service，调用者与 Service 之间没有关联，即使调用者退出，Service 服务依然运行。

(2) 绑定方式：通过 Context.bindService()启动 Service，调用者与 Service 之间绑定在一起，调用者一旦退出，Service 服务也就终止。

1. 启动方式

启动方式是通过调用 Context.startService()方法启动 Service，在服务未被创建时，系统会先调用服务的 onCreate()方法，接着调用 onStart()方法。如果在调用 startService()方法前服务已经被创建，系统则会直接调用 onStart()方法启动服务，此时不会调用 onCreate()方法多次创建服务。

启动方式的 Service 生命周期如图 6-1 所示。

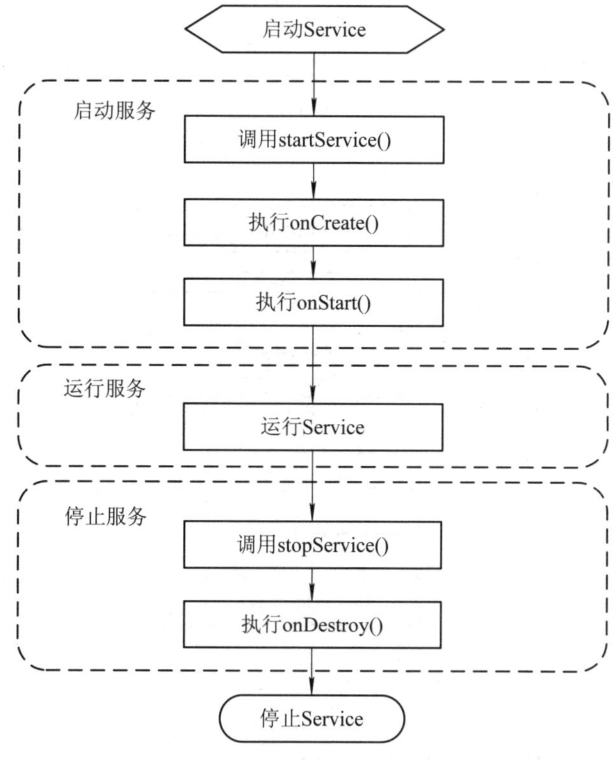

图 6-1　使用启动方式启动的 Service 生命周期

启动 Service 的代码如下：

```
//创建 Intent
Intent intent = new Intent();
//设置 Action 属性
intent.setAction("MY_SERVICE");
//启动该 Service
startService(intent);
```

上述代码先创建了一个 Intent 对象，并设置其 Action 属性值为 AndroidManifest.xml 配置文件中配置的 Service 名称，即通过 Intent 隐式方式找到相应的 Service；再调用 startService(intent)方法启动服务。其中，Service 的名称可以在调用 Intent 构造函数时指明，代码如下：

```
Intent intent = new Intent("MY_SERVICE");
```

2. 绑定方式

绑定方式是通过调用 Context.bindService()方法启动服务，和调用 startService()方法一样，如果 Service 还未创建，则调用 onCreate()方法来创建 Service，但是它不会调用 onStart()方法，而是调用 onBind()方法返回客户端一个 IBinder 接口。由于同一个 Service 可以绑定多个服务连接，因而通过捆绑方式就可以同时为多个不同的应用提供服务。绑定方式的生命周期如图 6-2 所示。

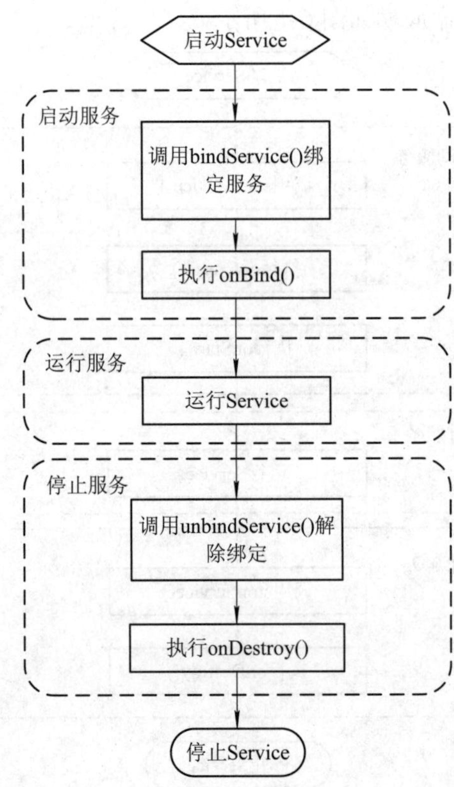

图 6-2　使用绑定方式启动的 Service 生命周期

调用 Context.bindService()方法绑定一个 Service 时需要三个参数：

(1) 第一个参数是 Intent 对象。

(2) 第二个参数是服务连接对象 ServiceConnection，通过实现其 onServiceConnected() 和 onServiceDisconnected()方法来判断服务是否连接成功或连接断开。

(3) 第三个参数是创建 Service 的方式，一般指定为在绑定时自动创建，即设置为 Service.BIND_AUTO_CREATE。

创建服务连接对象 ServiceConnection 的代码如下：

```
//连接对象
ServiceConnection conn = new ServiceConnection() {
        @Override
        public void onServiceConnected(ComponentName name, IBinder service) {
```

```
            Log.i("SERVICE", "连接成功！");
        }
        @Override
        public void onServiceDisconnected(ComponentName name) {
            Log.i("SERVICE", "断开连接！");
        }
};
```

绑定 Service 的代码如下：

```
//绑定 Service
Context.bindService(intent, conn, Service.BIND_AUTO_CREATE);
```

启动方式和绑定方式并不是完全独立的，而是可以混合使用。以 MP3 播放器为例，其功能主要分为启动音乐和暂停音乐两个部分，对于启动音乐功能，可通过 Context.startService()方法来播放相应的音频文件；然而对于暂停音乐功能，则可通过 Context.bindService()方法获取服务链接和 Service 对象，然后通过调用该 Service 对象来暂停音乐并保存相关信息。在这种情况下，如果只调用 Context.stopService()方法并不能够停止 Service，需要在所有的服务链接关闭后，才能够停止 Service。

6.3.3　停止 Service

当 Service 完成动作或处理后，应该调用相应的方法停止服务，释放服务所占用的资源。根据启动 Service 方式的不同，需采用不同的方法停止 Service。

(1) 使用 Context.startService()方法启动的 Service，通过调用 Context.stopService()或 Service.stopSelf()方法结束。

(2) 使用 Context.bindService()绑定的 Service，通过调用 Context.unbindservice()方法解除绑定。

与启动服务的过程类似，Context.stopService()或 Context.unbindservice()方法只能停止过程中的开始部分，系统最终会调用 onDestroy()方法销毁服务并释放资源。

注 意

stopService()方法和 stopSelf()方法不同，stopService()强行终止当前服务，而 stopSelf()直到 Intent 被处理完才停止服务。

6.3.4　Service 示例

由上一小节可知，服务 Service 的启动方式有两种，相应地，服务执行的生命周期方法和停止服务的方法也不同，下面结合具体示例来说明这种差别。

【示例 6.1】 在 Android Studio 中新建项目 Ch06_6D1_Service，演示 Service 的启动、绑定、停止以及解除绑定操作。

首先创建一个 MyService 类，该类继承 Service 并覆盖其生命周期中的各个方法，代码如下：

```
public class MyService extends Service{

    //可以返回null，通常返回一个Binder子类
    public IBinder onBind(Intent intent) {
        Log.i("SERVICE", "onBind.............");
        Toast.makeText(MyService.this, "onBind.............",
            Toast.LENGTH_LONG).show();
        return new MyBinder();
    }
    //Service 创建时调用
    public void onCreate() {
        Log.i("SERVICE", "onCreate.............");
        Toast.makeText(MyService.this, "onCreate.............",
            Toast.LENGTH_LONG).show();
    }
    //当客户端调用 startService()方法启动 Service 时，该方法被调用
    public void onStart(Intent intent, int startId) {
        Log.i("SERVICE", "onStart.............");
        Toast.makeText(MyService.this, "onStart.............",
            Toast.LENGTH_LONG).show();
    }
    //当 Service 不再使用时调用
    public void onDestroy() {
        Log.i("SERVICE", "onDestroy.............");
        Toast.makeText(MyService.this, "onDestroy.............",
            Toast.LENGTH_LONG).show();
    }
    public class MyBinder extends Binder {
        public MyService getService() {
            return MyService.this;
        }
    }
}
```

上述代码中：

(1) 在 Service 生命周期的各个方法中，使用 Log 输出日志，并使用 Toast 输出方法名称。

(2) 在 Service 中创建了一个内部类 MyBinder，此类为 Binder 的子类，其中的 getService()方法用于返回当前 Service 对象给调用者。

(3) 当该 Service 被绑定启动时，会自动调用生命周期方法 onBind()，此方法返回一个 MyBinder 对象，调用者通过此对象即可获取当前所绑定的 Service 对象。

创建一个 MainActivity，该界面布局中包含四个按钮，分别用来启动、停止、绑定和解除绑定 Service，代码如下：

```
/**
 *使用 Service 时需要采用隐式启动的方式，但是 Android 5.0 出来后，
 * 其中有个特性就是 Service Intent  must be explitict，
 * 也就是说从 Lollipop 开始，service 必须采用显示方式启动。
 */
public class MainActivity extends Activity {
        private Button startBtn;
    private Button stopBtn;
    private Button bindBtn;
    private Button unbindBtn;

    @Override
    protected void onCreate(Bundle savedInstanceState) {
        super.onCreate(savedInstanceState);
        setContentView(R.layout.activity_main);
        initView();
    }

    private void initView() {
        startBtn = (Button) findViewById(R.id.btn_start);
        stopBtn = (Button) findViewById(R.id.btn_stop);
        bindBtn = (Button) findViewById(R.id.btn_bind);
        unbindBtn = (Button) findViewById(R.id.btn_unbind);
        startBtn.setOnClickListener(this);
        stopBtn.setOnClickListener(this);
        bindBtn.setOnClickListener(this);
        unbindBtn.setOnClickListener(this);
    }

    @Override
    public void onClick(View view) {
        switch (view.getId()) {
            case R.id.btn_start://开启服务
                //创建 Intent
                Intent intent = new Intent();
```

```
            //设置 Action 属性
            intent.setAction("MYSERVICE");
            //设置应用的包名
            intent.setPackage(getPackageName());
            //启动该 Service
            startService(intent);
            break;
        case R.id.btn_stop://停止服务
            //创建 Intent
            Intent intent2 = new Intent();
            //设置 Action 属性
            intent2.setAction("MYSERVICE");
            //设置应用的包名
            intent2.setPackage(getPackageName());
            //启动该 Service
            stopService(intent2);
            break;
        case R.id.btn_bind://绑定服务
            //创建 Intent
            Intent intent3 = new Intent();
            //设置 Action 属性
            intent3.setAction("MYSERVICE");
            //设置应用的包名
            intent3.setPackage(getPackageName());
            //绑定 Service
            bindService(intent3, conn, Service.BIND_AUTO_CREATE);
            break;
        case R.id.btn_unbind://解绑服务
            //创建 Intent
            Intent intent4 = new Intent();
            //设置 Action 属性
            intent4.setAction("MYSERVICE");
            //设置应用的包名
            intent4.setPackage(getPackageName());
            //解除绑定 Service
            unbindService(conn);
            break;
        default:
            break;
```

```
            }
        }
        //连接对象
        private ServiceConnection conn = new ServiceConnection() {
            //成功绑定 Service，此回调方法被执行
            @Override
            public void onServiceConnected(ComponentName name, IBinder service) {
                Log.i("SERVICE", "连接成功！");
                Toast.makeText(MainActivity.this, "连接成功！", Toast.LENGTH_LONG).show();
                //获取 MyService 返回的 Binder 对象并强制转换为 MyBinder
                MyService.MyBinder binder=(MyService.MyBinder) service;
                //需要通过此 binder 获取所绑定的 MyService 对象
                MyService myService=binder.getService();
                //通过获取到的 myService 对象调用 MyService 中的成员变量或方法
            }

            @Override
            public void onServiceDisconnected(ComponentName name) {
                Log.i("SERVICE", "断开连接！");
                Toast.makeText(MainActivity.this, "断开连接！", Toast.LENGTH_LONG).show();
            }
        };
}
```

上述代码在四个按钮上分别添加了处理单击事件的监听器，当单击按钮时，会先创建一个 Intent 对象，然后将其 Action 属性设置为 MYSERVICE，再调用相应的方法启动、停止、绑定或解除绑定 Service；在以绑定方式启动 Service 时，需要传入一个参数 ServiceConnection 对象，这个对象就是绑定者与所绑定的 Service 对象之间的桥梁，可以通过此 ServiceConnection 对象获取所绑定的 Service 对象。

在配置文件 AndroidManifest.xml 中对 Service 进行配置，代码如下：

```
<?xml version="1.0" encoding="utf-8"?>
<manifest xmlns:android="http://schemas.android.com/apk/res/android"
    package="com.ugrow.ch06_6d1_service">

    <application
        android:allowBackup="true"
        android:icon="@mipmap/ic_launcher"
        android:label="@string/app_name"
        android:roundIcon="@mipmap/ic_launcher_round"
```

```
        android:supportsRtl="true"
        android:theme="@style/AppTheme">
        <activity android:name=".MainActivity">
            <intent-filter>
                <action android:name="android.intent.action.MAIN" />

                <category android:name="android.intent.category.LAUNCHER" />
            </intent-filter>
        </activity>
        <service android:name=".MyService">
            <intent-filter>
                <action android:name="MYSERVICE" />
            </intent-filter>
        </service>
    </application>

</manifest>
```

运行项目，当第一次单击【启动 Service】按钮时，界面显示字符串“onCreate..............”；再次单击【启动 Service】按钮时，则只会显示信息“onStart......”如图 6-3 所示。

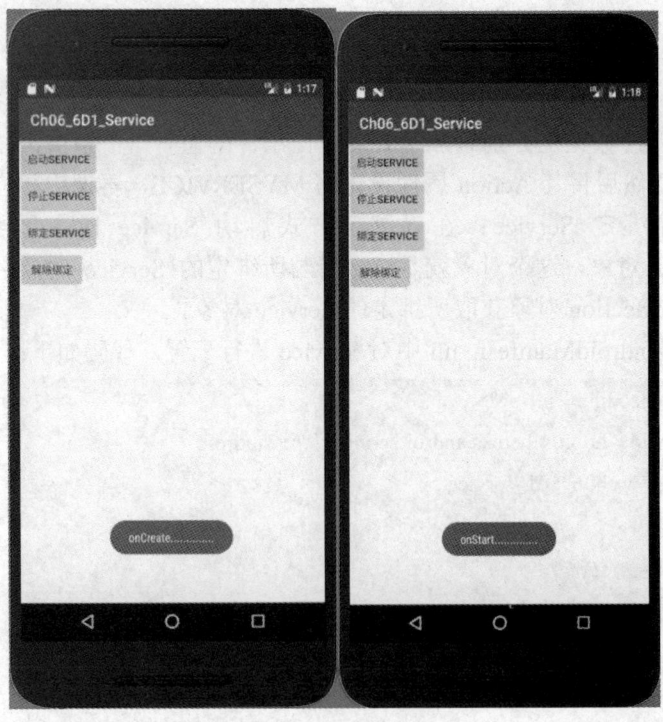

图 6-3　启动 Service

单击【绑定 Service】按钮时，会显示字符串“onBind...”(如果 Service 没创建，则会先显示“onCreate...”)；单击【停止 Service】和【解除绑定】按钮时，都会显示字符串“onDestroy...”，如图 6-4 所示。

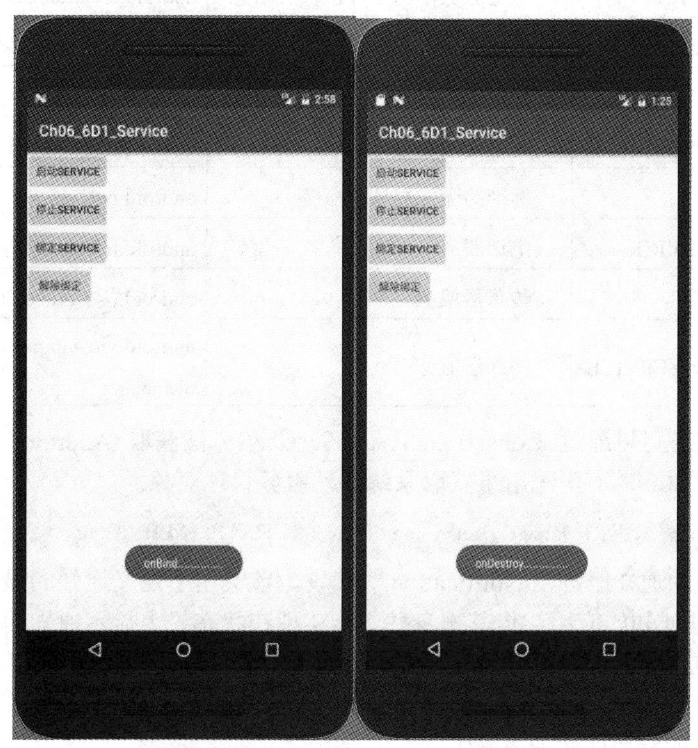

图 6-4　绑定和停止 Sevice

6.4　Android 系统服务

Android 中提供了大量的系统服务(Service)，这些 Service 用于完成不同的功能，如表 6-1 所示。

表 6-1　Andirod 系统服务

Service 名称	作　用	返回对象
WINDOW_SERVICE	窗口服务，例如获得屏幕的宽和高	android.view.WindowManager
LAYOUT_INFLATER_SERVICE	布局映射服务，根据 XML 布局文件来绘制视图(View)对象	android.view.LayoutInflater
ACTIVITY_SERVICE	活动服务，和全局系统状态一起使用	android.app.ActivityManager
NOTIFICATION_SERVICE	通知服务	android.app.NotificationManager
KEYGUARD_SERVICE	键盘锁的服务	android.app.KeyguardManager

续表

Service 名称	作　用	返回对象
LOCATION_SERVICE	位置服务，用于提供位置信息	android.location.LocationManager
SEARCH_SERVICE	本地查询服务	android.app.SearchManager
VEBRATOR_SERVICE	手机震动服务	android.os.Vibrator
CONNECTIVITY_SERVICE	网络连接服务	android.net.ConnectivityManager
WIFI_SERVICE	标准的无线局域网服务	android.net.wifi.WifiManager
TELEPHONY_SERVICE	电话服务	android.telephony.TelephonyManager
SENSOR_SERVICE	传感器服务	android.os.storage.StorageManager
INPUT_METHOD_SERVICE	输入法服务	android.view.inputmethod.InputMethodManager

这些系统服务可以通过 Context.getSystemService()方法获取 Android 系统所支持的服务管理对象。例如，以下代码用于获取系统活动服务管理对象：

```
ActivityManager am = (ActivityManager) getSystemService(ACTIVITY_SERVICE);
```

系统服务中的通知服务由 Notification 类实现，该类用于定义通知的显示(图片和标题等内容)以及处理通知的应用，但其本身并不能实现在状态栏上显示通知的功能，必须通过 NotificationManager 才能将 Notification 所定义的通知显示在手机中。otificationManager 类是系统的通知服务管理类，它能够将通知信息显示在状态栏上。

Notification 的常用属性如表 6-2 所示。

表 6-2　Notification 常用属性

属性名称	描　述
audioStreamType	设置 Notification 所用的音频流的类型
contentIntent	设置单击通知条目时所执行的 Intent
contentView	设置在状态条上显示通知时所显示的视图
defaults	设置默认值，例如：DEFAULT_LIGHTS(默认灯)、DEFAULT_SOUND(默认声音)、DEFAULT_VIBRATE(默认震动)、DEFAULT_ALL(以上默认)
deleteIntent	删除所有通知时被执行的 Intent
icon	设置状态栏上显示的图标
iconLevel	设置显示图标级别
ledARGB	设置 LED 的颜色
ledOffMS	设置关闭 LED 时的闪烁时间
ledOnMS	设置开启 LED 时的闪烁时间。ledOnMS 属性为 1，ledOffMS 属性为 0 来表示打开 LED；两者设置为 0，则表示关闭 LED

续表

属性名称	描　述
sound	设置一个音频文件作为 Notification，其值为一个 URI
tickerText	设置状态栏上显示的消息内容
vibrate.	设置 Notification 的振动模式，通常需要给 Notification 的 vibrate 属性设定一个时间数组，如 long[] vibrate = new long[] { 1000, 1000, 1000, 1000, 1000 }。注意使用震动之前，需要在配置文件中添加震动权限： <uses-permission android:name="android.permission.VIBRATE"/>
when	设置通知发生时的时间

【示例 6.2】 在 Android Studio 中新建项目 Ch06_6D2_Notification，通过 Notification 类在状态栏上显示天气信息，演示系统通知服务的应用。

显示天气信息的 Activity 代码 MainActivity.java 如下所示：

```
public class MainActivity extends AppCompatActivity implements View.OnClickListener{

    private Button mBtnClear;
    private Button mBtnCloud;
    private Button mBtnRain;
    private Button mBtnSunny;
    //声明 NotificationManager 对象，该对象用于管理 Notification
    private NotificationManager nm;


    @Override
    protected void onCreate(Bundle savedInstanceState) {
        super.onCreate(savedInstanceState);
        setContentView(R.layout.activity_main);
        //通过 getSystemService(Context.NOTIFICATION_SERVICE)获得 NotificationManager 对象。
        nm = (NotificationManager) getSystemService(NOTIFICATION_SERVICE);
        initView();
    }

    private void initView() {
        mBtnClear =(Button) findViewById(R.id.btn_clear);
        mBtnCloud = (Button)findViewById(R.id.btn_cloud);
        mBtnRain = (Button)findViewById(R.id.btn_rain);
        mBtnSunny =(Button) findViewById(R.id.btn_sunny);
```

```
        mBtnClear.setOnClickListener(this);
        mBtnCloud.setOnClickListener(this);
        mBtnRain.setOnClickListener(this);
        mBtnSunny.setOnClickListener(this);
    }

    @Override
    public void onClick(View view) {
        switch (view.getId()){
            case R.id.btn_clear:
                nm.cancel(R.layout.activity_main);
                break;
            case R.id.btn_cloud:
                displayWeather("阴", "天气预报", "阴云密布", R.mipmap.cloudy);
                break;
            case R.id.btn_rain:
                displayWeather("雨", "天气预报", "大雨连绵", R.mipmap.rain);
                break;
            case R.id.btn_sunny:
                displayWeather("晴", "天气预报", "晴空万里", R.mipmap.sun);
                break;
            default:
                break;
        }
    }

    //displayWeather()方法用于在状态栏上显示相应的天气信息
    private void displayWeather(String tickerText, String title, String content, int drawable) {
        Notification.Builder builder = new Notification.Builder(MainActivity.this);
        builder.setSmallIcon(drawable); //设置图标
        builder.setTicker(tickerText);
        builder.setContentTitle(title); //设置标题
        builder.setContentText(content); //消息内容
        builder.setWhen(System.currentTimeMillis()); //发送时间
        builder.setDefaults(Notification.DEFAULT_ALL); //设置默认的提示音，振动方式，灯光
        builder.setAutoCancel(true);//打开程序后图标消失
        Intent intent =new Intent (this,MainActivity.class);
```

```
        PendingIntent pendingIntent =PendingIntent.getActivity(MainActivity.this, 0, intent, 0);
        builder.setContentIntent(pendingIntent);
        Notification notification1 = builder.build();
        nm.notify(R.layout.activity_main, notification1); //通过通知管理器发送通知
    }
}
```

上述代码使用 Notification 实现了在状态栏上显示天气的功能，具体步骤如下：

(1) 通过 getSystemService(NOTIFICATION_SERVICE)方法创建 NotificationManager 对象，该对象用于管理 Notification。

(2) 使用 Notification 构造函数创建 Notification 对象，代码如下：

```
//创建一个 Notification 对象，该通知的图标为 drawable 对应的图像，标题为 tickerText 对应的文本，通知的发送时间为当前时间
Notification.Builder builder = new Notification.Builder(MainActivity.this);
builder.setSmallIcon(drawable); //设置图标
builder.setTicker(tickerText);
builder.setContentTitle(title); //设置标题
builder.setContentText(content); //消息内容
builder.setWhen(System.currentTimeMillis()); //发送时间
```

注意：除了通过 Notification 构造函数在创建的同时设置其在状态栏上的图标、内容以及显示的时间等之外，也可以单独通过属性设置这些显示内容，示例如下：

```
builder.setDefaults(Notification.DEFAULT_ALL); //设置默认的提示音、振动方式、灯光
builder.setAutoCancel(true);//设置打开程序后图标即消失
```

(3) 创建 Intent 对象，并指定其对应的 Notification 对象，代码如下：

```
Intent    intent=new Intent(notification,class);
```

(4) 根据创建的 Intent 对象创建 PendingIntent 对象，后者用于对前者的进一步封装，代码如下：

```
PendingIntent m_PendingIntent=PendingIntent.getActivity(NotificationName.this, 0, intent, 0)
```

(5) 调用 NotificationManager 的 notify()方法将通知发送到状态栏中，代码如下：

```
nm.notify(R.layout.main, notification);
```

(6) 使用 NotificationManager 的 cancel()方法删除 Notification，代码如下：

```
nm.cancel(R.layout.main);
```

在 MainActivity 的布局文件 activity_main.xml 中编写代码如下：

```
<?xml version="1.0" encoding="utf-8"?>
<LinearLayout xmlns:android="http://schemas.android.com/apk/res/android"
    xmlns:tools="http://schemas.android.com/tools"
```

```
    android:layout_width="match_parent"
    android:layout_height="match_parent"
    android:orientation="vertical"
    tools:context="com.ugrow.ch06_6d2_notification.MainActivity">

    <LinearLayout
        android:layout_width="match_parent"
        android:layout_height="wrap_content"
        android:orientation="vertical" >

        <Button
            android:id="@+id/btn_sunny"
            android:layout_width="wrap_content"
            android:layout_height="wrap_content"
            android:text="晴空万里" />

        <Button
            android:id="@+id/btn_cloud"
            android:layout_width="wrap_content"
            android:layout_height="wrap_content"
            android:text="阴云密布" />

        <Button
            android:id="@+id/btn_rain"
            android:layout_width="wrap_content"
            android:layout_height="wrap_content"
            android:text="大雨连绵" />
    </LinearLayout>

    <Button
        android:id="@+id/btn_clear"
        android:layout_width="wrap_content"
        android:layout_height="wrap_content"
        android:layout_marginTop="20dip"
        android:text="清除 Notification" />

</LinearLayout>
```

上述代码分别定义了【晴空万里】、【阴云密布】、【大雨连绵】、【清除 Notification】四个按钮，运行程序，单击界面上不同的按钮，状态栏上就会显示对应的图标，如图 6-5 所示。

图 6-5　在状态栏上显示天气信息

本 章 小 结

通过本章的学习，读者应当了解：

✧ Service 提供程序的后台服务，分为本地服务和远程服务两种类型。

✧ 定义一个 Service 子类需要继承 Service 类，并实现其生命周期中的方法。

✧ Service 必须在 AndroidManifest.xml 配置文件中通过<service>元素进行声明。

✧ Service 类创建后，可以通过两种方式启动 Service：启动方式和绑定方式。

✧ Service 的启动方式使用 Context.startService()方法来启动一个 Service，调用者与 Service 之间没有关联，即使调用者退出，Service 服务依然运行。

✧ Service 的绑定方式通过 Context.bindService()来启动一个 Service，调用者与 Service 之间绑定在一起，调用者一旦退出，Service 服务也就终止。

✧ 使用 Context.startService()方法启动的 Service，通过调用 Context.stopService()或 Service.stopSelf()方法结束服务。

✧ 使用 Context.bindService()绑定的 Service，通过调用 Context.unbindservice()解除绑定的服务。

✧ Android 提供大量的系统服务，这些系统服务用于完成不同的功能，通过 Context.getSystemService()方法可以获取不同服务管理对象。

✧ NotificationManager 类是系统的通知服务管理类，它能够将通知信息(Notification)显示在状态栏上。

本 章 练 习

1. 下列不是 Service 特点的是______。
   A. 没有用户界面，不与用户交互
   B. 长时间运行，不占程序控制权
   C. 比 Activity 的优先级低
   D. 可用于进程间通信
2. 关于启动、停止 Service 的说法，错误的是______。
   A. Context.startService()启动的 Service 可以调用 Context.stopService()结束
   B. Context.startService()启动的 Service 可以调用 Context.stopSelf()结束
   C. Context.bindService()启动的 Service 可以调用 Context.stopService()结束
   D. Context.bindService()启动的 Service 可以调用 Context.unbindservice()结束
3. Service 生命周期方法有 onCreate()、______和______。
4. 简述 Service 的生命周期。
5. 如果启动同一个 Service 类两次会怎样？请编写代码测试。
6. 编写程序，测试表 6-1 列出的系统服务。

第7章 数据存储

本章目标

- 了解 Android 系统中数据存储的方式
- 熟悉 SharedPreferences 工作原理，并会使用 SharedPreferences 存储数据
- 熟悉文件存储工作原理，并会使用文件存储数据
- 熟悉 SQLite 工作原理，并会使用 SQLite 存储数据
- 熟悉 ContentProvider 存储数据的原理

7.1　数据存储简介

程序可被理解为数据输入、输出以及数据处理的过程，程序执行中通常需读取处理数据并且将处理后的结果存放起来。存放数据需要使用数据存储机制，Android 提供了四种数据存储方式。

1．SharedPreferences 存储方式

SharedPreferences 采用“键-值”对方式组织和管理数据，其数据存储在 XML 文件中。相对于其他方式，它是一个轻量级的存储机制。该方式实现比较简单，适合简单数据的存储。

2．File 存储方式

文件存储的特点介于 SharedPreferences 与 SQLite 之间，比 SharedPreferences 方式更适合存储较大的数据；从存储结构化来看，这种方式不同于 SQLite，不适合结构化的数据存储。

3．SQLite 存储方式

SQLite 相对于 MySQL 数据库来说，是一个轻量级的数据库，适合移动设备中复杂数据的存储，Android 已经集成了 SQLite 数据库，通过这种方式能够很容易地对数据进行增加、插入、删除、更新等操作。相比 SharedPreferences 和文件存储，使用 SQLite 较为复杂。

4．网络存储方式

将数据通过 java.net.*和 android.net.*包中的类存储于网络，这将在后续的网络通信中涉及。

在开发过程中，可以根据程序的实际需要选择合适的存取方式。另外，在 Android 中各个应用程序组件之间是相互独立的，彼此间的数据不能直接访问。为此 Android 提供了内容提供器(ContentProvider)以达到应用程序之间数据共享的目的。ContentProvider 可以使用 SQLite 数据库或者文件作为存储方式(通常使用 SQLite 数据库)。

7.2　SharedPreferences 存储方式

SharedPreferences 是 Android 系统提供的一个通用的数据持久化框架，是一种轻量级的数据存储方式，将数据保存在一个 XML 文件中，用于存储和读取 Key-Value 类型的原始基本数据，目前支持 String、int、float、boolean 等基本类型数据的存储。

7.2.1　访问 SharedPreferences 的 API

使用 SharedPreferences 方式来存取数据时，需要用到 SharedPreferences 和 SharedPreferences.Editor 接口，这两个接口在 android.content 包中。SharedPreferences 提供了获得数据的方法，其常用的方法如表 7-1 所示。

表 7-1　SharedPreferences 常用方法

方　　法	功 能 描 述
contains (String key)	判断是否包含一个该键值
edit()	返回 SharedPreferencesEditor 对象
getAll ()	取得所有值 Map
getBoolean (String key, boolean defValue)	获取一个布尔值
getFloat (String key, float defValue)	获取一个 float 值
getInt (String key, int defValue)	获取一个 int 值
getString (String key, String defValue)	获取一个 String 值
getLong (String key, Long defValue)	获取一个 long 值
registerOnSharedPreferenceChangeListener()	注册 SharedPreferences 发生变化的监听函数
unregisterOnSharedPreferenceChangeListener()	注销一个之前注册的监听函数

SharedPreferences.Editor 编辑器提供保存数据的方法，如表 7-2 所示。

表 7-2　SharedPreferences.Editor 常用方法

方　　法	功 能 描 述
clear()	清除所有值
commit()	保存
getAll()	返回所有值 Map
putBoolean(String key, boolean value)	保存一个布尔值
putFloat(String key, float value)	保存一个 float 值
putInt(String key,int value)	保存一个 int 值
putLong(String key, long value)	保存一个 long 值
putString(String key, String value)	保存一个 String 值
remove(String key)	删除 key 所对应的值

从 SharedPreferences 及其 Editor 提供的方法可以看到，SharedPreferences 只支持简单数据类型(如 String、float、int 和 long)的存储操作。

注 意　使用 SharedPreferences.Editor 的 putXXX()方法保存数据时，需要根据数据类型调用相应的 putXXX()方法。例如调用 putString()方法为字符串建立键值对。

使用 SharedPreferences 存储数据比较简单，步骤如下：

(1) 首先使用 getSharedPreferences()得到 SharedPreferences 对象。调用 getSharedPreferences()方法时，需要指定以下两个参数：

- ✧ 一是存储数据的 XML 文件名，这个 XML 文件存储在“/data/data/包名/shared_prefs/”目录下，其文件名由该参数指定，注意文件名不需要指定后缀(.xml)，系统会在该文件名之后自动添加 xml 后缀并创建之。
- ✧ 二是操作模式，其取值有三种：MODE_WORLD_READABLE(可读)、MODE_WORLD_WRITEABLE(可写)和 MODE_PRIVATE(私有)。

(2) 使用 SharedPreferences.Editor 的 putXXX()方法保存数据。

(3) 使用 SharedPreferences.Editor 的 commit()方法将本次操作的数据写到 XML 文件中。

(4) 使用 SharedPreferences 的 getXXX()方法获取相应数据。

7.2.2 SharedPreferences 应用

SharedPreferences 主要用于存储系统的配置信息。例如上次登录的用户名，上次最后设置的配置信息(如：是否打开音效，是否使用振动，小游戏的玩家积分等)，当程序再次启动后依然保持原有设置。

【示例 7.1】 在 Android Studio 中新建项目 Ch07_7D1_SharedPreference，使用 SharedPreferences 将修改的音量值存储到 XML 文件中。

在 MainActivity 中编写代码如下：

```
public class MainActivity extends AppCompatActivity implements View.OnClickListener {

    //声明 3 个 Button 对象，这些对象用于调节音量
    private Button add_voice;
    private Button sub_voice;
    private Button mute_voice;
    //声明一个 SharedPreferences 对象
    private SharedPreferences sharedPreferences;
    /* 定义音量参数：cur_voice、MIN_VOICE 以及 MAX_VOICE */
    //cur_voice 记录了当前音量大小，初始设置为 0
    private static int cur_voice = 0;
    //MIN_VOICE 定义了最小音量值为 0
    private final static int MIN_VOICE = 0;
    //MAX_VOICE 定义了最大音量值为 8
    private final static int MAX_VOICE = 8;

    @Override
    protected void onCreate(Bundle savedInstanceState) {
        super.onCreate(savedInstanceState);
        setContentView(R.layout.activity_main);
        initView();
    }

    private void initView() {
        /*
         * 使用 getSharedPreferences()方法生成 SharedPreferences 对象，将数据存储到名为 preferences.xml
的文件中，该操作模式是私有的
         *     */
        sharedPreferences = getSharedPreferences("preferences", Context.MODE_PRIVATE);
        //根据 XML 定义生成 Button 对象
```

```
        add_voice = (Button) findViewById(R.id.add_voice);
        sub_voice = (Button) findViewById(R.id.sub_voice);
        mute_voice = (Button) findViewById(R.id.mute_voice);
        add_voice.setOnClickListener(this);
        sub_voice.setOnClickListener(this);
        mute_voice.setOnClickListener(this);
        //通过 getVoicevalue()方法获取 SharedPreferences 中存储的音量值，即最近一次设置的音量值
        cur_voice = getVoicevalue(sharedPreferences);
        //使用 Toast 显示上次音量值
        Toast.makeText(MainActivity.this, "上次设置音量：" + cur_voice, Toast.LENGTH_SHORT).show();
    }

    @Override
    public void onClick(View view) {
        switch (view.getId()) {
            case R.id.add_voice://增加音量，当单击该按钮时，当前音量值被增加并且使用 Toast 显示增加
后的音量
                //若音量值未达到最大值，则增加当前的音量值。否则，音量值不变。
                if (cur_voice < MAX_VOICE)
                    cur_voice = cur_voice + 1;
                //根据音量值构造音量显示文本，即每个“|”代表一个音量
                String voicetext = (String) generateVoice(cur_voice);
                //根据音量值显示一个 Toast 消息
                Toast.makeText(MainActivity.this, "音量" + cur_voice + "\n" + voicetext,
Toast.LENGTH_LONG).show();
                //将音量存储到 SharedPreferences 对象指定的 XML 文件中
                saveVoicevalue(cur_voice, sharedPreferences);
                break;
            case R.id.sub_voice://减小音量，当单击该按钮时，当前音量值被减少并且使用 Toast 显示减少
后的音量
                //若音量值未达到最小值，则减小当前的音量值。否则，音量值不变。
                if (cur_voice > MIN_VOICE)
                    cur_voice = cur_voice - 1;
                //根据音量值构造音量显示文本，即每个“|”代表一个音量
                String voicetext2 = (String) generateVoice(cur_voice);

                Toast.makeText(MainActivity.this, "音量" + cur_voice + "\n" + voicetext2,
Toast.LENGTH_LONG)
                        .show();
                //将音量存储到 SharedPreferences 对象指定的 XML 文件中
```

```
                saveVoicevalue(cur_voice, sharedPreferences);
                break;
            case R.id.mute_voice://静音，当单击该按钮时，当前音量值被置为 0 并且使用 Toast 显示音量
为 0
                //设置音量值为 0
                cur_voice = 0;
                String voicetext3 = (String) generateVoice(cur_voice);
                Toast.makeText(MainActivity.this, "音量" + cur_voice + "\n" + voicetext3,
Toast.LENGTH_LONG).show();
                //将音量存储到 SharedPreferences 对象指定的 XML 文件中
                saveVoicevalue(cur_voice, sharedPreferences);
                break;
            default:
                break;
        }
    }

    /* getVoicevalue()方法用于获取当前的音量值 */
    int getVoicevalue(SharedPreferences sharedPreferences) {
        //通过调用 SharedPreferences 对象的 getXXX 方法获取 SharedPreferences 中存储的值
        int ret = sharedPreferences.getInt("key", 0);
        //返回结果
        return ret;
    }

    /* generateVoice()方法根据音量值返回音量文本，该文本中的“l”的个数等于音量值。如音量值为 3
时，返回“lll” */
    private CharSequence generateVoice(int voice) {
        //声明并初始化 CharSequence 对象
        CharSequence str = "";
        /* 根据音量值构造 Toast 显示的文本 */
        while (voice > 0) {
            str = str + "l";
            voice--;
        }
        //返回 str
        return str;
    }

    /* saveVoicevalue()方法用于将当前的音量值存储到 SharedPreferences 中 */
```

```
    void saveVoicevalue(int voicevalue, SharedPreferences sharedPreferences) {
        //生成 SharedPreferences 编辑对象，通过该对象将数据放入到 SharedPreferences 中
        SharedPreferences.Editor editor = sharedPreferences.edit();
                /*
                 * 指定 XML 中存储 voicevalue 值的标签为 key，即 SharedPreferences 中包含以下信息：
<string
                 * name="key">$voicevalue</string>
                 */
        //将音量值放入到 SharedPreferences 中，该值通过 key 引用
        editor.putInt("key", voicevalue);
        //提交数据，将数据保存到 XML 文件中
        boolean ret = editor.commit();
        //使用 Toast 显示保存成功或失败提示信息
        if (ret == true)
            Toast.makeText(MainActivity.this, "保存成功", Toast.LENGTH_SHORT).show();
        else
            Toast.makeText(MainActivity.this, "保存失败", Toast.LENGTH_SHORT).show();
    }

}
```

上述代码声明了三个按钮并为其分别注册监听事件，分别实现音量的增加、减少以及静音功能，同时通过 saveVoicevalue()方法将当前的音量值存储到 SharedPreferences 中。getVoicevalue()方法用于获取当前的音量值；generateVoice()方法可以将数字的音量值转换成相应的音量文本，以便在显示时可不直接使用数字，而是使用相应的音量字符串进行提示。

在 activity_main.xml 文件中编写代码如下：

```
<?xml version="1.0" encoding="utf-8"?>
<LinearLayout xmlns:android="http://schemas.android.com/apk/res/android"
    xmlns:tools="http://schemas.android.com/tools"
    android:layout_width="match_parent"
    android:layout_height="match_parent"
    android:orientation="vertical"
    tools:context="com.ugrow.ch07_7d1_sharedpreferences.MainActivity">

    <!-- 音量增加按钮 -->
    <Button
        android:id="@+id/add_voice"
        android:layout_width="wrap_content"
        android:layout_height="wrap_content"
        android:layout_gravity="center"
        android:layout_x="0px"
```

```
        android:layout_y="20px"
        android:text="增加" />
    <!-- 音量降低按钮 -->

    <Button
        android:id="@+id/sub_voice"
        android:layout_width="wrap_content"
        android:layout_height="wrap_content"
        android:layout_gravity="center"
        android:layout_marginBottom="5dp"
        android:layout_marginTop="5dp"
        android:text="减少" />
    <!-- 静音按钮 -->

    <Button
        android:id="@+id/mute_voice"
        android:layout_width="wrap_content"
        android:layout_height="wrap_content"
        android:layout_gravity="center"
        android:text="静音" />
</LinearLayout>
```

运行程序，单击两次【增加】按钮，效果如图 7-1 所示。同理，单击【减少】按钮，音量会降低；单击【静音】按钮，音量会变为 0，此处不再演示。

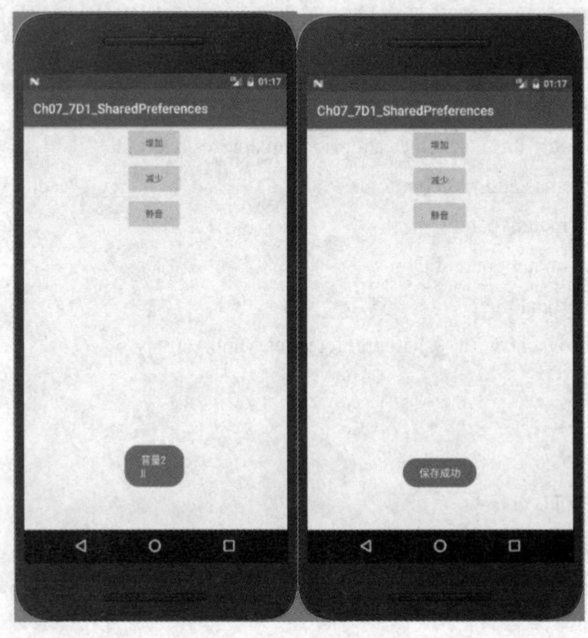

图 7-1　调节音量

可以通过 Android DDMS 查看 getSharedPreferences()方法第一个参数指定的 XML 文件：在 DDMS 的【File Explorer】选项卡下，展开目录/data/data/com.ugrow.ch07_7d1_sharedPreferences/ shared_prefs，可以看到在该目录下有一个文件 preferences.xml，如图 7-2 所示。

Name	Size	Date	Time	Permissions	Info
com.ugrow.ch07_7d1_sharedpreferences		2017-07-20	01:12	drwxr-x--x	
cache		2017-07-20	01:12	drwxrwx--x	
lib		2017-07-20	01:12	lrwxrwxrwx	-> /data/a...
shared_prefs		2017-07-20	01:14	drwxrwx--x	
preferences.xml	103	2017-07-20	01:14	-rw-rw----	

图 7-2　通过 DDMS 查看 preferences.xml 文件

单击按钮【】，可以将 preferences.xml 文件从模拟器中保存到指定计算机路径下，该文件的内容如下：

```
<?xml version='1.0' encoding='utf-8' standalone='yes' ?>
<map>
<int name="key" value="2" />
</map>
```

重新启动该程序，程序会获取最近一次保存的数据，如图 7-3 所示。

图 7-3　获取用户最近一次保存的音量调节数据

7.3　File 存储方式

不同于 SharedPreference 存储方式，File 存储方式不受类型限制，可以将一些数据直接以文件的形式保存在设备中，如文本文件、PDF、音频、图片等。如果需要存储复杂数据，可以使用文件进行存储。

7.3.1 File 操作

和在传统的 Java 中实现 I/O 的程序类似，通过 Context.openFileInput()方法可以获取标准的文件输入流(FileInputStream)，以读取设备上的文件；通过 Context.openFileOuput()方法可以获取标准的文件输出流(FileOutputStream)，将数据写到文件中。

读取文件的代码如下：

```
String file = "dh.txt";//定义文件名
//获取指定文件的文件输入流
FileInputStream fileInputStream = openFileInput(file);
//定义一个字节缓存数组
byte[] buffer=new byte[fileInputStream.available()];
//将数据读到缓存区
fileInputStream.read(buffer);
//关闭文件输入流
fileInputStream.close();
```

保存文件的代码如下：

```
//获取文件输出流，操作模式是私有
FileOutputStream fileOutputStream = openFileOutput(file,Context.MODE_PRIVATE);
//将内容写入文件
fileOutputStream.write(fileContent.getBytes());
```

openFileOutput()方法的第二个参数用于指定输出流的模式，Android 提供了四种输出模式，如表 7-3 所示。

表 7-3 Android 文件输出模式

模　式	功 能 描 述
Context.MODE_PRIVATE	私有模式，这种模式创建的文件是私有文件，因而创建的文件只能被应用本身访问。在该模式下，写入的内容会覆盖原文件的内容
Context.MODE_APPEND	附加模式，该模式会首先检查文件是否存在，若文件不存则创建新文件，否则在原文件中追加内容
Context.MODE_WORLD_WRITABLE	可写模式，该模式的文件可以被其他应用修改
Context.MODE_WORLD_READABLE	可读模式，该模式的文件可以被其他应用读取

7.3.2 File 应用

File 存储数据的功能更加强大，下面结合示例来演示 File 存储方式的具体使用方法。

【示例 7.2】 在 Android Studio 中新建项目 Ch07_7D2_File，使用 File 存储实现简单的文本编辑器。

在 activity_main.xml 中编写代码如下：

```
<?xml version="1.0" encoding="utf-8"?>
<LinearLayout xmlns:android="http://schemas.android.com/apk/res/android"
```

```
xmlns:tools="http://schemas.android.com/tools"
android:layout_width="match_parent"
android:layout_height="match_parent"
android:orientation="vertical"
tools:context="com.ugrow.ch07_7d2_file.MainActivity">

<TextView
    android:layout_width="match_parent"
    android:layout_height="wrap_content"
    android:text="文件名" />

<EditText
    android:id="@+id/et_name"
    android:layout_width="match_parent"
    android:layout_height="wrap_content" />

<TextView
    android:layout_width="match_parent"
    android:layout_height="wrap_content"
    android:text="内容" />

<EditText
    android:id="@+id/et_content"
    android:layout_width="match_parent"
    android:layout_height="wrap_content" />

<LinearLayout
    android:layout_width="match_parent"
    android:layout_height="match_parent"
    android:orientation="horizontal">

    <Button
        android:id="@+id/btn_save"
        android:layout_width="wrap_content"
        android:layout_height="wrap_content"
        android:text="保存数据" />

    <Button
        android:id="@+id/btn_read"
        android:layout_width="wrap_content"
```

```
            android:layout_height="wrap_content"
            android:text="读取数据" />
    </LinearLayout>

</LinearLayout>
```

在 MainActivity 中编写代码如下：

```
public class MainActivity extends AppCompatActivity implements View.OnClickListener{

    private EditText etName; //输入文件名的文本框
    private EditText etContent; //输入文件内容的文本框
    private Button btnRead; //读取文件的按钮
    private Button btnSave; //保存文件的按钮

    @Override
    protected void onCreate(Bundle savedInstanceState) {
        super.onCreate(savedInstanceState);
        setContentView(R.layout.activity_main);
        initView();
    }

    private void initView() {
        btnRead = (Button) findViewById(R.id.btn_read);
        btnSave = (Button) findViewById(R.id.btn_save);
        etName = (EditText) findViewById(R.id.et_name);
        etContent = (EditText) findViewById(R.id.et_content);
        btnRead.setOnClickListener(this);
        btnSave.setOnClickListener(this);
        etName.setOnClickListener(this);
        etContent.setOnClickListener(this);
    }
    @Override
    public void onClick(View view) {
        switch (view.getId()){
            case R.id.btn_save://单击该按钮时，保存当前编辑的文件
                //获取保存的文件名
                String str_file_name = etName.getText().toString();
                if (str_file_name == "")
                    Toast.makeText(MainActivity.this, "文件名为空", Toast.LENGTH_SHORT).show();
                else {
                    //获取 file_contentet 的内容
```

```
                String str_file_content = etContent.getText().toString();
                try {
                    //调用 save()方法保存文件
                    save(str_file_name, str_file_content);
                    //使用 Toast 显示保存成功
                    Toast.makeText(MainActivity.this, "保存成功", Toast.LENGTH_SHORT).show();
                } catch (Exception e) {
                    //产生异常，使用 Toast 显示保存失败
                    Toast.makeText(MainActivity.this, "保存失败", Toast.LENGTH_SHORT).show();
                }
            }
            break;
        case R.id.btn_read://单击该按钮时，根据文件名读取内容
            String file_name = etName.getText().toString();
            if (file_name == "")
                Toast.makeText(MainActivity.this, "文件名为空", Toast.LENGTH_SHORT).show();
            else {
                try {
                    String file_content = read(file_name);
                    etContent.setText(file_content);
                } catch (Exception e) {
                    Toast.makeText(MainActivity.this,"文件读写失败", Toast.LENGTH_SHORT).show();
                }
            }
            break;
        default:
            break;
    }
}

/* 方法 save()负责将指定的内容(fileContent)写入到文件(file) */
public void save(String file, String fileContent) throws Exception {
    //获取文件输出流，操作模式是私有
    FileOutputStream fileOutputStream = openFileOutput(file, Context.MODE_PRIVATE);
    //将内容写入文件
    fileOutputStream.write(fileContent.getBytes());
    //关闭文件输出流
    fileOutputStream.close();
}
```

```
    /* 方法 read 负责读取指定的文件(file) */
    public String read(String file) throws Exception {
        //获取指定文件的文件输入流
        FileInputStream fileInputStream = openFileInput(file);
        //定义一个字节缓存数组
        byte[] buffer = new byte[fileInputStream.available()];
        //将数据读到缓存区
        fileInputStream.read(buffer);
        //关闭文件输入流
        fileInputStream.close();
        //将缓冲区中的数据转换成字符串并返回
        return new String(buffer);
    }
}
```

上述代码基于文件存储方式实现了简单的文本编辑器。对于文本编辑器而言，主要的核心操作是保存和读取文件，而上述代码使用 save(String file, String fileContent)方法实现了保存操作，使用 read(String file)方法实现了文件的读取操作。

运行程序，效果如图 7-4 所示：输入文件名和内容，并单击【保存数据】按钮，清除输入文本后，再单击【读取数据】按钮，文件中保存的信息就会显示到内容文本框中。

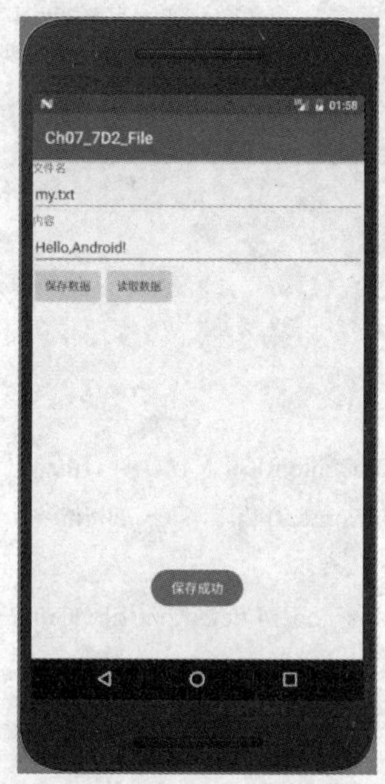

图 7-4　使用 File 存储方式建立文本编辑器

在 DDMS 的【File Explorer】选项卡下展开目录 /data/data/com.ugrow.ch07_7d2_file/files，在目录下可以看到保存数据的文件，如图 7-5 所示。

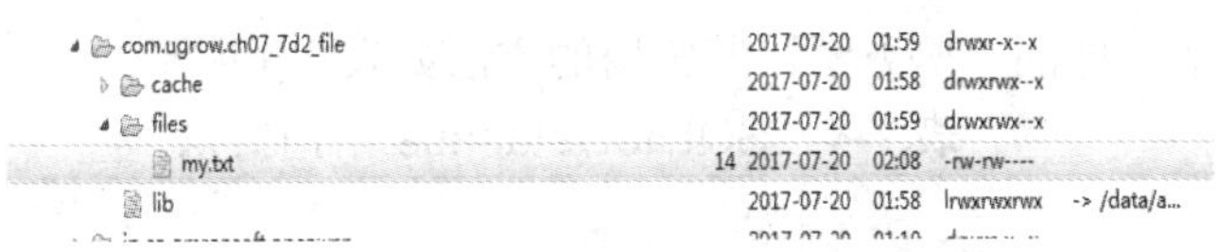

图 7-5　通过 DDMS 查看保存数据的文件

7.4　SQLite 存储方式

Android 中通过 SQLite 数据库实现结构化数据存储。SQLite 是一个嵌入式数据库引擎，目的在于为内存等资源有限的设备，如手机、PDA 等，在数据的增、删、改、查等操作上提供一种高效的方法。

7.4.1　SQLite 简介

Android 使用 SQLite 作为存储数据库，SQLite 数据库是一种免费开源的且底层无关的数据库。SQLite 是基于 C 语言设计开发的开源数据库，最大支持 2048G 数据。它具有如下特征：

(1) 轻量级：大多数数据库的读写模型是基于 C/S 架构设计的，该架构下的数据库分为客户端和服务器端。C/S 架构数据库是重量型的数据库，系统功能复杂且尺寸较大。SQLite 和 C/S 模式的数据库软件不同，它不使用分布式架构作为数据引擎。SQLite 数据库功能简单且尺寸较小，一般只需要带上一个 DDL 就可使用 SQLite 数据库。

(2) 独立：SQLite 与底层操作系统无关，其核心引擎既不需要安装，也不依赖任何第三方软件，SQLite 几乎能在所有的操作系统上运行，具有较高的独立性。

(3) 便于管理和维护：SQLite 数据库具有较强的数据隔离性。SQLite 的一个文件包含了数据库的所有信息(比如表、视图、触发器)，有利于数据的管理和维护。

(4) 可移植性：SQLite 数据库应用可快速无缝移植到大部分操作系统，如 Android、Windows Mobile、Symbian、Palm 等。

(5) 语言无关：SQLite 数据库与语言无关，支持很多语言比如 Python、.Net、C/C++、Java、Ruby、Perl 等。

(6) 事务性：SQLite 数据库采用独立事务处理机制，SQLite 遵守 ACID(Atomicity、Consistency、Isolation、Durability)原则，使用数据库的独占性和共享锁处理事务。这种方式规定必须在获得该共享锁后，才能执行写操作，因而 SQLite 既允许数据库被多个进程并发读取，又保证最多只有一个进程写数据。这种方式可有效防止读脏数据、不可重复读、丢失修改等异常。

SQLite 不需要系统提供太大的资源，占用不到 1M 的内存空间就可运行，因此广泛地应用在小型的嵌入式设备中。SQLite 操作简单，且数据库功能强大，提供了基本数据库、表以及记录的操作，包括数据库创建、数据库删除、表创建、表删除、记录插入、记录删除、记录更新、记录查询。

7.4.2　SQLite 数据库操作

Android 提供了创建和使用 SQLite 数据库的 API。SQLiteDatabase 代表一个数据库对象，提供了操作数据库的一些方法，其常用的方法如表 7-4 所示。

表 7-4 SQLiteDatabase 常用方法

方 法	方 法 描 述
openOrCreateDatabase(String path,SQLiteDatabase.CursorFactory factory)	打开或创建数据库
openDatabase(String path, SQLiteDatabase.CursorFactory factory, int flags)	打开指定的数据库
close()	关闭数据库
insert(String table,String nullColumnHack,ContentValues values)	插入一条记录
delete(String table,String whereClause,String[] whereArgs)	删除一条记录
query (boolean distinct, String table, String[] columns, String selection, String[] selectionArgs, String groupBy, String having, String orderBy, String limit)	查询记录
update(String table,ContentValues value,String whereClause, String[] whereArgs)	修改记录
execSQL(String sql)	执行一条 SQL 语句

1．数据库操作

(1) 创建或打开数据库。

openDatabase()方法用于打开指定的数据库，该方法有三个参数，其中：

- ✧ path：用于指定数据库的路径，若指定数据库不存在，则抛出 FileNotFoundException 异常。
- ✧ factory：用于构造查询时的游标，若 factory 为 null，则表示使用默认的 factory 构造游标。
- ✧ flags：指定了数据库打开的模式，SQLite 定义了四种数据库打开模式。这四种模式分别为：
  - ➢ OPEN_READONLY：以只读的方式打开数据库。
  - ➢ OPEN_READWRITE：以可读可写的方式打开数据库。
  - ➢ CREATE_IF_NECESSARY：检查数据库是否存在，若不存在则创建数据库。
  - ➢ NO_LOCALIZED_COLLATORS：打开数据库时，不按照本地化语言对数据进行排序。

数据库打开模式可以同时指定多个，中间使用“|”进行分隔即可。

使用 openOrCreateDatabase()创建打开数据库时，数据库默认不按照本地化语言对数据进行排序，其作用同 openDatabase(path,factory,CREATE_IF_NECESSARY)一样。因为创建 SQLite 数据库也就是在文件系统中创建一个 SQLite 数据库的文件，所以应用程序必须对创建数据库的目录有可写的权限，否则会抛出 SQLiteException 异常。

使用 openDatabase()方法打开指定的数据库，代码如下：

```
SQLiteDatabase sqliteDatabase = SQLiteDatabase
        .openDatabase("qdu_Student.db", null, NO_LOCALIZED_COLLATORS);
```

使用 openOrCreateDatabase()方法打开或创建指定数据库，代码如下：

```
SQLiteDatabase sqliteDatabase = SQLiteDatabase
```

```
        .openOrCreateDatabase ("qdu_Student.db", null);
```

(2) 删除数据库。

android.content.Context.deleteDatabase()方法用于删除指定的数据库，例如，可以在 Activity 中使用以下代码删除数据库：

```
deleteDatabase("qdu_Student.db");  //删除数据库 qdu_Student.db
```

(3) 关闭数据库，代码如下：

```
sqliteDatabase.close(); //关闭数据库
```

2. 表操作

(1) 创建表。

数据库包含多个表，每个表可存储多条记录。数据库创建后，下一步需要创建表。SQLite 没有提供专门的方法创建表，可使用 execSQL()方法并指定 SQL 语句来创建表，格式如下：

```
String SQL_CT = "CREATE TABLE student (ID INTEGER PRIMARY KEY, "
      + "age INTEGER,name TEXT)";//创建表的 SQL 语句
sqliteDatabase.execSQL(SQL_CT);//执行该 SQL 语句创建表
```

(2) 删除表。

使用 execSQL()方法并指定 SQL 语句来删除表，代码如下：

```
String SQL_DROP_TABLE= "DROP TABLE student";//删除表的 SQL 语句
sqliteDatabase.execSQL(SQL_DROP_TABLE); //删除表 student
```

3. 记录操作

(1) 插入记录。

向表中插入记录有两种实现方式：insert()方法和 execSQL()方法。

✧ insert()方法。

可使用 SQLiteDatabase 的 insert()方法向 SQLite 数据库的表中插入数据，格式如下：

```
insert(String table,String nullColumnHack,ContentValues values)
```

其中：

- 第 1 个参数是要插入数据的表名称。
- 第 2 个参数是空列的默认值。
- 第 3 个参数是 android.content.ContentValues 类型的对象，它是一个封装了列名称和列值的 Map，代表一条记录信息。

使用 insert()方法插入记录，代码如下：

```
ContentValues contentValues = new ContentValues();//创建 ContentValues 对象
contentValues.put("ID", 1);//将 ID、age 和 name 放入 contentValues
contentValues.put("age", 26);
contentValues.put("name", "StudentA");
//调用 insert()方法将 contentValues 对象封装的数据插入到 student 表中
sqliteDatabase.insert("student" , null, contentValues);
```

✧ execSQL()方法。

使用 execSQL()方法向数据库中插入数据时，需要先编写插入数据的 SQL 语句，然后再执行 execSQL()方法，格式如下：

```
String SQL_INSERT= "INSERT INTO student (ID,age,name) "
    + "values (1, 26, 'StudentA')";//定义插入 SQL 语句
//调用 execSQL()方法执行 SQL 语句，将数据插入到 student 表中
sqliteDatabase.execSQL(SQL_INSERT);
```

(2) 更新记录。

与插入记录类似，更新记录也有两种实现方式：update()方法和 execSQL()方法。

✧ upadate()方法。

可使用 SQLiteDatabasede 的 update()方法对数据库表中的数据进行更新，格式如下：

```
update(String table,ContentValues value,String whereClause,
    String[] whereArgs)
```

其中：

➢ 第 1 个参数是要更新表数据的名称。
➢ 第 2 个参数是更新的记录信息 ContentValues 对象。
➢ 第 3 个参数是更新条件(where 子句)。
➢ 第 4 个参数是更新条件值数组。

以修改 StudentA 的年龄为例，使用 update()更新记录的代码如下：

```
ContentValues contentValues = new ContentValues();//创建 ContentValues 对象
contentValues.put("ID", 1);
contentValues.put("age", 25);
contentValues.put("name", "StudentA");
//调用 update()方法更新 student 表中名为 StudentA 记录数据
sqliteDatabase.update("student", contentValues, "name=StudentA", null);
```

✧ execSQL()方法。

使用 execSQL()方法更新数据时，需先编写更新数据的 SQL 语句，然后再执行 execSQL()方法来更新一条记录，格式如下：

```
//定义更新 SQL 语句
String   SQL_UPDATE= "UPDATE student SET age=25 where name='StudentA'";
//调用 execSQL()方法执行 SQL 语句更新 student 表中的记录
sqliteDatabase.execSQL(SQL_UPDATE);
```

(3) 查询记录。

使用 SQLiteDatabase 的 query()方法查询记录。SQLiteDatabase 提供了 6 种 query()方法用于不同方式的查询，其中常用的 query()方法如下：

```
public Cursor query (boolean distinct, String table, String[] columns, String selection, String[] selectionArgs,
String groupBy, String having, String orderBy, String limit);
```

该方法的参数说明如下：

- distinct：一个可选的布尔值，用来区分返回的记录是否只包含唯一的值。
- table：表名称。
- columns：列名称数组。
- selection：条件子句，可以包含通配符“?”，可被参数数组中的值替换。
- selectionArgs：参数数组，替换 where 子句中的“?”。
- groupBy：分组列。
- having：分组条件。
- orderBy：排序列。
- limit：一个可选的字符串，用来定义返回的行数限制，即分页查询限制。

query()方法的返回值是一个 Cursor 游标对象，相当于结果集 ResultSet。游标提供了一种对从表中检索出的数据进行操作的灵活手段，它实际上是一种能从多条数据记录的结果集中每次提取一条记录的机制。游标总是与一条 SQL 选择语句相关联，因为游标由结果集(可以是零条、一条或多条记录)和结果集中指向特定记录的游标位置组成。当对结果集进行处理时，必须声明一个指向该结果集的游标。Cursor 游标常用的方法如表 7-5 所示。

表 7-5　Cursor 游标常用方法

方　法	功 能 描 述
move(int offset)	以当前的位置为基准，将 Cursor 移动到偏移量为 offset 的位置。若移动成功返回 true，失败返回 false。注意 offset 为正值时，游标向前移动；负值时，向后移动
moveToPosition(int position)	将 Cursor 移动到绝对位置 position 位置，若移动成功返回 true，失败返回 false。注意 moveToPosition 移动到一个绝对位置，而 move 移动以当前位置为基准
moveToNext()	将 Cursor 向前移动一个位置，成功返回 true，失败返回 false。其功能等同于 move (1)
moveToLast()	将 Cursor 移动到最后一条记录，成功返回 true，失败返回 false。若当前记录数为 count，则其功能等同于 moveToPosition (count)
moveToFisrt()	将 Cursor 移动到第一条记录，成功返回 true，失败返回 false。其功能等同于 moveToPosition (1)
isBeforeFirst()	判断 Cursor 是否指向第一项数据之前。若指向第一项数据之前，返回 true；否则返回 false
isAfterLast()	判断 Cursor 是否指向最后一项数据之后。若指向最后一项数据之后，返回 true；否则返回 false
isClosed()	判断 Cursor 是否关闭。若 Cursor 关闭，返回 true；否则返回 false
isFirst()	判断 Cursor 是否指向第一条记录
isLast()	判断 Cursor 是否指向最后一条记录
isNull(int columnIndex)	判断指定的位置 columnIndex 的记录是否存在
getCount()	获取当前表的行数即记录总数
getInt(int columnIndex)	获取指定列索引的 int 类型值
getString(int columnIndex)	获取指定列索引的 String 类型值

以查询 StudentA 为例，使用 query()方法查询记录的代码如下：

```
//查询获得游标
Cursor cursor=sqliteDatabase.query(true, "student", null, "name=StudentA", null, null, null,null,null);
//将游标移动到第一条记录，并判断
if(cursor.moveToFirst()){
        int id=cursor.getInt(0);//获得列信息
        int age=cursor.getInt(1);
        String name=cursor.getString(3);
        System.out.println(id+":"+age+":"+name);//输出
}
```

(4) 删除记录。

与插入、修改记录相同，删除记录也有两种实现方式：

✧ delete()方法。

可使用 SQLiteDatabasede 的 delete()方法删除数据库表中的数据，格式如下：

```
delete(String table,String whereClause,String[] whereArgs)
```

其中：第 1 个参数是表名称；第 2 个参数是删除条件；第 3 个参数是删除条件的参数数组。

以删除 StudentA 的记录为例，使用 delete()方法的代码如下：

```
sqliteDatabase.delete("student","name=?",new String[]{"StudentA"});
```

✧ execSQL()方法。

使用 execSQL()方法删除表数据需要先编写 SQL 语句，然后再执行 execSQL()方法，格式如下：

```
//定义删除 SQL 语句
String SQL_DELETE= "DELETE FORM student where name='StudentA'";
//调用 execSQL()方法执行 SQL 语句删除 student 表中的记录
sqliteDatabase.execSQL(SQL_DELETE);
```

7.4.3 SQLiteOpenHelper

SQLiteOpenHelper 是 SQLiteDatabase 的一个帮助类，用来管理数据库的创建和版本更新。通过实现 SQLiteOpenHelper，可以隐藏那些用于决定一个数据库在被打开之前是否需要被创建或者升级的逻辑。一般用法是定义一个类继承 SQLiteOpenHelper，并实现 onCreate()和 onUpgrade()两个方法。SQLiteOpenHelper 类的常用方法如表 7-6 所示。

表 7-6 SQLiteOpenHelper 常用方法

方 法	功 能 描 述
SQLiteOpenHelper(Context context,String name, SQLiteDatabase.CursorFactory,int version)	构造函数，第二个参数是数据库名称
onCreate(SQLiteDatabase db)	创建数据库时调用
onUpgrade(SQLiteDatabase db,int oldVersion,int newVersion)	版本更新时调用
getReadableDatabase()	创建或打开一个只读数据库
getWritableDatabase()	创建或打开一个读写数据库

【示例 7.3】 在 Android Studio 中新建项目 Ch07_7D3_SQLite，使用 SQLiteOpenHelper 实现音乐播放列表的添加、删除和浏览功能。

(1) 创建一个数据库工具类 DBHelper，该类继承 SQLiteOpenHelper，重写 onCreate() 和 onUpgrade()方法，并添加 insert()、delete()、query()方法，分别实现数据的添加、删除和查询功能。DBHelper 类的代码如下：

```
public class DBHelper extends SQLiteOpenHelper {
    private static final String DB_NAME = "music.db";//数据库名称
    private static final String TBL_NAME = "MusicTbl";//表名
    private SQLiteDatabase db;//声明 SQLiteDatabase 对象
    //构造函数
    DBHelper(Context c) {
        super(c, DB_NAME, null, 2);
    }
    @Override
    public void onCreate(SQLiteDatabase db) {
        //获取 SQLiteDatabase 对象
        this.db = db;
        //创建表
        String CREATE_TBL = "create table MusicTbl(_id integer primary " +
        "key autoincrement,name text,singer text) ";
        db.execSQL(CREATE_TBL);
    }
    //插入
    public void insert(ContentValues values) {
        SQLiteDatabase db = getWritableDatabase();
        db.insert(TBL_NAME, null, values);
        db.close();
    }
    //查询
    public Cursor query() {
        SQLiteDatabase db = getWritableDatabase();
        Cursor c = db.query(TBL_NAME, null, null, null, null, null,null);
        return c;
    }
    //删除
    public void del(int id) {
        if (db == null)
            db = getWritableDatabase();
        db.delete(TBL_NAME, "_id=?",new String[]{ String.valueOf(id) });
```

```
    }
    //关闭数据库
    public void close() {
        if (db != null)
            db.close();
    }
    @Override
    public void onUpgrade(SQLiteDatabase db, int oldVersion, int newVersion) {
    }
}
```

(2) 创建添加音乐的 MainActivity，添加界面中提供两个文本框和一个按钮，用于输入音乐名和歌手名，当单击【添加】按钮时，将数据插入到表中，代码如下：

```
public class MainActivity extends AppCompatActivity {

    private EditText mEtName;
    private EditText mEtSinger;
    private Button mBtnAdd;

    @Override
    protected void onCreate(Bundle savedInstanceState) {
        super.onCreate(savedInstanceState);
        setContentView(R.layout.activity_main);
        initView();
    }

    private void initView() {
        mEtName = (EditText) findViewById(R.id.et_name);
        mEtSinger = (EditText)findViewById(R.id.et_singer);
        mBtnAdd = (Button) findViewById(R.id.btn_add);
        mBtnAdd.setOnClickListener(new View.OnClickListener() {
            @Override
            public void onClick(View view) {
                //获取用户输入的文本信息
                String name = mEtName.getText().toString();
                String singer = mEtSinger.getText().toString();
                //创建 ContentValues 对象，封装记录信息
                ContentValues values = new ContentValues();
                values.put("name", name);
                values.put("singer", singer);
```

```
                //创建数据库工具类 DBHelper
                DBHelper helper = new DBHelper(getApplicationContext());
                //调用 insert()方法插入数据
                helper.insert(values);
                //跳转到 QueryActivity，显示音乐列表
                Intent intent = new Intent(MainActivity.this, QueryActivity.class);
                startActivity(intent);
            }
        });
    }
}
```

当在上述代码中单击“添加”按钮时，先将用户输入的音乐名和歌手信息封装到 ContentValues 对象中，再调用 DBHelper 的 insert()方法将记录插入到数据库中，然后跳转到 QueryActivity 来显示音乐列表。

(3) 创建显示音乐列表的 QueryActivity，代码如下：

```
public class QueryActivity extends AppCompatActivity {

    private ListView mLv;

    @Override
    protected void onCreate(Bundle savedInstanceState) {
        super.onCreate(savedInstanceState);
        setContentView(R.layout.activity_query);
        initView();
    }

    private void initView() {
        setTitle("浏览音乐列信息");
        mLv = (ListView) findViewById(R.id.lv);
        final DBHelper helpter = new DBHelper(this);
        //查询数据，获取游标
        Cursor c = helpter.query();
        //列表项数组
        String[] from = {"_id", "name", "singer"};
        //列表项 ID
        int[] to = {R.id.text0, R.id.text1, R.id.text2};
        //适配器
        SimpleCursorAdapter adapter = new SimpleCursorAdapter(this,
                R.layout.item_list, c, from, to);
```

```
        //设置适配器
        mLv.setAdapter(adapter);

        //提示对话框
        final AlertDialog.Builder builder = new AlertDialog.Builder(this);

        //设置 ListView 单击监听器
        mLv.setOnItemClickListener(new AdapterView.OnItemClickListener() {
            @Override
            public void onItemClick(AdapterView<?> arg0, View arg1, int arg2,
                                long arg3) {
                final long temp = arg3;
                builder.setMessage("真的要删除该记录吗？")
                        .setPositiveButton("是",
                                new DialogInterface.OnClickListener() {
                                    public void onClick(DialogInterface dialog, int which) {
                                        //删除数据
                                        helpter.del((int) temp);
                                        //重新查询数据
                                        Cursor c = helpter.query();
                                        String[] from = {"_id", "name", "singer"};
                                        int[] to = {R.id.text0, R.id.text1, R.id.text2};
                                        SimpleCursorAdapter adapter = new SimpleCursorAdapter(
                                                getApplicationContext(), R.layout.item_list, c, from, to);
                                        mLv.setAdapter(adapter);
                                    }
                                })
                        .setNegativeButton("否", null);
                AlertDialog ad = builder.create();
                ad.show();
            }
        });
        helpter.close();
    }
}
```

上述代码中调用 DBHelper 的 query()方法查询数据库，并返回一个 Cursor 游标，然后使用 SimpleCursorAdapter 适配器将数据绑定到 ListView 控件上，并在 ListView 控件上注册单击监听器，使在单击一条记录时，会显示一个警告对话框提示是否删除记录，如果单击【是】按钮，则调用 DBHelper 的 del()方法删除指定记录。

运行程序，添加音乐记录，如图 7-6 所示。

图 7-6　添加音乐记录

在音乐列表页面中单击一条记录，然后通过弹出的警告对话框删除一条记录，如图 7-7 所示。

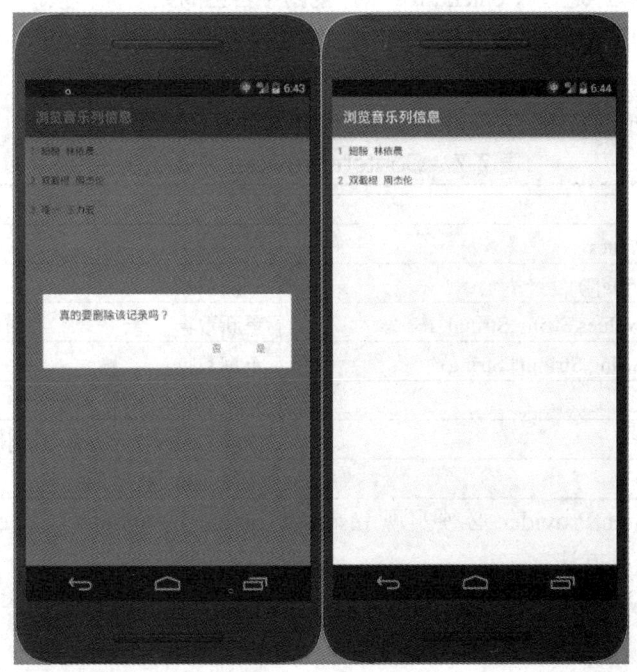

图 7-7　删除音乐记录

7.5 数据共享 ContentProvider

数据库在 Android 当中是私有的，不能将数据库设为 WORLD_READABLE，每个数据库都只能由创建它的包访问，这就意味着只有创建这个数据库的应用程序才能访问它，也就是说数据的访问不能跨越进程和包的边界，直接访问别的应用程序的数据库。那么，如何在应用程序间交换数据呢？

可以使用 ContentProvider(内容提供器)实现进程间的数据传递。ContentProvider 是所有应用程序之间数据存储和检索的一个桥梁，其作用是使各个应用程序之间能共享数据。

7.5.1 ContentProvider

ContentProvider 是 Android 提供的应用组件，定义在 android.content 包下面，通过这个组件可访问 Android 提供的应用数据，如联系人列表。作为应用程序之间唯一的数据共享途径，ContentProvider 主要功能是存储、检索数据并向应用程序提供访问数据的接口。Android 系统为一些常见的应用(如音乐、视频、图像、联系人列表等)定义了相应的 ContentProvider，它们被定义在 android.provider 包下。需要特别指出的是，只有在 AndroidManifest.xml 配置文件中添加许可，才能访问 ContentProvider 中的数据。

当一个应用程序需要把自己的数据暴露给其他应用程序时，该应用程序可以通过提供的 ContentProvider 来实现，而其他应用程序需要使用这些数据时，可以通过 ContentResolver 来操作 ContentProvider 暴露的数据。

一旦某个应用程序通过 ContentProvider 暴露了自己的数据操作接口，那么不管该应用程序是否启动，其他应用程序都可以通过该接口来操作被暴露的内部数据，包括增加数据、删除数据、修改数据、查询数据等。

ContentProvider 的常用方法如表 7-7 所示。

表 7-7　ContentProvider 常用方法

方　法	功能描述
insert(Uri,ContentValues)	插入数据
delete(Uri,String,String[])	删除数据
update(Uri,ContentValues,String,String[])	更新数据
query(Uri,String[],String,String[],String)	查询数据
getType(Uri)	获得 MIME 数据类型
onCreate()	创建 ContentProvider 时调用
getContext()	获得 Context 对象

定义一个 ContentProvider 必须实现 insert()、delete()、update()、query()和 getType()这几个操作数据的抽象方法。

在 ContentProvider 中，数据模型和 URI 是两个重要概念，详细内容如下所述。

1. 数据模型

ContentProvider 将其存储的数据以数据表的形式提供给访问者。在数据表中每一行为一条记录，而每一列为具有特定类型和意义的字段。每一条数据记录都包括一个“_ID”

数据列，该字段唯一标识一个记录。

2. URI

每一个 ContentProvider 都对外提供一个自身数据集的唯一标识，这个唯一标识就是 URI。若一个 ContentProvider 管理多个数据集，这个 ContentProvider 将会为每个数据集分配一个独立且唯一的 URI。所有的 ContentProvider 的 URI 都以“content://”开头，其中“content:”是用来标识 ContentProvider 管理的 schema。

URI 一般格式如下：

```
content://数据路径/标识 ID(可选)
```

例如：

```
content://media/internal/images （该 URI 返回设备上存储的所有图片）
content://contacts/people/5        （该 URI 返回 ID 为 5 的联系人信息）
```

Android 中使用 Uri 类来定义 URI，例如：

```
Uri uri = Uri.parse("content://media/internal/images");
Uri uri = Uri.parse("content://contacts/people/5");
```

URI 后面可以加上记录的 ID 值，Android 提供了两种方法用于在 URI 后扩展一个记录的 ID：

(1) withAppendedId()：该方法是 ContentUris 类的方法。除了 withAppendedId()方法之外，ContentUris 还提供了 parseId()方法用于从路径中获取 ID 部分。使用 ContentUris 的方法来添加和获取 ID 的代码如下：

```
Uri uri = Uri.parse("content://qdu.edu/student");
Uri resultUri = ContentUris.withAppendedId(uri, 3);
//生成后的 Uri 为：content://qdu.edu/student/3
Uri uri = Uri.parse("content://qdu.edu/student/3") ;
long personid = ContentUris.parseId(uri); //获取的结果为:3
```

(2) withAppendedPath()：该方法是 Uri 类的方法。通过该方法可以很简单地在 URI 后扩展一个 ID。比如要在联系人数据库中查找记录 41，代码如下：

```
Uri uri= Uri.withAppendedPath(Contacts.CONTENT_URI, 41);
```

几乎所有的 ContentProvider 的操作都会用到 URI，因此通常将 URI 定义为常量，例如 android.provider.ContactsContract.Contacts.CONTENT_URI 就是联系人列表的 CONTENT_URI 常量，这样在简化开发的同时也提高了代码的可维护性。

7.5.2 ContentResolver

应用程序不能直接访问 ContentProvider 中的数据，但可通过 Android 系统提供的 ContentResolver(内容解析器)来间接访问。ContentResolver 提供了对 ContentProvider 数据进行查询、插入、修改和删除等操作的方法，在开发过程中是通过间接操作 ContentResolver 来操作 ContentProvider 的。通常 ContentProvider 是单实例的，但可以有多个 ContentResolver 在不同的应用程序和不同的进程之间与 ContentProvider 进行交互。

每个应用程序的上下文都有一个 ContentResolver 实例，可以使用 getContentResolver()

方法获取该实例对象，代码如下：

```
ContentResolver cr=getContentResolver();
```

1．查询

使用 ContentResolver 的 query()方法查询数据与 SQLite 查询一样，返回一个指向结果集的游标 Cursor，代码如下：

```
ContentResolver resolver = getContentResolver();//获取 ContentResolver 对象
Cursor cursor=resolver.query(Contacts.CONTENT_URI,null, null, null, null);
```

使用 Activity 的 managedQuery()方法也可以查询 ContentProvider 中的数据，与 ContentResolver 的 query()方法类似，它们的第一个参数都是 ContentProvider 的 CONTENT_URI 常量。这个常量用来标识某个特定的 ContentProvider 和数据集。query()和 managedQuery()方法都返回一个 Cursor 对象。两者之间的唯一区别是：Activity 可使用 managedQuery()方法来管理 Cursor 的生命周期，而 ContentResolver 却无法通过 query() 方法来管理 Cursor 的生命周期。

2．插入

使用 ContentResolver.insert()方法向 ContentProvider 中增加一个新的记录时，需要先将新记录的数据封装到 ContentValues 对象中，然后调用 ContentResolver.insert()方法。insert()方法将返回一个 URI，该 URI 内容是由 ContentProvider 的 URI 加上该新记录的扩展 ID 得到的，可以通过该 URI 对该记录做进一步的操作，代码如下：

```
ContentValues contentValues = new ContentValues();
values.put(Contacts._ID, 1);//联系人 ID
contentValues .put(Contacts.DISPLAY_NAME, "zhangsan");//联系人名
//获取 ContentResolver 对象
ContentResolver resolver = getContentResolver();
Uri uri = resolver.insert(Contacts.CONTENT_URI, contentValues);//插入
```

3．删除

如果要删除单个记录，可以调用 ContentResolver.delete()方法，通过给该方法传递一个特定行的 URI 参数来实现删除操作。如果要对多行记录执行删除操作，就需要给 delete()方法传递需要被删除的记录类型的 URI 以及一个 where 子句来实现多行删除，代码如下：

```
//获取 ContentResolver 对象
ContentResolver resolver = getContentResolver();
//删除单个记录
resolver.delete(Uri.withAppendedPath(Contacts.CONTENT_URI,41),null, null);
//删除前 5 行记录
resolver.delete(Contacts.CONTENT_URI,"_id<5", null);
```

4．更新

使用 ContentResolver.update()方法实现记录的更新操作，代码如下：

```
ContentValues contentValues = new ContentValues();//创建一个新值
```

```
contentValues .put(Contacts.DISPLAY_NAME, "zhangsan");
//获取 ContentResolver 对象
ContentResolver resolver = getContentResolver();
//更新
resolver.update(Contacts.CONTENT_URI,contentValues, "_id=5",null);
```

7.5.3　ContentProvider 应用

ContentProvider 为应用间的数据交互提供了一个安全的环境，允许应用把自己的数据根据需求开放给其他应用，进行增删改查，而不需要担心直接开放数据权限带来的安全问题。

【示例 7.4】 在 Android Studio 中新建项目 Ch07_7D4_ContentProvider，使用 ContentProvider 访问联系人信息。

在 MainActivity 中编写代码如下：

```
public class MainActivity extends AppCompatActivity {

    //联系人的 URI
    Uri contact_uri = ContactsContract.Contacts.CONTENT_URI;
    TextView textview; //声明 TextView 对象
    int textcolor = Color.BLACK; //定义文本颜色

    @Override
    protected void onCreate(Bundle savedInstanceState) {
        super.onCreate(savedInstanceState);
        setContentView(R.layout.activity_main);
        textview = (TextView) findViewById(R.id.tv); //创建 textview 对象
        //调用 getContactInfo()方法获取联系人的信息
        String result = getContactInfo();
        textview.setTextColor(textcolor); //设置文本框的颜色
        textview.setTextSize(20.0f); //定义字体大小
        textview.setText("记录\t 名字\n" + result); //设置文本框的文本
    }

    /**
     * getContactInfo()获取联系人列表的信息，返回 String 对象
     */
    public String getContactInfo() {
        String result = "";
        //获取 ContentResolver 对象
        ContentResolver resolver = getContentResolver();
        //查询联系人
```

```
        Cursor cursor = resolver.query(contact_uri, null, null, null, null);
        //获得_ID 字段的索引
        int idIndex = cursor.getColumnIndex(ContactsContract.Contacts._ID);
        //获得 Name 字段的索引
        int nameIndex = cursor.getColumnIndex(ContactsContract.Contacts.DISPLAY_NAME);
        //遍历 Cursor 提取数据
        cursor.moveToFirst();
        for (; !cursor.isAfterLast(); cursor.moveToNext()) {
            result = result + cursor.getString(idIndex) + "\t\t\t";
            result = result + cursor.getString(nameIndex) + "\t\n";
        }
        //使用 close 方法关闭游标
        cursor.close();
        //返回结果
        return result;
    }
}
```

上述实例使用 ContentProvider 访问联系人信息。在 Android 中，用户可使用 Contacts 来添加、修改、编辑以及删除联系人。在桌面上有联系人应用图标，如图 7-8 所示；单击该应用图标，则进入联系人列表页面，在联系人页面中可以添加新联系人，图 7-9 中就添加了 3 个联系人信息。

图 7-8　联系人应用图标

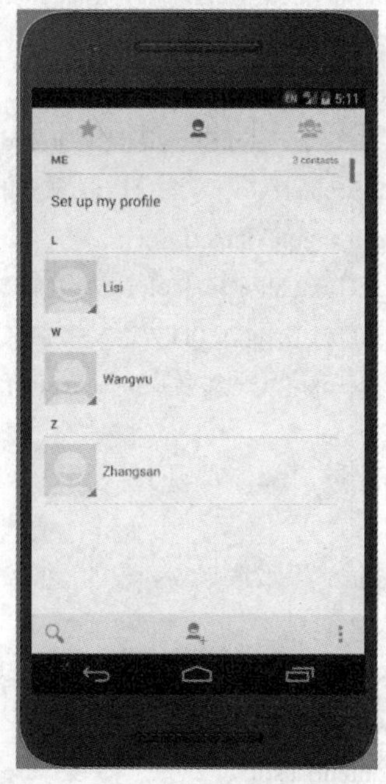

图 7-9　联系人列表

Contacts 对应一个 SQLite 数据库，用于存放联系人信息，因而对 Contact 的操作结果最终要存储到一个 SQLite 数据库中，该数据库位于/data/data/com.android.provides.

contacts/databases 目录下，如图 7-10 所示。

名称	大小	日期	时间	权限
◢ com.android.providers.contacts		2017-07-20	06:26	drwxr-x--x
▷ cache		2017-07-20	06:26	drwxrwx--x
◢ databases		2017-07-21	05:10	drwxrwx--x
contacts2.db	331776	2017-07-21	05:10	-rw-rw----
contacts2.db-journal	0	2017-07-21	05:10	-rw-------
profile.db	331776	2017-07-20	06:26	-rw-rw----
profile.db-journal	16928	2017-07-20	06:26	-rw-------

图 7-10　Contact 对应的 SQLite 数据库

本实例使用 ContentResolver 来实现联系人的查询，ContentResolver 提供了 query()方法来查询 ContentProvider 中的数据。执行完 query()方法后，会返回一个游标对象，通过该对象获取表对应字段的索引，然后使用游标的 getXXX()方法获取该索引对应的值。注意要让程序能够读取 Contacts 信息，必须在配置文件 AndroidManifest.xml 中添加如下权限：

```
<uses-permission android:name="android.permission.READ_CONTACTS" />
```

运行程序，记录手机中所有联系人的 ID 以及名字，并在屏幕上显示，效果如图 7-11 所示。

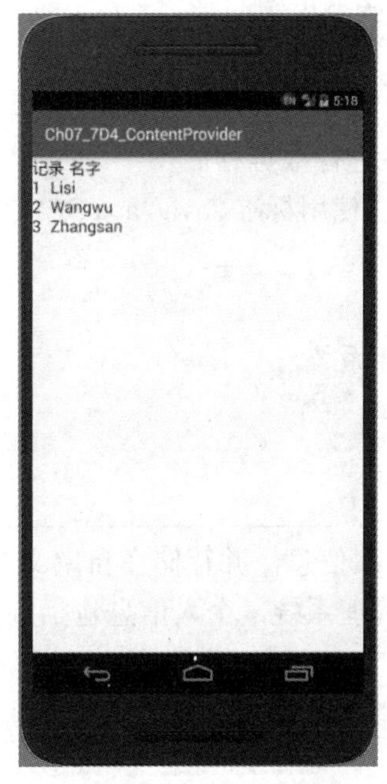

图 7-11　在屏幕上显示所有联系人信息

本 章 小 结

通过本章的学习，读者应当了解：

✧ SharedPreference 提供了一种轻量级的数据存储方式，以“key-value”方式将数据保存在一个 XML 配置文件中。

✧ 通过 Context.openFileInput() 方法可以获取标准的文件输入流(FileInputStream)，以读取设备上的文件；通过 Context.openFileOuput()方法

可以获取标准的文件输出流(FileOutputStream)。

- ✧ Android 中通过 SQLite 数据库实现结构化数据存储。
- ✧ SQLiteOpenHelper 是 SQLiteDatabase 的一个帮助类，用来管理数据库的创建和版本更新。
- ✧ ContentProvider(内容提供器)是所有应用程序之间数据存储和检索的一个桥梁，其作用就是使得各个应用程序之间实现数据共享。
- ✧ Android 系统为一些常见的应用(如音乐、视频、图像、联系人列表等)定义了相应的 ContentProvider，它们被定义在 android.provider 包下。

本 章 练 习

1. 适合结构化数据存储的是______。

   A. SharedPreference　　B. 文件方式　　C. SQLite　　D. 网络

2. 可以存储为 XML 文件的存储方式是______。

   A. SharedPreference　　B. 文件方式　　C. SQLite　　D. 网络

3. 关于 Android 数据存储的说法不正确的是______。

   A. SharedPreference 适合小数据量的存储

   B. 文件方式适合大文件存储

   C. SQLite 适合嵌入式设备进行数据存储

   D. Android 的文件存储无法使用标准 Java(Java SE)中的 IO 机制

4. 关于 SQLite 的说法不正确的是______。

   A. SQLite 不支持事务

   B. SQLite 只能用于 Android 系统

   C. SQLite 不支持完整的 SQL 规范

   D. SQLite 允许网络访问

5. Android 的四种数据存储机制是______、______、______、______。
6. 编写代码，读取所有的联系人信息，并存储在自定义的 SQLite 表中。
7. 使用 SQLite 实现个人信息管理系统，个人信息包括姓名、年龄、性别以及学历。

第 8 章　片段(Fragment)

本章目标

- 了解片段 Fragment 的作用及特点
- 掌握 Fragment 的生命周期
- 掌握 Fragment 的创建方法
- 掌握 Fragment 的切换方法

8.1 Fragment 简介

Fragment(片段)是一种可以嵌套在 Activity 当中的 UI 片段，它能让程序界面更加充分合理地利用屏幕空间，因此在设计手机或平板应用程序时被广泛采用。Fragment 的用法和 Activity 类似，同样都包含布局，同样都有自己的生命周期。

8.1.1 Fragment 的作用

随着移动设备的迅速发展，手机和平板电脑已经成为当今的生活必需品，两者最重要的区别之一就在于屏幕的大小。而如果屏幕尺寸差距过大，同样的应用程序界面在视觉效果上会产生较大的差异，比如在手机屏幕上十分美观的界面，放到更大的平板屏幕上后，由于相关控件被过度拉伸，极易导致效果不再美观。为了解决这一问题，Android 自 3.0 版本开始，引入了 Fragment 的概念。

Fragment 可以充分利用不同尺寸的屏幕空间。以设计可同时支持平板电脑设备和手机的新闻应用为例，使用 Fragment 实现的界面效果如图 8-1 所示。

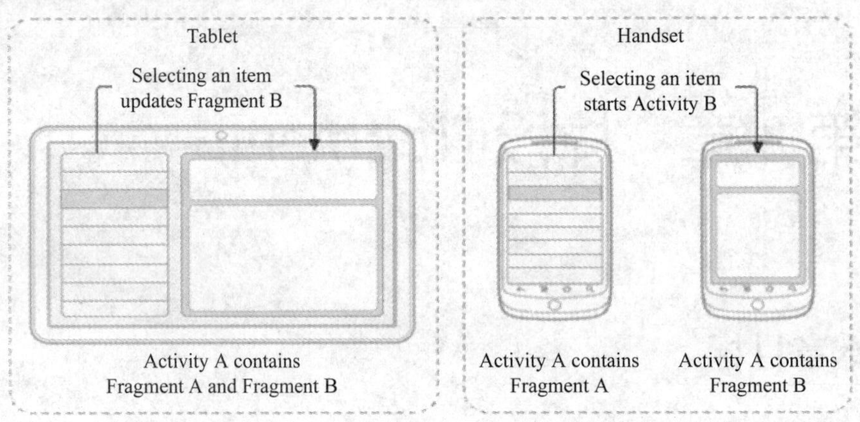

图 8-1 使用 Fragment 实现的平板和手机应用界面

在左图 Tablet 所示的平板电脑设备上运行时，可以在 Activity A 中嵌入两个 Fragment，用一个 Fragment 在左侧显示文章列表，另一个在右侧显示相应文章，两个 Fragment 并排显示在一个 Activity 中，每个 Fragment 都具有自己的一套生命周期回调方法，并各自处理自己的用户输入事件。这样，用户就不需要使用一个 Activity 来选择文章，然后使用另一个 Activity 来阅读文章，而是可以在同一个 Activity 内选择文章，并进行阅读。

而在右图 Handset 所示的手机屏幕上没有足以储存两个 Fragment 的空间，无法在同一 Activity 内储存多个 Fragment，因此可以使用单独 Fragment 来实现单窗格 UI。此时，Activity A 只包括用于显示文章列表的 Fragment，当用户选择文章时，它会启动 Activity B，其中包括用于阅读文章的第二个 Fragment。这样，应用就可通过重复使用不同组合的 Fragment 来同时支持平板电脑和手机。

8.1.2　Fragment 的特点

Fragment 作为一个嵌套到 Activity 中的 UI 片段，是 Activity 界面的一部分，它拥有如下特点：

(1) Fragment 不能独立存在或使用，而是必须要嵌套到 Activity 中才能使用。

(2) 在同一个 Activity 中可以嵌套一个或多个 Fragment，同一个 Fragment 也可以同时被多个 Activity 重用。

(3) Fragment 拥有自己的生命周期，包含 11 个生命周期方法。

(4) Fragment 的生命周期直接受所在 Activity 生命周期的影响。

(5) Fragment 可以创建动态灵活的 UI 设计，可以适应不同的屏幕尺寸，从手机到平板。

8.1.3　Fragment 生命周期

Fragment 的生命周期如图 8-2 所示。

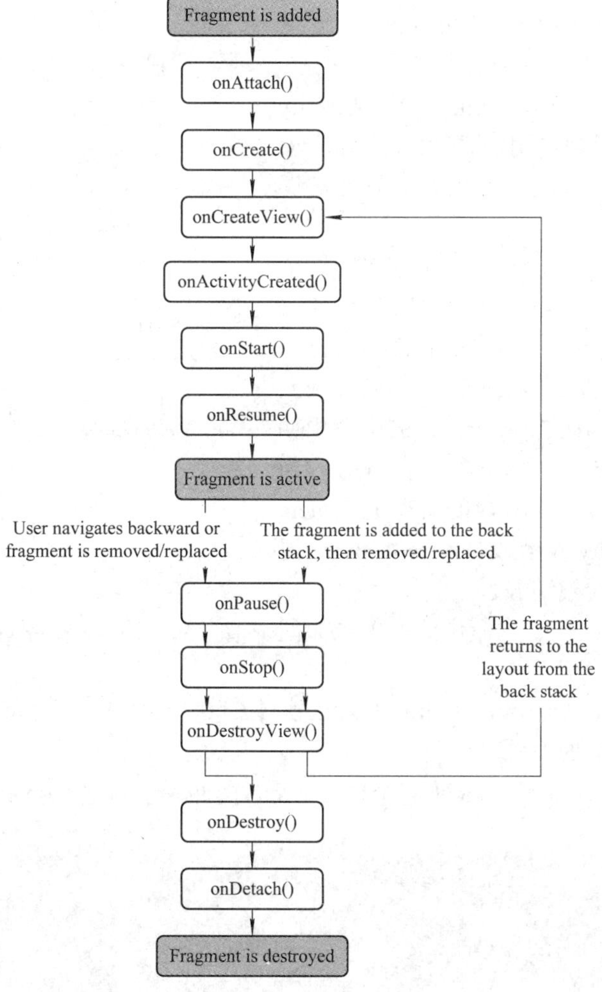

图 8-2　Fragment 的生命周期

Fragment 的生命周期包括 11 个生命周期方法：

(1) onAttach()：当 Fragment 被添加到 Activity 时被调用，该方法只会被调用一次。

(2) onCreate()：当创建 Fragment 时被调用，该方法只会被调用一次。

(3) onCreateView()：每次创建、绘制 Fragment 的 View 组件时被调用，Fragment 中将会显示该方法返回的 View 组件。

(4) onActivityCreated()：当 Fragment 的宿主 Activity 启动完成后被调用。

(5) onStart()：启动 Fragment 时被调用。

(6) onResume()：恢复 Fragment 时被调用，调用 onStart()方法后一定会调用该方法。

(7) onPause()：暂停 Fragment 时被调用。

(8) onStop()：停止 Fragment 时被调用。

(9) onDestroyView()：销毁 Fragment 所包含的 View 组件时调用。

(10) onDestroy()：销毁 Fragment 时被调用。该方法只会被调用一次。

(11) onDetach()：将 Fragment 从 Activity 中删除、替换完成时被调用，调用 onDestroy()方法后一定会调用该方法，且该方法只会被调用一次。

在同一个 Activity 中可以包含多个 Fragment，同一个 Fragment 也可以被多个 Activity 使用。虽然每个 Fragment 都有自己的布局和生命周期，但因为 Fragment 必须被嵌入到 Activity 中使用，因此 Fragment 的生命周期是受其宿主 Activity 的生命周期限制的。当 Activity 暂停时，该 Activtiy 内的所有 Fragment 都会暂停；当 Activity 被销毁时，该 Activity 内的所有 Fragment 都会被销毁。

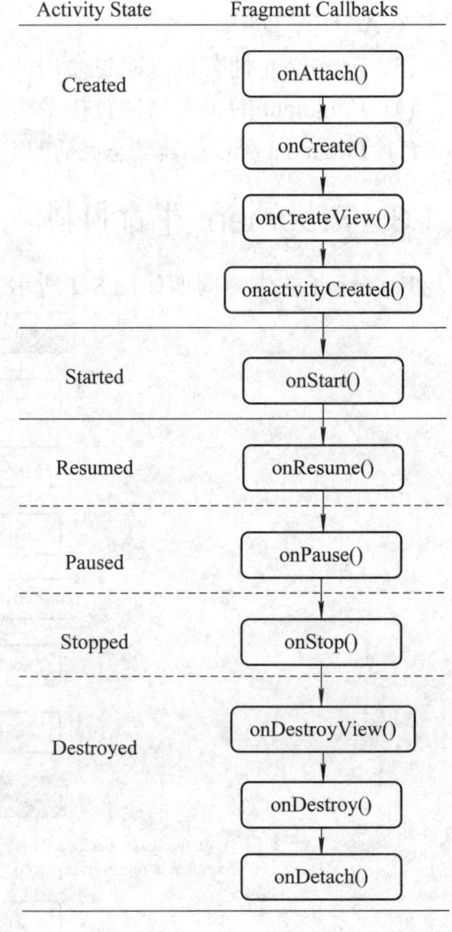

图 8-3　Activity 和 Fragment 生命周期的关系

Activity 和 Fragment 生命周期的关系如图 8-3 所示。

【示例 8.1】 在 Android Studio 中新建项目 Ch08_8D1_LifeCycle，用于观察 Activity 和 Fragment 生命周期的关系。

在 src/main/java/com.ugrow.ch08_8d1_lifecycle 文件夹下新建 ContentFragment，在其中编写以下代码：

```
public class ContentFragment extends Fragment {

    @Override
    public void onAttach(Activity activity) {
        super.onAttach(activity);
```

```
        Log.i("TAG", "fragment----->onAttach 执行");
    }

    @Override
    public void onCreate(Bundle savedInstanceState) {
        Log.i("TAG", "fragment----->onCreate 执行");
        super.onCreate(savedInstanceState);
    }

    @Override
    public View onCreateView(LayoutInflaterinflater, ViewGroup container,
                             Bundle savedInstanceState) {
        View view = inflater.inflate(R.layout.content_fragment, null);
        Log.i("TAG", "fragment----->onCreateView 执行");
        return view;
    }

    @Override
    public void onActivityCreated(Bundle savedInstanceState) {
        Log.i("TAG", "fragment----->onActivityCreated 执行");
        super.onActivityCreated(savedInstanceState);
    }

    @Override
    public void onStart() {
        Log.i("TAG", "fragment----->onStart 执行");
        super.onStart();
    }

    @Override
    public void onResume() {
        Log.i("TAG", "fragment----->onResume 执行");
        super.onResume();
    }

    @Override
    public void onPause() {
        Log.i("TAG", "fragment----->onPause 执行");
        super.onPause();
    }
```

```
    @Override
    public void onStop() {
        Log.i("TAG", "fragment----->onStop 执行");
        super.onStop();
    }

    @Override
    public void onDestroyView() {
        Log.i("TAG", "fragment----->onDestroyView 执行");
        super.onDestroyView();
    }

    @Override
    public void onDestroy() {
        Log.i("TAG", "fragment----->onDestroy 执行");
        super.onDestroy();
    }

    @Override
    public void onDetach() {
        Log.i("TAG", "fragment----->onDetach 执行");
        super.onDetach();
    }

}
```

上述代码中，ContentFragment 继承了 App 包下的 Fragment，同时重写了 Fragment 的 11 个生命周期方法，然后在每个生命周期方法中使用 Log 打印对应的生命周期方法名称。

编写 MainActivity，代码如下：

```
public class MainActivity extends AppCompatActivity {

    @Override
    protected void onCreate(Bundle savedInstanceState) {
        super.onCreate(savedInstanceState);
        setContentView(R.layout.activity_main);
        FragmentManager fragmentManager = getFragmentManager();
        fragmentManager.beginTransaction().add(R.id.fl, new ContentFragment()).commit();
        Log.e("TAG", "Activity----->onCreate 执行");
    }
```

```
    @Override
    protected void onStart() {
        Log.e("TAG", "Activity----->onStart 执行");
        super.onStart();
    }

    @Override
    protected void onResume() {
        Log.e("TAG", "Activity----->onResume 执行");
        super.onResume();
    }
    @Override
    protected void onRestart() {
        Log.e("TAG", "Activity----->onRestart 执行");
        super.onRestart();
    }
    @Override
    protected void onPause() {
        Log.e("TAG", "Activity----->onPause 执行");
        super.onPause();
    }
    @Override
    protected void onStop() {
        Log.e("TAG", "Activity----->onStop 执行");
        super.onStop();
    }

    @Override
    protected void onDestroy() {
        Log.e("TAG", "Activity----->onDestroy 执行");
        super.onDestroy();
    }
}
```

上述代码重写了 Activity 的 7 个生命周期方法，并通过 Log 打印对应方法名称。

在模拟器中安装运行应用程序，然后打开 LogCat，可以看到以下日志信息：

```
07-27 03:06:00.538 2876-2876/com.ugrow.ch08_8d1_lifecycle E/TAG: Activity----->onCreate 执行
07-27 03:06:00.538 2876-2876/com.ugrow.ch08_8d1_lifecycle I/TAG: fragment----->onAttach 执行
07-27 03:06:00.538 2876-2876/com.ugrow.ch08_8d1_lifecycle I/TAG: fragment----->onCreate 执行
07-27 03:06:00.543 2876-2876/com.ugrow.ch08_8d1_lifecycle I/TAG: fragment----->onCreateView 执行
07-27 03:06:00.543 2876-2876/com.ugrow.ch08_8d1_lifecycle I/TAG: fragment----->onActivityCreated 执行
```

```
07-27 03:06:00.543 2876-2876/com.ugrow.ch08_8d1_lifecycle E/TAG: Activity----->onStart 执行
07-27 03:06:00.543 2876-2876/com.ugrow.ch08_8d1_lifecycle I/TAG: fragment----->onStart 执行
07-27 03:06:00.550 2876-2876/com.ugrow.ch08_8d1_lifecycle E/TAG: Activity----->onResume 执行
07-27 03:06:00.550 2876-2876/com.ugrow.ch08_8d1_lifecycle I/TAG: fragment----->onResume 执行
```

从上面的日志信息中可以看到：当第一次运行应用程序时，首先会执行 Activity 的 onCreate()方法，然后依次执行 Fragment 的 onAttach()、onCreate()、onCreateView()、onActivityCreated()方法，之后再依次执行 Activity 的 onStart()方法、Fragment 的 onStart()方法、Activity 的 onResume()方法与 Fragment 的 onResume()方法。

此时，单击模拟器中的 Home 键，在 LogCat 中可以看到以下日志信息：

```
07-27 03:07:39.037 2876-2876/com.ugrow.ch08_8d1_lifecycle I/TAG: fragment----->onPause 执行
07-27 03:07:39.037 2876-2876/com.ugrow.ch08_8d1_lifecycle E/TAG: Activity----->onPause 执行
07-27 03:07:39.317 2876-2876/com.ugrow.ch08_8d1_lifecycle I/TAG: fragment----->onStop 执行
07-27 03:07:39.317 2876-2876/com.ugrow.ch08_8d1_lifecycle E/TAG: Activity----->onStop 执行
```

然后，在模拟器中单击该应用程序的图标，再次打开应用程序，在 LogCat 中可以看到以下日志信息：

```
07-27 03:08:08.313 2876-2876/com.ugrow.ch08_8d1_lifecycle E/TAG: Activity----->onRestart 执行
07-27 03:08:08.313 2876-2876/com.ugrow.ch08_8d1_lifecycle E/TAG: Activity----->onStart 执行
07-27 03:08:08.313 2876-2876/com.ugrow.ch08_8d1_lifecycle I/TAG: fragment----->onStart 执行
07-27 03:08:08.314 2876-2876/com.ugrow.ch08_8d1_lifecycle E/TAG: Activity----->onResume 执行
07-27 03:08:08.314 2876-2876/com.ugrow.ch08_8d1_lifecycle I/TAG: fragment----->onResume 执行
```

最后，单击模拟器的退出按钮，在 LogCat 中可以看到以下日志信息：

```
07-27 03:08:47.660 2876-2876/com.ugrow.ch08_8d1_lifecycle I/TAG: fragment----->onPause 执行
07-27 03:08:47.660 2876-2876/com.ugrow.ch08_8d1_lifecycle E/TAG: Activity----->onPause 执行
07-27 03:08:48.501 2876-2876/com.ugrow.ch08_8d1_lifecycle I/TAG: fragment----->onStop 执行
07-27 03:08:48.501 2876-2876/com.ugrow.ch08_8d1_lifecycle E/TAG: Activity----->onStop 执行
07-27 03:08:48.502 2876-2876/com.ugrow.ch08_8d1_lifecycle I/TAG: fragment----->onDestroyView 执行
07-27 03:08:48.502 2876-2876/com.ugrow.ch08_8d1_lifecycle I/TAG: fragment----->onDestroy 执行
07-27 03:08:48.502 2876-2876/com.ugrow.ch08_8d1_lifecycle I/TAG: fragment----->onDetach 执行
07-27 03:08:48.502 2876-2876/com.ugrow.ch08_8d1_lifecycle E/TAG: Activity----->onDestroy 执行
```

通过以上打印出来的日志信息，可以看出，Fragment 的生命周期是被宿主 Activity 的生命周期所控制的。

8.2 创建 Fragment

Fragment 作为一个嵌套在 Activity 中的 UI 片段，在使用前要先进行创建。有两种方式可以创建 Fragment：静态创建和动态创建。

8.2.1 静态创建

静态创建方式是 Fragment 最简单的一种创建方式，即把 Fragment 当作控件，使用

<fragment>标签在布局中添加 Fragment，不同的是，这里还需要通过 android:name 属性来指定要添加的 Fragment 类名，注意一定要将类的包名也加上，同时需要声明该 fragment 控件的 ID。

【示例 8.2】 在 Android Studio 中新建项目 Ch08_8D2_Fragment，使用静态方式创建 Fragment。

在 src/main/java/com.ugrow.ch08_8d2_fragment 文件夹下创建 MyFragment，继承 App 包下的 Fragment，代码如下：

```
public class MyFragment extends Fragment {

    @Nullable
    @Override
    public View onCreateView(LayoutInflater inflater, @Nullable ViewGroup container, Bundle savedInstanceState) {
        View view = inflater.inflate(R.layout.my_fragment, null);
        return view;
    }
}
```

上述代码重写了 Fragment 的 onCreateView()方法，然后使用 LayoutInflater(布局加载器)的 inflate()方法将所创建的 MyFragment 布局加载进来。

在布局 layout 下，新建 my_fragment.xml 文件，在其中编写以下代码，创建 MyFragment 布局：

```
<?xml version="1.0" encoding="utf-8"?>
<LinearLayout xmlns:android="http://schemas.android.com/apk/res/android"
android:layout_width="match_parent"
android:layout_height="match_parent"
android:orientation="vertical">

<TextView
    android:layout_width="match_parent"
    android:layout_height="match_parent"
    android:gravity="center"
    android:text="静态创建Fragment"
    android:textColor="#000"/>


</LinearLayout>
```

编写 activity_main.xml，代码如下：

```
<?xml version="1.0" encoding="utf-8"?>
<LinearLayout xmlns:android="http://schemas.android.com/apk/res/android"
    xmlns:tools="http://schemas.android.com/tools"
    android:layout_width="match_parent"
    android:layout_height="match_parent"
```

```
    tools:context="com.ugrow.ch08_8d2_fragment.MainActivity">

    <fragment
    android:id="@+id/fragment"
    android:layout_width="0dp"
    android:layout_weight="1"
    android:layout_height="match_parent"
    android:name="com.ugrow.ch08_8d2_fragment.MyFragment"/>

</LinearLayout>
```

编写上述代码时，需要注意一点：要使用 name 属性将新建的 MyFragment 与 MainActivity 关联起来。

在新建 MyFragment 类并让其继承 Fragment 时，可能会有两个不同包下的 Fragment 供选择，一个是系统内置的 android.app.Fragment，一个是 support-v4 库中的 android.support.v4.app.Fragment。这里需要选择 support-v4 库中的 Fragment，因为它可以让 Fragment 在所有 Android 系统版本中保持一致。

至此，Fragment 的静态创建就完成了，运行应用程序，效果如图 8-4 所示。

图 8-4　静态创建 Fragment

8.2.2　动态创建

上文讲解了在布局文件中静态创建 Fragment 的方法，不过 Fragment 真正的强大之处在于：可以在程序运行时，根据具体情况将 Fragment 动态地创建并添加到 Activity 中。

动态创建 Fragment 的步骤如下：

(1) 创建待添加的 Fragment 实例，继承 Fragment 类，并重写 onCreateView()方法。

(2) 获取 FragmentManager，可以在 Activity 中通过 getSupportFragmentManager()方法实现。

(3) 调用 beginTransation()方法开启一个事务，得到一个 transaction 对象。

(4) 向 Activity 内添加或替换 Fragment，一般通过 transaction 对象的 replace()方法实现，需要向其中传入容器的 ID 和待添加的 Fragment 实例。

(5) 调用 commit()方法提交事务。

其中，第(4)步中的操作可以根据需求替换为以下常用方法：

(1) transaction.add()：向 Activity 中添加一个 Fragment。

(2) transaction.remove()：从 Activity 中移除一个 Fragment，如果被移除的 Fragment 没有添加到回退栈，则这个 Fragment 实例会被销毁。

(3) transaction.replace()：使用另一个 Fragment 替换当前的，实际等同于 remove()方法与 add()方法的合体。

(4) transaction.hide()：隐藏当前的 Fragment，注意仅仅是设为不可见，并不会销毁。

(5) transaction.show()：显示之前隐藏的 Fragment。

(6) transaction.detach()：将 view 从 UI 中移除，和 remove()不同，此时 Fragment 的状态依然由 FragmentManager 维护。

(7) transaction.attach()：重建 view 视图，将其附加到 UI 上并显示。

(8) transatcion.commit()：提交一个事务。

【示例 8.3】 在 Android Studio 中新建项目 Ch08_8D3_Fragment，动态添加 Fragment，并实现 Fragment 的切换。

编写 activity_main.xml，代码如下：

```
<?xml version="1.0" encoding="utf-8"?>
<LinearLayout xmlns:android="http://schemas.android.com/apk/res/android"
    xmlns:tools="http://schemas.android.com/tools"
    android:layout_width="match_parent"
    android:layout_height="match_parent"
    android:orientation="vertical"
    tools:context="com.ugrow.ch08_8d3_fragment.MainActivity">

    <FrameLayout
        android:id="@+id/fl"
        android:layout_width="match_parent"
        android:layout_height="0dp"
        android:layout_weight="1" />

    <RadioGroup
        android:id="@+id/rg"
        android:layout_width="match_parent"
```

```
        android:layout_height="56dp"
        android:orientation="horizontal"
        android:paddingTop="8dp">

    <RadioButton
        android:id="@+id/rb_home"
        android:layout_width="0dp"
        android:layout_height="match_parent"
        android:layout_weight="1"
        android:button="@null"
        android:drawableTop="@drawable/tab_home"
        android:gravity="center_horizontal"
        android:text="首页"
        android:textColor="@drawable/text_color"/>

    <RadioButton
        android:id="@+id/rb_list"
        android:layout_width="0dp"
        android:layout_height="match_parent"
        android:layout_weight="1"
        android:button="@null"
        android:drawableTop="@drawable/tab_list"
        android:gravity="center_horizontal"
        android:text="列表"
        android:textColor="@drawable/text_color"/>

    <RadioButton
        android:id="@+id/rb_mine"
        android:layout_width="0dp"
        android:layout_height="match_parent"
        android:layout_weight="1"
        android:button="@null"
        android:drawableTop="@drawable/tab_mine"
        android:gravity="center_horizontal"
        android:text="我的"
        android:textColor="@drawable/text_color"/>
</RadioGroup>

</LinearLayout>
```

上述代码中创建了一个线性布局 LinearLayout，包括一个 FrameLayout 帧布局和三个

单选按钮 RadioButton。

在 src 文件夹下新建三个 Fragment，分别名为 HomeFragment、ListFragment 和 MineFragment，然后在其中编写代码。以 HomeFragment 为例，首先在 HomeFragment 中编写以下代码：

```
public class HomeFragment extends Fragment {

    public static String TAG = HomeFragment.class.getSimpleName();

    public static HomeFragment newInstance() {
        Bundle args = new Bundle();
        HomeFragment fragment = new HomeFragment();
        fragment.setArguments(args);
        return fragment;
    }
    @Override
    public View onCreateView(LayoutInflater inflater, ViewGroup container,
                Bundle savedInstanceState) {
        View view = inflater.inflate(R.layout.home_fragment, null);
        return view;
    }
}
```

然后在 home_fragment.xml 中编写以下代码：

```
<?xml version="1.0" encoding="utf-8"?>
<LinearLayout xmlns:android="http://schemas.android.com/apk/res/android"
    android:layout_width="match_parent"
    android:layout_height="match_parent"
    android:orientation="vertical" >

    <TextView
        android:layout_width="match_parent"
        android:layout_height="match_parent"
        android:gravity="center"
        android:text="首页"
        android:textColor="#d81e06"/>

</LinearLayout>
```

使用相同方法，创建 ListFragment 与 MineFragment，两者的代码与 HomeFragment 基本相同，唯一不同之处在于：<TextView>标签的 android:text 属性值不同，分别为 android:text="列表"和 android:text="个人中心"。

三个 Fragment 的代码都编写完毕后，编写 MainActivity，代码如下：

```
public class MainActivity extends AppCompatActivity implements RadioGroup.OnCheckedChangeListener{

private FrameLayout mFl;
private RadioGroup mRg;
private RadioButton mRbHome;
private RadioButton mRbList;
private RadioButton mRbMine;

    @Override
    protected void onCreate(Bundle savedInstanceState) {
        super.onCreate(savedInstanceState);
        setContentView(R.layout.activity_main);
        initView();
    }

    private void initView() {
        mFl = (FrameLayout) findViewById(R.id.fl);
        mRg = (RadioGroup) findViewById(R.id.rg);
        mRbHome =(RadioButton) findViewById(R.id.rb_home);
        mRbList =(RadioButton) findViewById(R.id.rb_list);
        mRbMine = (RadioButton)findViewById(R.id.rb_mine);
        mRg.setOnCheckedChangeListener(this);
        mRbHome.setChecked(true);
    }
    /**
     * 切换Fragment
     *
     * @param fragmentTag 要切换Fragment的Tag
     */
    private void replaceFragment(String fragmentTag){
        Fragment fragment = getSupportFragmentManager().findFragmentByTag(fragmentTag);
        if (fragment == null) {
            if (fragmentTag.equals(HomeFragment.TAG)) {
                fragment = HomeFragment.newInstance();
            } else if (fragmentTag.equals(ListFragment.TAG)) {
                fragment = ListFragment.newInstance();
            }else if (fragmentTag.equals(MineFragment.TAG)){
                fragment = MineFragment.newInstance();
            } else {
            return;
            }
```

```
        }
            getSupportFragmentManager()
                .beginTransaction()
                .replace(R.id.fl, fragment, fragmentTag)
                .addToBackStack(fragmentTag)
                .commitAllowingStateLoss();
    }
    @Override
    public void onCheckedChanged(RadioGroup radioGroup, @IdRes int checkedId) {
        switch (checkedId) {
            case R.id.rb_home:
                replaceFragment(HomeFragment.TAG);
                break;
            case R.id.rb_list:
                replaceFragment(ListFragment.TAG);

                break;
            case R.id.rb_mine:
                replaceFragment(MineFragment.TAG);
                break;
            default:
                break;
        }
    }
}
```

为使切换底部单选按钮时的效果更加明显，可以在 res/drawable 文件夹下新建一个图片选择器和一个文字颜色选择器。

新建 tab_home.xml，写入图片选择器的代码：

```
<?xml version="1.0" encoding="utf-8"?>
<selector xmlns:android="http://schemas.android.com/apk/res/android">
    <item android:drawable="@mipmap/home" android:state_checked="false"/>
    <item android:drawable="@mipmap/home_selected"
    android:state_checked="true"/>
</selector>
```

新建 text_color.xml，写入文字颜色选择器的代码：

```
<?xml version="1.0" encoding="utf-8"?>
<selector xmlns:android="http://schemas.android.com/apk/res/android">
    <item android:state_checked="true" android:color="#d81e06"></item>
    <item android:state_checked="false" android:color="#8a8a8a"></item>
</selector>
```

运行应用程序，其效果如图 8-5 所示：当程序打开时，底部按钮【首页】被选中，页面显示 HomeFragment；当单击底部按钮【列表】时，页面切换到 ListFragment；当单击底部按钮【我的】时，页面切换到 MineFragment。

图 8-5　动态创建并切换 Fragment

本 章 小 结

通过本章的学习，读者应当了解：

✧ 为合理利用屏幕空间，Android 3.0 引入了片段(Fragment)的概念。

✧ 作为一个嵌套到 Activity 中的 UI 片段，Fragment 不能独立存在或使用，而是必须要嵌套到 Activity 中使用。

✧ Fragment 拥有自己的生命周期方法，共有 11 个生命周期方法。

✧ Fragment 的创建方式有两种：静态创建和动态创建。

本 章 练 习

1. Fragment 的生命周期方法共______。

   A．7 个　　B．9 个　　C．10 个　　D．11 个

2. Activity 的生命周期方法共______。

   A．7 个　　B．9 个　　C．10 个　　D．11 个

3. 将 Fragment 动态加载到 Activity 中使用的方法是______。

   A．remove()　　B．replace()　　C．hide()　　D．show()

4. Fragment 的创建方式有______和______两种。

5. 简述 Fragment 的特点。

6. 编写代码，使用静态创建方式和动态创建方式分别创建 Fragment。

第 9 章　网络通信

本章目标

- 了解 Android 中网络通信的方式
- 理解 Socket、ServerSocket 的原理及常用方法
- 掌握使用 Socket 和 ServerSocket 进行网络通信的方法
- 掌握 HttpURLConnection 的使用方法
- 掌握 HttpClient 的使用方法
- 熟悉 WebKit 的组成及原理
- 掌握使用 WebView 组件浏览网页的方法
- 掌握解析 JSON 的方法以及使用 AsyncTask 获取网络数据的方法

9.1 网络通信简介

无线网络的产生为人们的生活提供了诸多方便，比如高速的无线宽带上网、视频通话、无线搜索、手机音乐、网游等。无线网络的迅猛发展，使人们不必受时间和空间的限制，可以随时随地进行数据交换、浏览 Internet，并使用网络来处理一些事务，因此在 Android 中，掌握网络通信便可以开发一些优秀的网络应用程序。

网络通信包含三部分的内容：发送方、接收方以及协议栈。其中，发送方和接收方是参与通信的主体，协议栈是发送方和接收方进行通信的规约。按照服务类型，网络通信可分为面向连接方式和无连接方式：面向连接方式是在通信前建立通信链路，通信结束后释放该链路；无连接方式则不需要在通信前建立通信连接，但不能保证传输的质量。

Android 提供了多种网络通信的方式，Java 中的网络编程功能为 Android 中的网络通信提供了支持。Android 中常用的网络编程方式如下：

(1) 针对 TCP/IP 协议的 Socket 和 ServerSocket。

(2) 针对 HTTP 协议的网络编程方式，如 HttpURLConnection 和 HttpClient。

(3) 直接使用 WebKit 访问网络。

9.2 Socket 通信

Socket 通信是指双方采用 Socket 机制交换数据。Socket 是比较低层的网络编程方式，其他的高级协议，如 HTTP 都是建立在此基础之上的，而且 Socket 是跨平台的编程，可以在异构语言之间进行通信，所以掌握 Socket 网络编程是掌握其他网络编程方式的基础。

9.2.1 Socket 和 ServerSocket

Socket 通常也称作“套接字”，用来描述通信链的句柄(包括 IP 地址和端口)。通过套接字，应用程序之间可传输信息。常见的网络通信协议有 TCP 和 UDP 两种，TCP 协议是可靠的、面向连接的协议，这种方式需要在通信前建立通信双方的连接链路，而通信结束后又释放该链路；UDP 数据报协议是不可靠的、无连接的协议，这种协议不需要在通信前建立通信双方的连接，相当于使用可靠性来换取传输开销，其传输开销比 TCP 小。

java.net 包中提供了两个 Socket 类：

(1) java.net.Socket 是客户端的 Socket 对应的类。

(2) java.net.ServerSocket 是服务器端的 Socket 对应的类，这个类表示一个等待客户端连接的服务器端套接字。

1. Socket 类

Socket 类的常用方法如表 9-1 所示。

表 9-1　Socket 类的常用方法

方　　法	功 能 描 述
Socket(String host ,int port)	Socket 的构造方法。该构造方法带两个参数，用于创建一个到主机 host、端口号为 port 的套接字，并连接到远程主机
bind(SocketAddress localAddr)	将该 Socket 与参数 localAddr 指定的地址和端口绑定
InetAddress getInetAddress()	获取该 Socket 连接的目标主机的 IP 地址
synchronized int getReceiveBufferSize()	获取该 Socket 的接受缓冲区的尺寸
synchronized void close()	关闭 Socket
InputStream getInputStream()	获取该 Socket 的输入流，这个输入流用来读取数据
boolean isConnected()	判断该 Socket 是否连接
boolean isOutputShutdown()	判断该 Socket 的输出管道是否关闭
boolean isInputShutdown()	判断该 Socket 的输入管道是否关闭
SocketAddress getLocalSocketAddress()	获取该 Socket 的本地地址和端口
int getPort()	获取端口号

一般情况下，Socket 类的工作步骤如下：

(1) 根据指定地址和端口创建一个 Socket 对象。

(2) 调用 getInputStream()方法或 getOutputStream()方法打开连接到 Socket 的输入/输出流。

(3) 客户端与服务器根据一定的协议交互，直到关闭连接。

(4) 关闭客户端的 Socket。

以下是一个典型的创建客户端 Socket 的代码片段：

```
try {
        Socket socket = new Socket("127.0.0.1", 4700);
} catch (IOException ioe)
{
        System.out.println("Error:" + ioe);
} catch(UnknownHostException uhe)
{
        System.out.println("Error:" + uhe);
}
```

2. ServerSocket 类

ServerSocket 类用于监听在特定端口的 TCP 连接。当客户端的 Socket 试图与服务器指定端口建立连接时，服务器就会被激活，判断客户程序是否连接，一旦客户端与服务器建立了连接，两者之间就可以相互传送数据。ServerSocket 类的常用方法如表 9-2 所示。

表 9-2　ServerSocket 类的常用方法

方　法	功 能 描 述
ServerSocket(int port)	ServerSocket 的构造方法
Socket accept()	等待客户端的连接，当客户端请求连接时，返回一个 Socket
void close()	关闭服务器 Socket
SocketAddress getLocalSocketAddress()	获取该 Socket 的本地地址和端口
int getLocalPort()	获取端口号
InetAddress getInetAddress()	获取该 Socket 的 IP 地址
boolean isClosed()	判断连接是否关闭
void setSoTimeout(int timeout)	设置 accpet()的超时时间

一般情况下，ServerSocket 类的工作步骤如下：

(1) 根据指定端口创建一个新的 ServerSocket 对象。

(2) 调用 ServerSocket 的 accept()方法，在指定的端口监听到来的连接。accept()方法一直处于阻塞状态，直到有客户端试图建立连接，这时 accept()方法会返回连接客户端与服务器的 Socket 对象。

(3) 调用 getInputStream()方法或 getOutputStream()方法建立与客户端交互的输入/输出流。

(4) 服务器与客户端根据一定的协议交互，直到关闭连接。

(5) 关闭服务器端的 Socket。

(6) 回到第 2 步，继续监听下一次的连接。

以下是一个典型的创建服务器端 ServerSocket 的代码片段：

```
ServerSocket server = null;
try {
        //创建一个 ServerSocket 在端口 4700 监听客户请求
        server = new ServerSocket(4700);
} catch (IOException e) {
        System.out.println("can not listen to :" + e);
}
Socket socket = null;
try {
        //accept()是一个阻塞方法，一旦有客户请求，它就会返回一个 Socket 对象用于同客户进行交互

        socket = server.accept();
} catch (IOException e) {
        System.out.println("Error:" + e);
}
```

9.2.2　Socket 应用

Socket 利用客户/服务器模式巧妙地解决了进程之间建立通信的问题，下面通过示例

讲解如何使用 Socket 实现通信。

【示例 9.1】 在 Android Studio 中新建项目 Ch09_9D1_Socket，使用 Socket 和 ServerSocket 实现一个简易聊天室。

使用 Socket 进行通信至少要实现服务器端和客户端两部分功能。

服务器端的代码如下：

```
public class Server {
    private int ServerPort = 9898; // 定义端口
    private ServerSocket serversocket = null; // 声明服务器套接字
    private OutputStream outputStream = null; // 声明输出流
    private InputStream inputStream = null; // 声明输入流
    // 声明 PrintWriter 对象，用于将数据发送给对方
    private PrintWriter printWriter = null;
    private Socket socket = null; // 声明套接字，注意同服务器套接字不同
    // 声明 BufferedReader 对象，用于读取接受的数据
    private BufferedReader reader = null;

    /* Server 类的构造方法 */
    public Server() {
        try {
            // 根据指定的端口号，创建套接字
            serversocket = new ServerSocket(ServerPort);
            System.out.println("服务启动...");
            // 用 accept 方法等待客户端的连接
            socket = serversocket.accept();
            System.out.println("客户已连接...\n");
        } catch (Exception ex) {
            ex.printStackTrace(); // 打印异常信息
        }

        try {
            // 获取套接字输出流
            outputStream = socket.getOutputStream();
            // 获取套接字输入流
            inputStream = socket.getInputStream();
            // 根据 outputStream 创建 PrintWriter 对象
            printWriter = new PrintWriter(outputStream, true);
            // 根据 inputStream 创建 BufferedReader 对象
            reader = new BufferedReader(newInputStreamReader(inputStream));
            // 根据 System.in 创建 BufferedReader 对象
            BufferedReader in = new BufferedReader(new InputStreamReader(System.in));
```

```
                while (true) {
                        // 读客户端的传输信息
                        String message = reader.readLine();
                        // 将接受的信息打印出来
                        System.out.println("Client: " + message);
                        // 若消息为 Bye 或者 bye，则结束通信
                        if (message.equals("Bye") || message.equals("bye"))
                                break;
                        System.out.print("Service：");
                        message = in.readLine();// 接收键盘输入
                        printWriter.println(message); // 将输入的信息向客户端输出
                }
                outputStream.close(); // 关闭输出流
                inputStream.close(); // 关闭输入流
                socket.close(); // 关闭套接字
                serversocket.close(); // 关闭服务器套接字
                System.out.println("Client is disconnected");
        } catch (Exception e) {
                e.printStackTrace(); // 打印异常信息
        } finally {

        }
    }

    /* 程序入口，程序从 main 方法开始执行 */
    public static void main(String[] args) {
        new Server();
    }
}
```

上述代码实现服务器端等待客户端的连接，此程序运行在 Windows 系统中，而不是 Android 系统中。客户端的 MainActivity 代码如下：

```
public class MainActivity extends AppCompatActivity {

    // 声明文本视图 chatmessage，用于显示聊天记录
    private TextView chatmessage = null;
    // 声明编辑框 sendmessage，用于用户输入短信内容
    private EditText sendmessage = null;
    // 声明按钮 send_button，用于发送短信按钮
    private Button send_button = null;
    private static final String HOST = "192.168.1.46"; // 服务器的 IP 地址
```

```
private static final int PORT = 9898; // 服务器端口号
private Socket socket = null; // 声明套接字类，用于传输数据
private BufferedReader bufferedReader = null; // 声明 BufferedReader 类，用于读取接受的数据
private PrintWriter printWriter = null; // 声明 printWriter 类，用于将数据发送给对方
private String string = ""; // 声明字符串变量

@Override
protected void onCreate(Bundle savedInstanceState) {
    super.onCreate(savedInstanceState);
    setContentView(R.layout.activity_main);
    initView();
    initSocket();
}

private void initView() {
    chatmessage = (TextView) this.findViewById(R.id.chatmessage);
    sendmessage = (EditText) this.findViewById(R.id.sendmessage);
    send_button = (Button) this.findViewById(R.id.SendButton);
    send_button.setOnClickListener(new Button.OnClickListener() {
        public void onClick(View view) {
            // 获取输入框的内容
            String message = sendmessage.getText().toString();
            // 清空 sendmessage 的内容以便下次输入
            sendmessage.setText("");
            chatmessage.setText(chatmessage.getText().toString() + "\n"
                    + "Client: " + message);
            // 发送消息
            sendMsg(message);

        }
    });
}

/**
 * 发送消息
 *
 * @param message
 */
protected void sendMsg(final String message) {
    new Thread(new Runnable() {
```

```
            @Override
            public void run() {
                // 判断 Socket 是否连接
                if (socket.isConnected()) {
                    if (!socket.isOutputShutdown()) {
                        // 将输入框内容发送到服务器
                        printWriter.println(message);
                    }
                }
            }
        }).start();
    }

    /**
     * 初始化 Socket
     */
    private void initSocket() {
        new Thread(new Runnable() {

            @Override
            public void run() {
                try {
                    // 指定 IP 和端口号创建套接字
                    socket = new Socket(HOST, PORT);
                    // 使用套接字的输入流构造 BufferedReader 对象
                    bufferedReader = new BufferedReader(new InputStreamReader(
                            socket.getInputStream()));
                    // 使用套接字的输出流构造 PrintWriter 对象
                    printWriter = new PrintWriter(new BufferedWriter(
                            new OutputStreamWriter(socket.getOutputStream())),
                            true);
                    // 连接成功后，启动客户端监听
                    if (socket != null) {
                        while (true) {
                            // 若套接字同服务器的链接存在且输入流存在，则发送消息
                            if (socket.isConnected()) {
                                if (!socket.isInputShutdown()) {
                                    if ((string = bufferedReader.readLine()) != null) {
                                        Log.i("TAG", "++ " + string);
```

```
                        string += " ";
                        handler.sendEmptyMessage(1);
                    } else {
                        // TODO
                    }
                }
            }

        }
    }

            } catch (Exception e) {
                e.printStackTrace(); // 打印异常
                CreateDialog(e.getMessage()); // 调用 CreateDialog()方法生成对话框
            }
        }
    }).start();

}

/* 声明 CreateDialog()方法，用于创建对话框*/
public void CreateDialog(String msmessage) {
    AlertDialog.Builder builder = new AlertDialog.Builder(this);
    // 首先获取 AlertDialog 的 Builder 类，该 Builder 对象用于构造对话框
    builder.setTitle("异常"); // 指定对话框的标题
    builder.setMessage(msmessage); // 设置显示的信息
    // 设置 PositiveButton 的名称以及监听器
    builder.setPositiveButton("Yes", null);
    // 设置 NegativeButton 的名称以及监听器
    builder.show(); // 显示对话框
}

public Handler handler = new Handler() {
    @Override
    public void handleMessage(Message msg) {
        if (msg.what == 1) {

            Log.i("TAG", "-- " + msg);
            chatmessage.setText(chatmessage.getText().toString() + "\n"
                    + "Server: " + string);
```

```
            }
        }
    };
}
```

上述代码作为客户端运行在 Android 系统中，客户端通过套接字绑定服务器端的 IP 地址和端口号。注意：这里的 IP 地址是服务器端的 IP 地址，即使服务器端和 Android 的模拟器在同一机器上运行，也不能使用回环地址(127.0.0.1)作为服务器的 IP 地址，否则若指定回环地址，程序会出现拒绝连接的错误。

使用 Handler 消息处理机制处理服务器端的消息，这种处理机制是异步的。Handler 对于发送和接受信息有不同的处理方式：向消息队列发送消息时会立即返回，而从消息队列中接受消息时会阻塞。当从消息队列中读取消息时，会执行 Handler 中的 handleMessage(Message msg)方法，因此在创建 Handler 时应重写该方法，在其中写入读取消息后的操作，并使用 Handler 的 obtainMessage()来获得消息对象。

要让客户端能够访问服务器，必须在 AndroidManifest.xml 配置文件中增加如下权限：

```
<uses-permission android:name="android.permission.INTERNET" />
```

先启动 Server 服务器，再运行客户端项目 ClientActivity。在客户端界面的文本框中输入信息并单击【发送】按钮，信息会发送给服务器；然后在服务器端通过键盘输入返回信息，就实现了服务器与客户端的相互通信。

客户端的输出效果如图 9-1 所示。

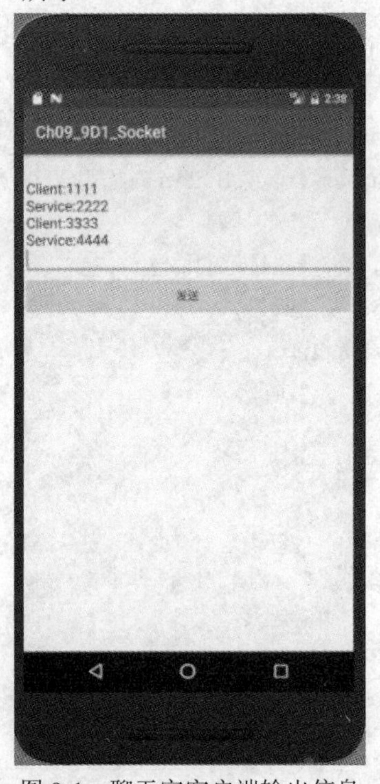

图 9-1　聊天室客户端输出信息

服务器端的输出信息如下：

```
服务启动...
```

```
客户已连接...

Client: 1111
Service：2222
Client: 3333
Service：4444
```

注 意

这里需要特别注意：(1) Android 中所有访问网络的操作必须在新的线程中执行，不能直接在主线程(UI 线程)中执行；(2) 不能在主线程之外的其他线程更新 UI，但可以通过 Handler 来更新 UI。

9.3 HTTP 网络编程

HTTP 协议是 Internet 上使用最为广泛的通信协议，随着移动互联网时代的来临，基于 HTTP 协议的手机等移动终端的应用也会更加广泛。

在 Android 中使用 HTTP 协议进行网络通信的方式有以下两种：

(1) HttpURLConnection 方式。

(2) Apache HTTP 客户端组件 HttpClient 方式。

9.3.1 HttpURLConnection

如果网上某个资源 URL 是基于 HTTP 协议的，则可以使用 java.net.HttpURLConnection 包进行请求和响应。每个 HttpURLConnection 实例都可用于生成单个请求，并可以透明地共享连接到 HTTP 服务器的基础网络。HttpURLConnection 类的常用方法如表 9-3 所示。

表 9-3 HttpURLConnection 类的常用方法

方 法	功 能 描 述
InputStream getInputStream()	返回从此打开的连接读取的输入流
OutputStream getOutputStream()	返回写入到此连接的输出流
String getRequestMethod()	获取请求方法
int getResponseCode()	获取状态码，如 HTTP_OK、HTTP_UNAUTHORIZED
void setRequestMethod(Stringmethod)	设置 URL 请求的方法
void setDoInput(booleandoinput)	设置输入流。如果使用 URL 连接进行输入，则将 DoInput 标志设置为 true(默认值)；如果不打算使用，则设置为 false
void setDoOutput(booleandooutput)	设置输出流。如果使用 URL 连接进行输出，则将 DoOutput 标志设置为 true；如果不打算使用，则设置为 false(默认值)
void setUseCaches(booleanusecaches)	设置连接是否使用任何可用的缓存
void disconnect()	关闭连接

HttpURLConnection 是一个抽象类，无法直接实例化，其实例主要通过 URL 的 openConnection()方法获得。例如，以下代码用于获取一个 HttpURLConnection 连接：

```
//创建 URL
URL url=new URL("http://www.baidu.com/");
//获取 HttpURLConnection 连接
HttpURLConnection urlConn=(HttpURLConnection)url.openConnection();
```

在进行连接操作之前，可以对 HttpURLConnection 的一些属性进行设置，示例如下：

```
//设置输出、输入流
urlConn.setDoOutput(true);
urlConn.setDoInput(true);
//设置方式为 POST
urlConn.setRequestMethod("POST");
//请求不能使用缓存
urlConn.setUseCaches(false);
```

连接完成之后，可以关闭连接，代码如下：

```
urlConn.disconnect();
```

【示例 9.2】 在 Android Studio 中新建项目 Ch09_9D2_HttpUrlConnection，使用 HttpURLConnection 访问 Servlet，实现用户登录功能。

(1) 创建 Android 项目，设计登录界面布局，并编写 MainActivity，代码如下：

```
public class MainActivity extends AppCompatActivity {

    private EditText mEtName;
    private EditText mEtCode;
    private Button mBtnCancel;
    private Button mBtnLogin;

    @Override
    protected void onCreate(Bundle savedInstanceState) {
        super.onCreate(savedInstanceState);
        setContentView(R.layout.activity_main);
        initView();
    }

    private void initView() {
        /**
         * 实例化视图组件
         */
        mEtName = (EditText) findViewById(R.id.et_name);
        mEtCode = (EditText)findViewById(R.id.et_code);
```

```
        mBtnCancel = (Button) findViewById(R.id.btn_cancel);
        mBtnLogin = (Button)findViewById(R.id.btn_login);
        /**
         * 设置取消按钮监听事件
         */
        mBtnCancel.setOnClickListener(new View.OnClickListener() {
            @Override
            public void onClick(View view) {
                finish();
            }
        });
        /**
         * 设置登录按钮监听事件
         */
        mBtnLogin.setOnClickListener(new View.OnClickListener() {
            @Override
            public void onClick(View view) {
                String name = mEtName.getText().toString();
                String code = mEtCode.getText().toString();
                login(name,code);
            }
        });
    }

    /**
     * 通过用户名称和密码进行查询，发送请求，获得响应结果
     */
    private void login(final String name, final String code) {
        new Thread(new Runnable() {

            @Override
            public void run() {
                // URL 地址，用于访问指定网站的 Servlet
                String urlStr = "http://192.168.1.46:8080/ch09_9D2_Server/LoginServlet?";
                // 请求参数，用于传递用户名和密码值
                String queryString = "username=" + name + "&password="+ code;
                urlStr += queryString;
                try {
                    // 根据地址创建 URL 对象
                    URL url = new URL(urlStr);
```

```
                // 获取 HttpURLConnection 连接
                HttpURLConnection conn = (HttpURLConnection) url
                        .openConnection();

                // 获取状态码并判断其值是不是 HTTP_OK
                if (conn.getResponseCode() == HttpURLConnection.HTTP_OK) {

                    // 获取输入流
                    InputStream in = conn.getInputStream();
                    // 创建一个缓冲字节数组
                    byte[] buffer = new byte[in.available()];
                    // 从输入流中读取数据并存放到缓冲字节数组中
                    in.read(buffer);
                    // 将字节数据转换成字符串
                    String str = new String(buffer);
                    // 关闭输入流
                    in.close();
                    Message msg = new Message();
                    msg.what = 1;
                    msg.obj = str;
                    handler.sendMessage(msg);
                }
                // 关闭连接
                conn.disconnect();
            } catch (Exception e) {
                showDialog(e.getMessage());
            }
        }
    }).start();
}
Handler handler = new Handler() {
    @Override
    public void handleMessage(Message msg) {
        if (msg.what == 1) {
            String msgStr = (String) msg.obj;
            showDialog(msgStr);
        }
    }
};
/*
```

```
     * 定义一个显示提示信息的对话框
     */
    private void showDialog(String msg) {
        AlertDialog.Builder builder = new AlertDialog.Builder(this);
        builder.setMessage(msg).setCancelable(false)
                .setPositiveButton("确定", null);
        AlertDialog alert = builder.create();
        alert.show();
    }
}
```

上述代码中，在新创建的线程中执行网络操作，返回的登录结果通过 Handler 显示。

(2) 在 AndroidManifest.xml 配置文件中添加网络访问权限，代码如下：

```
<uses-permission android:name="android.permission.INTERNET" />
```

(3) 创建一个动态 Web 项目，并添加一个名为 LoginServlet 的 Servlet，代码如下：

```
public class LoginServlet extends HttpServlet {
    protected void doGet(HttpServletRequest request,
                HttpServletResponse response) throws ServletException, IOException {
            String username = request.getParameter("username");
            String password = request.getParameter("password");
            System.out.println(username + ":" + password);
            response.setContentType("text/html");
            response.setCharacterEncoding("utf-8");
            PrintWriter out = response.getWriter();
            String msg = null;
            if (username != null && username.equals("admin") && password !=
                    null&& password.equals("1")) {
                    msg = "登录成功!";
            } else {
                    msg = "登录失败!";
            }
            out.print(msg);
            out.flush();
            out.close();
    }
    protected void doPost(HttpServletRequest request,
            HttpServletResponse response) throws ServletException,
                    IOException {
                    doGet(request, response);
    }
```

```
}
```

在上述代码中实现了 Servlet 的 doGet()和 doPost()方法。在 doGet()方法中首先调用 request 请求对象的 getParameter()方法获取客户端传来的用户名和密码信息；然后调用 response 响应对象的 getWriter()方法获得打印输出流；再验证用户名和密码是否正确，并将验证结果通过打印输出流向客户端输出返回信息。

(4) 先启动 Web 项目，再运行 Android 客户端应用，运行结果如图 9-2 所示。

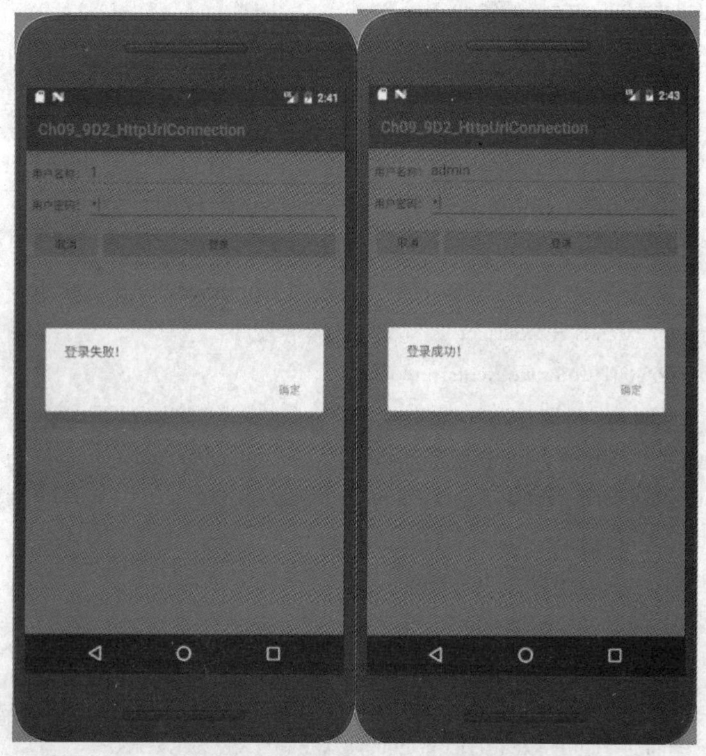

图 9-2　使用 HttpURLConnection 登录客户端

9.3.2　HttpClient

Apache 提供了 HTTP 客户端组件 HttpClient(API 23 之后已删除，如果使用需要在 gradle 文件中添加 HttpClient 依赖)，它对 java.net 中的类进行封装和抽象，更适合在 Android 上开发网络应用，使得针对 HTTP 编程更加方便、高效。HttpClient 本身不是一个浏览器，而是一个客户端的 HTTP 传输库，其目的是为了让 HttpClient 接收和发送 HTTP 消息。HttpClient 最重要的作用是执行 HTTP 方法。执行一个 HTTP 方法涉及到一个或几个 HTTP 请求或响应的交互，根据请求方法的不同会用到 HttpGet 和 HttpPost 两个对象，而 HttpClient 负责将该项请求转送到目标服务器并返回一个相应的响应对象，因此 HttpClient API 的主要部分是定义上面功能的 HttpClient 接口。HttpClient 接口代表了最基本的 HTTP 请求执行规约。它没有在请求执行的过程上强加任何限制或特定的具体细节，不关心连接管理细节、状态管理细节、认证和重定向处理个别的实现，这使得使用额外的

功能实现接口变得更容易。

通常使用 HttpClient 的子类 DefaultHttpClient 进行操作。DefaultHttpClient 是 HttpClient 的默认实现类，用来负责处理 HTTP 协议的某一方面功能，如重定向或认证处理、关于保持连接和保活时间的决策，这使得用户可以选择性地使用特定的应用替换默认的功能。

以下代码是一个使用 HttpClient 请求执行过程的简单例子。

```
//使用 DefaultHttpClient 生成一个 HttpClient 对象
HttpClient httpclient = new DefaultHttpClient();
//定义一个 URL 地址
String uri = "http://test/";
//定义一个以 Get 方式提交的 HttpGet 请求对象
HttpGet httpget = new HttpGet(uri);
//执行 HttpClient 对象的 execute()方法，即将请求对象提交给服务器，并返回一个响应对象
HttpResponse httpesponse = httpclient.execute(httpget);
//获取响应信息
HttpEntity httpentity = httpresponse.getEntity();
......
```

【示例 9.3】 对示例 9.2 进行修改，使用 HttpClient 访问 Servlet，实现用户登录功能。

服务器 Servlet 无需变动，只需将 MainActivity 中的 login()方法中的代码进行修改，修改后的代码如下：

```
private void login(final String username, final String password) {
        new Thread(new Runnable() {
                @Override
                public void run() {
                // URL 地址，访问指定网站的 Servlet
                String urlStr =
                "http://192.168.2.152:8080/ch09_9D2_Server/LoginServlet";
                // 定义一个以 Post 方式提交的 HttpPost 请求对象
                HttpPost request = new HttpPost(urlStr);
                // 如果传递参数个数比较多的话，可以对传递的参数进行封装
                List<NameValuePair> params = new ArrayList<NameValuePair>();
                // 添加用户名和密码参数
                params.add(new BasicNameValuePair("username", username));
                params.add(new BasicNameValuePair("password", password));
                try {
                // 设置请求参数项及其编码字符集
                request.setEntity(new UrlEncodedFormEntity(params, HTTP.UTF_8));
                // 使用 DefaultHttpClient 生成一个 HttpClient 对象
                HttpClient httpclient = new DefaultHttpClient();
```

```
            // 执行 HttpClient 对象的 execute()方法
            //将请求对象提交给服务器，并返回一个响应对象
            HttpResponse response=httpclient.execute(request);
            // 判断请求是否成功
            if (response.getStatusLine().getStatusCode() ==
                    HttpStatus.SC_OK) {
                    // 调用响应对象的 getEntity()方法获取响应信息，并转换成字符串
                    String str=EntityUtils.toString(response.getEntity());
                    Message msg=new Message();
                    msg.what=1;
                    msg.obj=str;
                    handler.sendMessage(msg);
                }
            } catch (Exception e) {
            e.printStackTrace();
            }
        }
    }).start();
}
```

上述代码中使用 HttpPost 封装请求，Post 方式要比 Get 方式复杂，需要使用 NameValuePair 来保存要传递的参数，此处先使用 BasicNameValuePair 来构造一个传递的参数，然后通过 add()方法将其添加到 NameValuePair 列表中，代码如下：

```
// 如果传递参数个数比较多的话，可以对传递的参数进行封装
List<NameValuePair> params = new ArrayList<NameValuePair>();
//添加用户名和密码参数
params.add(new BasicNameValuePair("username", username));
params.add(new BasicNameValuePair("password", password));
```

Post 方式需要设置所用的编码字符集，并调用 HttpPost 请求对象的 setEntity()方法设置参数项，即将参数列表添加到请求对象中，代码如下：

```
request.setEntity( new UrlEncodedFormEntity(params,HTTP.UTF_8));
```

与 Get 方式一样调用 HttpClient 对象的 execute()方法执行请求并返回一个响应对象，根据响应对象的状态码判读请求是否成功，如果成功则调用 getEntity()获取响应信息。运行结果同示例 9.2。

9.4 WebKit

Android 提供了内置的浏览器，该浏览器使用了开源的 WebKit 引擎。相比其他浏览器引擎，WebKit 引擎不仅具有较好的渲染效果，而且兼容 Web 标准，可扩展性好，因而被多种手机操作系统所采纳。

9.4.1　WebKit 介绍

在 Android 平台中，WebKit 引擎可分为 Java 引擎库和 WebCore 引擎库两个部分：

(1) Java 引擎使用 JavaScript 实现，该引擎负责与 Android 应用程序进行通信。

(2) WebCore 引擎库负责处理实际的网页生成与版面元素。

这两个引擎之间使用 JNI 和 Bridge 相互调用。图 9-3 显示了 WebKit 引擎子模块的调用关系。

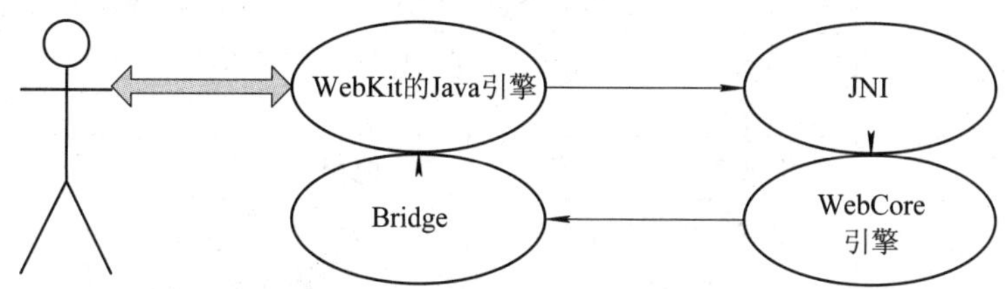

图 9-3　WebKit 引擎子模块的调用关系

Android 对 WebKit 进行封装，为开发者提供丰富的接口和类，定义在 android.webkit 包中。通过 Webkit 包中提供的类，可实现 Internet 的访问。WebKit 引擎中主要的类及其功能如表 9-4 所示。

表 9-4　WebKit 引擎中主要的类及其功能

类　　名	功　　能
BrowserFrame	BrowserFrame 类是一个封装器，封装了 WebCore 库中的 Frame 类。使用该类可创建 WebCore 中定义的 Frame，并且创建该对象的事件监听方法
WebView	WebView 类是 WebKit 模块 Java 层的视图类，所有需要使用 Web 浏览功能的 Android 应用程序都要创建该视图对象显示和处理请求的网络资源
HttpAuthHandler	Http 认证处理类
DownloadManagerCore	DownloadManagerCore 类负责管理下载网络资源，所有的网络资源的下载均由该类管理。该类实例运行在 WebKit 线程当中，通过调用 CallbackProxy 对象实现与 UI 线程的交互
CacheLoader	缓存加载器，用于加载缓存内容
DataLoader	数据加载器，用于载入网页数据
FileLoader	文件加载器，用于将文件数据加载到 Frame 中
WebViewDatabase	Web 视图数据库类，该类封装了 WebKit 引擎对 SQLiteDatabase 的操作
CacheSyncManager	Cache 同步管理器，该类负责同步浏览器缓存数据
Network	该类封装了网络连接逻辑，为开发人员提供更为高级的网络连接
CallbackProxy	回调代理类，该类用于 UI 和 WebCore 之间的交互。与用户相关的通知方法是在该类中定义的。当 WebCore 完成相应的数据处理时，则会调用 CallbackProxy 类中对应的方法，这些方法通过消息方式间接调用相应处理对象的处理方法
SslErrorHandler	SSL 错误处理器，该类提供了处理 SSL 错误的方法

续表

类　名	功　能
JsResult	Js 结果类，该类用于用户交互
WebChromeClient	WebChrome 客户端，该类定义了一系列的事件，这些事件与浏览窗口修饰相关，例如接收到 Title、进度变化等
CellList	CellList 类定义了图片集合中的 Cell，该类用于管理 Cell 图片的绘制、状态变化以及索引
LoadListener	加载监听器。当有下载事件时，该类的 DownloadFileMethod 会被调用
WebViewClient	Web 视图客户端类定义了页面载入、资源载入、页面访问错误等情况发生时的处理方法
DragClient	拖拽客户端类定义了与页面拖拽相关的处理
StreamLoader	包含三种类型的载入器： ✧ CacheLoader(缓存加载器)：加载缓存内容； ✧ DataLoader(数据加载器)：用于载入网页数据； ✧ FileLoader(文件加载器)：将文件数据加载到 Frame 中

9.4.2 WebView 视图组件

WebView 是 WebKit 中专门用来浏览网页的视图组件，它作为应用程序的 UI 接口，为用户提供了一系列的网页浏览和用户交互接口，通过这些接口显示和处理请求的网络资源。WebView 具有以下几个优点：

(1) 功能强大，支持 CSS、JavaScript 和 HTML，并能很好地融入布局，使页面更加美观。

(2) 能够对浏览器控件进行详细的设置，例如字体、背景颜色、滚动条样式。

(3) 能够捕捉到所有浏览器操作，例如单击、打开或关闭 URL。

(4) 能够显示和渲染 Web 页面。

(5) 能够直接使用 html 文件(网络上或本地 assets 中)作为布局。

(6) 能够和 JavaScript 交互调用。

WebView 提供了一些常用的浏览器方法，如表 9-5 所示。

表 9-5　WebView 常用方法

方　法	功能描述
loadUrl()	打开一个指定的 Web 资源页面
loadData()	显示 HTML 格式的网页内容
getSettings()	获取 WebView 的设置对象
addJavascriptInterface()	将一个对象添加到 JavaScript 的全局对象 Window 中，这样可以通过 window.XXX 进行调用，与 JavaScript 进行交互
clearCache()	清除缓存
destory()	销毁 WebView

【示例 9.4】 在 Android Studio 中新建项目 Ch09_9D4_WebView，使用 WebView 视图组件来浏览网页。

首先，在布局文件 activity_main.xml 中添加一个 WebView 组件，代码如下：

```
<LinearLayout xmlns:android="http://schemas.android.com/apk/res/android"
    android:layout_width="match_parent"
    android:layout_height="match_parent"
    android:background="#ededed"
    android:orientation="vertical" >
<WebView
        android:id="@+id/mv"
        android:layout_width="match_parent"
        android:layout_height="match_parent" />
</LinearLayout>
```

然后，在 AndroidManifest.xml 配置文件中添加能够访问网络的权限，代码如下：

```
<uses-permission android:name="android.permission.INTERNET" />
```

最后，创建一个 Activity，名为 MainActivity，代码如下：

```
public class MainActivity extends AppCompatActivity {

    private WebView mMv;
    private String mUrl = "http://www.baidu.com";

    @Override
    protected void onCreate(Bundle savedInstanceState) {
        super.onCreate(savedInstanceState);
        setContentView(R.layout.activity_main);
        mMv = (WebView) findViewById(R.id.wv);
        WebSettings settings = mMv.getSettings();
        //如果访问的页面中要与 Javascript 交互，则 Webview 必须设置为支持 Javascript
        settings.setJavaScriptEnabled(true);
        //将页面调整到适合 Webview 的大小
        settings.setUseWideViewPort(true);
        // 将页面缩放至屏幕的大小
        settings.setLoadWithOverviewMode(true);
        mMv.loadUrl(mUrl);
        mMv.setWebViewClient(new WebViewClient() {
            @Override
            public boolean shouldOverrideUrlLoading(WebView view, String url) {
                mMv.loadUrl(url);
                return true;
```

```
            }
        });
    }

    /**
     * 单击返回后，是网页回退而不是退出浏览器
     */
    @Override
    public boolean onKeyDown(int keyCode, KeyEvent event) {
        if ((keyCode == KEYCODE_BACK) && mMv.canGoBack()) {
            mMv.goBack();
            return true;
        }
        return super.onKeyDown(keyCode, event);
    }
}
```

上述代码使用 WebView 组件的 loadUrl()方法直接打开 Web 网页，运行结果如图 9-4 所示。

图 9-4　使用 WebView 控件浏览网页

也可以使用 WebView 组件的 loadData()方法显示 HTML 页面，代码如下：

```
String html = "";
html += "<html>";
html += "<body>";
html += "<a href=http://www.baidu.com>Baidu Home</a>";
html += "</body>";
html += "</html>";
webView.loadData(html, "text/html", "utf-8");
```

上述代码中，WebView 组件的 loadData()方法有三个参数，其中：

(1) 第 1 个参数是 html 内容。

(2) 第 2 个参数是 MIME 类型，即指明文本类型是 HTML 格式。

(3) 第 3 个参数是编码字符集。

9.5　JSON 数据

JSON 是一种轻量级的数据交换格式，是基于 ECMAScript 的一个子集，它采用完全独立于编程语言的文本格式来存储和传递数据。

JSON 是取代 xml 的数据结构，与 xml 相比更加小巧，但描述能力却不差，简洁清晰

的层次结构使得 JSON 成为理想的数据交换语言，其易于阅读和编写，也易于机器解析和生成，在传输数据时可以使用更少的网络流量，从而可以有效地提升数据的网络传输效率。

本质上，JSON 就是一串字符串，只不过其中的元素会使用特定的符号标注：{}表示对象；[]表示数组；“”内是属性或值；“:”表示后者是前者的值(这个值可以是字符串、数字、也可以是另一个数组或对象)。

例如：

✧ {“name”：“zhangsan”}可以理解为是一个包含的 name 为“zhangsan”的对象。

✧ [{“name”：“zhangsan”},{“name”：“lisi”}]可以理解为是包含两个对象的数组或集合。

在解析 JSON 数据时，需要理解 JSON 数据的格式(对象或是集合)。在 Android 实际项目开发中经常会使用两种 JSON 数据的解析方式：原生解析和 GSON 解析。下面逐一进行讲解。

9.5.1 原生解析

Android 中的 JSON 数据主要有以下两类，其原生解析 API 存放于包 org.json 下。

1．JSONObject

可以看做一个 JSON 数据对象，是系统中有关 JSON 数据定义的基本单元。一个 JSONObject 实例包含一对(key/value)数值，其中的 key 和 value 被冒号“:”分隔(例如{"JSON": "Hello, World"})。

2．JSONArray

代表一组有序的 JSON 数据集合。将其转换为 String 并输出(使用 toString()方法)的形式是用方括号包裹，数据间以逗号“,”分隔(例如[value1,value2,value3]，其中每个 value 都代表着一个 JSONObject 对象。

【示例 9.5】 在 Android Studio 中新建项目 Ch09_9D5_Json，使用原生解析方式解析 JSON 数据，并将解析出的数据在 ListView 中显示。

新建一个实体类对象 News，代码如下：

```
public class News {
    private int id;
    private int seq;
    private String description;
    private String keywords;
    private String name;
    private String title;

    public int getId() {
        return id;
```

```
    }

    public void setId(int id) {
        this.id = id;
    }

    public int getSeq() {
        return seq;
    }

    public void setSeq(int seq) {
        this.seq = seq;
    }

    public String getDescription() {
        return description;
    }

    public void setDescription(String description) {
        this.description = description;
    }

    public String getKeywords() {
        return keywords;
    }

    public void setKeywords(String keywords) {
        this.keywords = keywords;
    }

    public String getName() {
        return name;
    }

    public void setName(String name) {
        this.name = name;
    }

    public String getTitle() {
        return title;
    }
```

```
    public void setTitle(String title) {
        this.title = title;
    }
}
```

上述代码声明了对象 News.class 具有的属性，同时生成对应的 get 和 set 方法。

在项目包下新建一个工具类 JsonUtil，代码如下：

```
public class JsonUtil {

    public static List<News> parseJsonString(String json) throws Exception{
        List<News> mList = new ArrayList<News>();
        JSONObject object = new JSONObject(json);
        boolean status = object.getBoolean("status");
        if (status) {
            JSONArray array = object.getJSONArray("news");
            for (int i = 0; i < array.length(); i++) {
                JSONObject jsonObject = array.getJSONObject(i);
                News news = new News();
                news.setId(jsonObject.getInt("id"));
                news.setName(jsonObject.getString("name"));
                news.setDescription(jsonObject.getString("description"));
                news.setTitle(jsonObject.getString("title"));
                news.setKeywords(jsonObject.getString("keywords"));
                mList.add(news);
            }
        }

        return mList;
    }
}
```

上述代码按照 JSON 数据格式编写，最外层是一个对象，对象中又包含了一个数组，因此在解析的时候要一层一层地解析。

在 MainActivity 中编写代码如下：

```
public class MainActivity extends AppCompatActivity {
    private ListView mLv;
    private String urlJson = Content.JSON;
    private List<News> mlist;

    @Override
    protected void onCreate(Bundle savedInstanceState) {
        super.onCreate(savedInstanceState);
```

```
        setContentView(R.layout.activity_main);
        mLv = (ListView) findViewById(R.id.lv);

        try {
            /**
             * 原生解析
             */
            mlist = JsonUtil.parseJsonString(urlJson);
        } catch (Exception e) {
            e.printStackTrace();
        }
        //设置适配器
        mLv.setAdapter(new MyAdapter());

    }

    class MyAdapter extends BaseAdapter {

        @Override
        public int getCount() {
            return mlist.size();
        }

        @Override
        public Object getItem(int position) {
            return mlist.get(position);
        }

        @Override
        public long getItemId(int position) {
            return position;
        }

        @Override
        public View getView(int position, View convertView, ViewGroup parent) {
            ViewHolder holder;
            if (convertView == null) {
                holder = new ViewHolder();
                convertView = LayoutInflater.from(MainActivity.this).inflate(R.layout.item_list, null);
                holder.tvTitle = (TextView) convertView.findViewById(R.id.tv_title);
                holder.tvDes = (TextView) convertView.findViewById(R.id.tv_des);
                convertView.setTag(holder);
```

```
            } else {
                holder = (ViewHolder) convertView.getTag();
            }
            //加载数据
            News news = mlist.get(position);
            if (news != null) {
                holder.tvTitle.setText(news.getTitle());
                holder.tvDes.setText(news.getDescription());
            }

            return convertView;
        }

        class ViewHolder {
            TextView tvTitle;
            TextView tvDes;
        }
    }
}
```

运行项目，效果如图 9-5 所示。

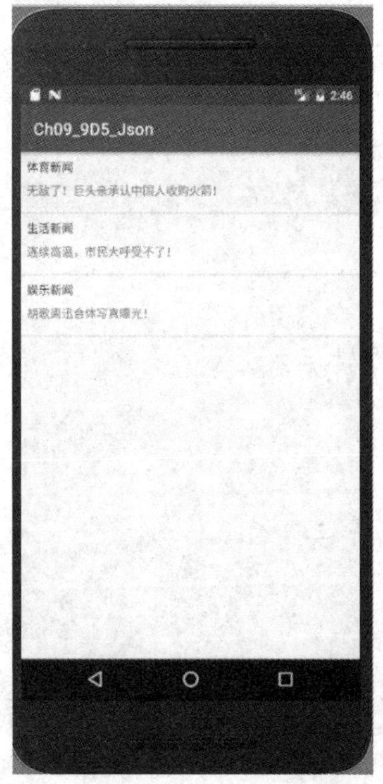

图 9-5　使用原生方式解析 JSON 数据

9.5.2 GSON 解析

GSON 是 Google 提供的用于在 Java 对象和 JSON 数据之间进行映射的 Java 类库，GSON 可以很轻松地将一串 JSON 数据转换成一个 Java 对象，或是将一个 Java 对象转换为相应的 JSON 数据。

如果使用 ADT 开发工具，在使用 GSON 之前需要导入 jar 包；在 Android Studio 中，则需要在 build.gradle 中添加 GSON 所依赖的库，示例代码如下：

```
dependencies {
    compile fileTree(include: ['*.jar'], dir: 'libs')
    androidTestCompile('com.android.support.test.espresso:espresso-core:2.2.2', {
        exclude group: 'com.android.support', module: 'support-annotations'
    })
    compile 'com.android.support:appcompat-v7:25.2.0'
    compile 'com.android.support.constraint:constraint-layout:1.0.2'
    testCompile 'junit:junit:4.12'
    compile 'com.google.code.gson:gson:2.8.1'
}
```

新建一个实体类，实体类中的属性要与 JSON 数据中的名称匹配，否则无法解析数据，代码如下：

```
public class NewsList {
    private boolean status;
    private List<News> news;

    public boolean isStatus() {
        return status;
    }

    public void setStatus(boolean status) {
        this.status = status;
    }

    public List<News> getNews() {
        return news;
    }

    public void setNews(List<News> news) {
        this.news = news;
    }
}
```

将示例 9.5 中 MainActivity 的代码修改如下：

```
public class MainActivity extends AppCompatActivity {
    private ListView mLv;
    private String urlJson = Content.JSON;
    private List<News> mlist;

    @Override
    protected void onCreate(Bundle savedInstanceState) {
        super.onCreate(savedInstanceState);
        setContentView(R.layout.activity_main);
        mLv = (ListView) findViewById(R.id.lv);

        try {

            /**
             *使用 GSON 解析
             */
            NewsList newsList = new Gson().fromJson(Content.JSON, NewsList.class);
            mlist = newsList.getNews();
        } catch (Exception e) {
            e.printStackTrace();
        }
        //设置适配器
        mLv.setAdapter(new MyAdapter());

    }


    class MyAdapter extends BaseAdapter {

        @Override
        public int getCount() {
            return mlist.size();
        }

        @Override
        public Object getItem(int position) {
            return mlist.get(position);
        }
```

```
        @Override
        public long getItemId(int position) {
            return position;
        }

        @Override
        public View getView(int position, View convertView, ViewGroup parent) {
            ViewHolder holder;
            if (convertView == null) {
                holder = new ViewHolder();
                convertView = LayoutInflater.from(MainActivity.this).inflate(R.layout.item_list, null);
                holder.tvTitle = (TextView) convertView.findViewById(R.id.tv_title);
                holder.tvDes = (TextView) convertView.findViewById(R.id.tv_des);
                convertView.setTag(holder);
            } else {
                holder = (ViewHolder) convertView.getTag();
            }
            //加载数据
            News news = mlist.get(position);
            if (news != null) {
                holder.tvTitle.setText(news.getTitle());
                holder.tvDes.setText(news.getDescription());
            }

            return convertView;
        }

        class ViewHolder {
            TextView tvTitle;
            TextView tvDes;
        }
    }
}
```

运行项目，效果与图 9-5 所示相同。

9.6 异步任务 AsyncTask

在 Android 应用程序开发过程中，需要时刻注意确保应用程序的稳定和 UI 操作响应的及时，不稳定或响应不及时会导致不好的用户体验。

在 Android 程序开始运行的时候会单独启动一个进程，默认情况下，该程序所有的操

作都在这个进程中进行。一个 Android 程序默认情况下只有一个进程，但这一个进程可以有多个线程。在这些线程中，有一个线程叫作 UI 线程(也叫 Main Thread)，除了 UI 线程外的线程都叫作子线程(Worker Thread)。UI 线程主要负责控制 UI 界面的显示、更新、交互等，因此，UI 线程中的操作延迟越短越好(流畅)，把一些耗时的操作(网络请求、数据库操作、逻辑计算等)放到单独的线程中，可以避免主线程阻塞。

为解决线程的阻塞问题，Android 提供了一种轻量级的异步任务类 AsyncTask，该类可以实现异步操作，并提供了接口，用于反馈当前任务的异步执行结果及进度，这些接口有的直接运行在主线程中(如 onPostExecute()，onPreExcute()等)。

AsyncTask 是一个抽象类，其中包括了三个参数——Params、Progress 和 Result。

(1) Params：启动任务执行的输入参数，比如 HTTP 请求的 URL，一般是 String 类型。

(2) Progress：后台任务执行的百分比，一般使用 Integer 类型。

(3) Result：后台执行任务的返回结果，一般使用 byte[]或者 String 类型。

AsyncTask 的使用步骤如下所示：

(1) 定义一个 AsyncTask 的子类。

(2) 实现 AsyncTask 中定义的方法，可以全部实现，也可以只实现其中一部分。

- ✧ onPreExecute()：该方法将在执行实际的后台操作前被 Main Thread 调用，可以在该方法中做一些准备工作，如在界面上显示一个进度条。
- ✧ doInBackgroud(Params...)：该方法将在 onPreExecute()方法执行后立刻执行，运行在后台线程中，主要负责执行那些很耗时的操作。
- ✧ onProgressUpdate(Progress...)：主线程调用该方法在界面上展示任务的进展情况。
- ✧ onPostExecute(Result)：在 doInBackgroud(Params...)方法执行完毕后，该方法将被 UI 主线程调用，将执行的结果传递到 UI 主线程中。

【示例 9.6】 在 Android Studio 中新建项目 Ch09_9D6_AsyncTask，使用异步任务下载图片，并显示下载进度。

在 activity_main.xml 中编写代码如下：

```
<?xml version="1.0" encoding="utf-8"?>
<LinearLayout xmlns:android="http://schemas.android.com/apk/res/android"
    xmlns:tools="http://schemas.android.com/tools"
    android:layout_width="match_parent"
    android:layout_height="match_parent"
    android:orientation="vertical"
    tools:context="com.ugrow.ch09_9d6_asynctask.MainActivity">

    <ImageView
        android:id="@+id/iv"
        android:layout_width="match_parent"
        android:layout_height="200dp"
        android:scaleType="centerCrop"/>
```

```
    <Button
        android:id="@+id/btn_download"
        android:layout_width="wrap_content"
        android:layout_height="wrap_content"
        android:layout_gravity="center_horizontal"
        android:layout_marginTop="16dp"
        android:text="点击下载网络图片" />

</LinearLayout>
```

在 MainActivity 中编写代码如下：

```
public class MainActivity extends AppCompatActivity {

    private ImageView mIv;
    private Button mBtnDownload;
    private String urlString = "http://img1.imgtn.bdimg.com/it/u=2227804654,860253351&fm=26&gp=0.jpg";

    @Override
    protected void onCreate(Bundle savedInstanceState) {
        super.onCreate(savedInstanceState);
        setContentView(R.layout.activity_main);
        mIv = (ImageView) findViewById(R.id.iv);
        mBtnDownload = (Button) findViewById(R.id.btn_download);
        mBtnDownload.setOnClickListener(new View.OnClickListener() {
            @Override
            public void onClick(View view) {
                new DownloadImageTask(MainActivity.this).execute(urlString);
            }
        });
    }
    /*
        * Params：输入任务的参数：doInBackground()方法的参数、execute()方法的参数
        * Progress：任务执行的进度指示：onProgressUpdate()方法的参数、publishProgress()方法的参数
        * Result：任务执行的结果：doInBackground()方法的返回值、onPostExecute()方法的参数
        */
    class DownloadImageTask extends AsyncTask<String, Integer, Bitmap> {

        private ProgressDialog progressDialog;

        public DownloadImageTask(Context context) {
            progressDialog = new ProgressDialog(context);
```

```
        progressDialog.setIcon(R.mipmap.ic_launcher);
        progressDialog.setTitle("提示信息");
        progressDialog.setMessage("正在努力为您加载...");
        progressDialog.setProgressStyle(ProgressDialog.STYLE_HORIZONTAL);
        progressDialog.setMax(100);
    }

    //在主线程中调用 onPreExecute()方法
    @Override
    protected void onPreExecute() {
        progressDialog.show();
        super.onPreExecute();
    }

    //在子线程中调用 doInBackground()方法
    @Override
    protected Bitmap doInBackground(String... params) {
        InputStream inStream = null;
        ByteArrayOutputStream outStream = new ByteArrayOutputStream();
        try {
            URL url = new URL(params[0]);
            HttpURLConnection conn = (HttpURLConnection) url.openConnection();
            conn.setReadTimeout(5000);
            conn.setDoInput(true);
            conn.setRequestMethod("GET");
            conn.connect();
            if (conn.getResponseCode() == 200) {
                inStream = conn.getInputStream();
                byte[] buffer = new byte[128];
                int length = 0;

                // 要下载文件的总长度
                int fileLength = conn.getContentLength();

                // 已经下载的文件长度
                int downloadLength = 0;


                while ((length = inStream.read(buffer)) != -1) {
                    outStream.write(buffer, 0, length);
```

```
                    // 计算当前的进度值
                    downloadLength += length;
                    int progress = (int) ((downloadLength / (float) fileLength) * 100);

                    // 发布进度，引起 onProgressUpdate()方法的调用
                    publishProgress(progress);

                    // 为了让加载进度的效果更明显
                    SystemClock.sleep(100);
                }
                byte[] imgData = outStream.toByteArray();
                Bitmap imgBitmap = BitmapFactory.decodeByteArray(imgData, 0, imgData.length);
                return imgBitmap;
            }
        } catch (Exception e) {
            e.printStackTrace();
        } finally {
            if (inStream != null) {
                try {
                    inStream.close();
                } catch (IOException e) {
                    e.printStackTrace();
                }
            }
        }
        return null;
    }

    //在主线程中调用 onProgressUpdate()方法
    @Override
    protected void onProgressUpdate(Integer... values) {

        // 使用 ProgressDialog 显示进度值
        progressDialog.setProgress(values[0]);
        super.onProgressUpdate(values);
    }

    //在主线程中调用 onPostExecute()方法
    @Override
```

```
        protected void onPostExecute(Bitmap result) {
            progressDialog.dismiss();
            if(result != null) {
                mIv.setImageBitmap(result);
            } else {
                Toast.makeText(MainActivity.this, "获取图片失败", Toast.LENGTH_SHORT).show();
            }
            super.onPostExecute(result);
        }

    }
}
```

运行项目，效果如图 9-6 所示。

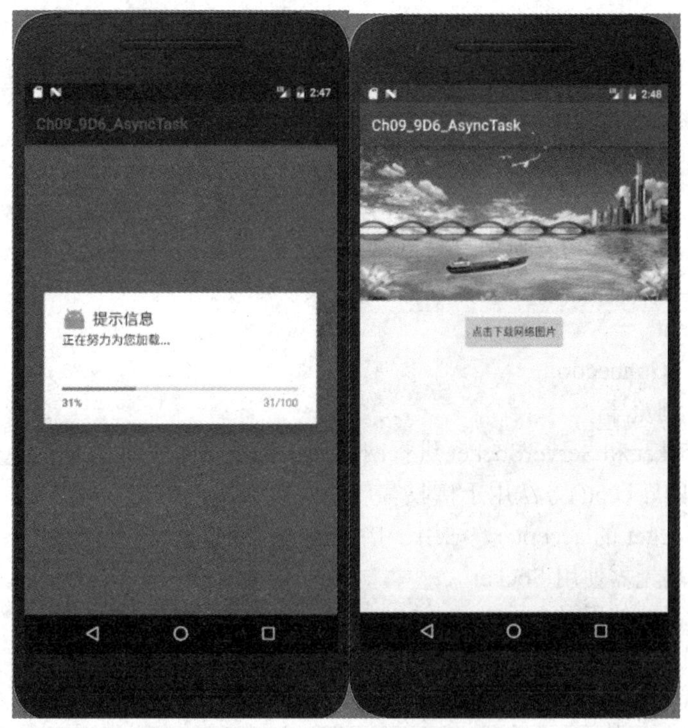

图 9-6　使用 AsyncTask 下载网络图片

本 章 小 结

通过本章的学习，读者应当了解：

✧ Socket 通常也称作“套接字”，用来描述通信链的句柄：IP 地址和端口。

✧ ServerSocket 用于监听特定端口的 TCP 连接，其 accept()方法用于监听到来的连接，并返回一个 Socket。

✧ HttpURLConnection 用于连接基于 HTTP 协议的 URL 网络资源。

- ✧ HttpURLConnection 是一个抽象类，无法直接实例化，其对象主要通过 URL 的 openConnection()方法获得。
- ✧ Apache 提供了 HTTP 客户端组件 HttpClient，它对 java.net 中的类进行封装和抽象，更适合在 Android 上开发网络应用。
- ✧ HttpClient 类默认使用 DefaultHttpClient 生成 HttpClient 对象。
- ✧ 根据请求方法的不同会用到 HttpGet 和 HttpPost 两个对象，HttpGet 封装是以 Get 方式提交的请求，而 HttpPost 封装是以 Post 方式提交的请求。
- ✧ Android 提供了内置的浏览器，该浏览器使用了开源的 WebKit 引擎。
- ✧ WebView 是 WebKit 中专门用来浏览网页的视图组件，它作为应用程序的 UI 接口，为用户提供了一系列的网页浏览、用户交互接口，通过这些接口显示和处理请求的网络资源。
- ✧ Android 中所有访问网络的操作必须在新的线程中执行，不能直接在主线程(UI 线程)中执行。
- ✧ 不能在主线程之外的其他线程更新 UI，可以通过 Handler 来更新 UI。
- ✧ 解析 JSON 数据的方式有两种：原生解析和 GSON 解析。
- ✧ 异步任务 AsyncTask 通过 doInBackground()方法执行耗时操作。

本 章 练 习

1. 下列不属于 Android 内置网络支持的是______。
   A. Socket
   B. HttpClient
   C. HttpURLConnection
   D. Firefox 浏览器
2. 关于使用 Socket 和 ServerSocket 进行网络通信的说法不正确的是______。
   A. Socket 的 accept()方法用于接收客户端连接
   B. ServerSocket 的 accept()方法用于接收客户端连接
   C. 服务器端无需使用 Socket
   D. 客户端无需使用 ServerSocket
3. 关于 HttpClient 的说法不正确的是______。
   A. HttpClient 是 Apache 的开源项目
   B. HttpClient 是针对 HTTP 的开发包
   C. HttpClient 提供了 HTML 的渲染支持
   D. HttpClient 提供了 JavaScript 的解释器
4. 在 Android 中，针对 HTTP 进行网络通信有______和______两种方式。
5. 编写代码，使用 HttpClient 上传文件。
6. 编写代码，使用 AsyncTask 获取网络数据，并使用 ListView 显示出来。

第 10 章　消息处理机制

本章目标

- 掌握 Android 消息处理机制的原理
- 掌握 Looper 循环器的工作原理
- 掌握 Handle 异步处理机制
- 掌握封装任务 Message

10.1 消息处理机制简介

Android 消息处理机制是 Android 提供的一套异步消息处理的机制。

在开发 Android 应用时必须遵守单线程模型的原则，因为 Android UI 操作是线程不安全的，所以这些操作必须在主线程中执行。

单线程模型的原则有以下两条：

(1) 不要阻塞主线程。

(2) 确保只在主线程中访问 Android UI 控件。

但是，当需要执行一些耗时操作，比如发起一个网络请求时，考虑到网络环境等其他原因，服务器未必会立刻响应请求，如果不将这类耗时操作放在子线程中去进行，就会导致主线程被阻塞，程序崩溃，从而影响软件的正常使用。

Android 消息处理机制的出现，完美地解决了必须将耗时操作放在子线程中进行，然而在子线程中又无法更新 UI 的问题。

10.1.1 子线程开启方式

子线程的开启方式有三种：继承方式、Runnable 接口方式、匿名内部类方式。

1. 继承方式

在 Android 中的多线程编程与 Java 中的多线程编程使用的语法基本相同。例如，定义一个子线程时，只需要新建一个类继承 Thread，然后重写父类 Thread 的 run()方法，并在里面编写耗时操作的逻辑代码即可，如下所示：

```
class MyThread extends Thread{
        @Override
        public void run() {
            //处理具体逻辑
        }
    }
```

子线程定义好之后，只需要新建一个 MyThread 的实例，然后调用它的 start()方法，就可以开启一个子线程，同时 run()方法中的代码就会在该子线程中执行，如下所示：

```
new MyThread().start();
```

2. Runnable 接口方式

由于继承方式的耦合度较高，所以更多时候会使用实现 Runnable 接口的方式来定义一个子线程，如下所示：

```
class MyThread implements Runnable{
        @Override
        public void run() {
//处理具体逻辑
```

```
        }
    }
```

同时，开启子线程的方法需要进行相应改变，如下所示：

```
MyThread myThread = new MyThread();
new Thread(myThread).start();
```

可以看到，新建的 MyThread 实例是一个实现了 Runnable 接口的对象，因此可以直接把新建的实例 myThread 作为参数传入到 Thread 的构造方法中，然后调用 start()方法，即可开启一个子线程。

3. 匿名内部类方式

如果不想专门定义一个类去实现 Runnable 接口，也可以使用匿名内部类的方式开启一个线程，这种方式是最常见的，如下所示：

```
new Thread(new Runnable() {
            @Override
            public void run() {
                //处理具体逻辑


            }
    }).start();
```

可以看到，这种方式把代码整合到一起编写，使用更加方便。

10.1.2　消息处理机制示例

消息处理机制通常用于解决不能在子线程中更新 UI 的问题，在上一小节中介绍了子线程的三种开启方式，下面以匿名内部类开启方式为例，讲解消息处理机制的应用。

【示例 10.1】 在 Android Studio 中新建项目 Ch10_10D1_Thread，尝试在子线程中更新 UI。

在 activity_main.xml 中编写以下代码：

```
<?xml version="1.0" encoding="utf-8"?>
<RelativeLayout xmlns:android="http://schemas.android.com/apk/res/android"
    xmlns:tools="http://schemas.android.com/tools"
    android:layout_width="match_parent"
    android:layout_height="match_parent"
    tools:context="com.ugrow.ch10_10d1_thread.MainActivity">
<Button
        android:id="@+id/btn"
        android:layout_width="match_parent"
        android:layout_height="wrap_content"
        android:text="按钮"
        android:textSize="24sp"/>
```

```
    <TextView
        android:id="@+id/tv_info"
        android:layout_width="wrap_content"
        android:layout_height="wrap_content"
        android:layout_centerInParent="true"
        android:text="Hello World"
        android:textSize="24sp"
        android:textColor="#000"/>
</RelativeLayout>
```

上述代码在布局文件中定义了一个文本控件 TextView，用于在屏幕中间显示一个字符串；然后定义了一个按钮 Button，单击该按钮可以改变 TextView 上显示的字符串。

在 MainActivity 中编写如下代码：

```
public class MainActivity extends AppCompatActivity {

    private Button mBtn;
    private TextView mTv;

    @Override
    protected void onCreate(Bundle savedInstanceState) {
        super.onCreate(savedInstanceState);
        setContentView(R.layout.activity_main);
        initView();
    }

    private void initView() {
        mBtn = (Button) findViewById(R.id.btn);
        mTv =(TextView) findViewById(R.id.tv_info);
        mBtn.setOnClickListener(new View.OnClickListener() {
            @Override
            public void onClick(View view) {
                new Thread(new Runnable() {
                    @Override
                    public void run() {

                        mTv.setText("Welcome to Android Thread!");
                    }
                }).start();
            }
        });
    }
```

```
}
```

上述代码给前面定义的 Button 绑定了一个监听事件：在 onClick()方法中开启一个子线程，并调用 TextView 的 setText()方法给 TextView 赋值。

编写完毕，单击 Android Studio 运行按钮，运行程序，会弹出程序崩溃提示，如图 10-1 所示。

图 10-1　程序崩溃提示

然后打开 LogCat，在其中可看到以下信息：

```
FATAL EXCEPTION: Thread-4
Process: com.ugrow.ch10_10d1_thread, PID: 3027
android.view.ViewRootImpl$CalledFromWrongThreadException: Only the original thread that created a view
hierarchy can touch its views.
```

显而易见，Android 中不能在子线程中进行 UI 操作，否则程序有可能崩溃。此时，就可以采用 Android 消息处理机制来解决这个问题。比如，可将 MainActivity 中的代码修改如下：

```
public class MainActivity extends AppCompatActivity {


    private Button mBtn;
    private TextView mTv;
    public static final int UPDATE = 1;
    private Handler handler = new Handler(){
        @Override
        public void handleMessage(Message msg) {
```

```
            super.handleMessage(msg);
            switch (msg.what){
                case UPDATE:
                    //在这里操作UI
                    mTv.setText("Welcome to Android Thread!");
                    break;
                default:
                    break;
            }
        }
    };

    @Override
    protected void onCreate(Bundle savedInstanceState) {
        super.onCreate(savedInstanceState);
        setContentView(R.layout.activity_main);
        initView();
    }

    private void initView() {
        mBtn = (Button) findViewById(R.id.btn);
        mTv =(TextView) findViewById(R.id.tv_info);
        mBtn.setOnClickListener(new View.OnClickListener() {
            @Override
            public void onClick(View view) {
                new Thread(new Runnable() {
                    @Override
                    public void run() {

                        Message message = new Message();
                        message.what =UPDATE;
                        handler.sendMessage(message);
                    }
                }).start();
            }
        });
    }
}
```

上述代码定义了一个整型常量 UPDATE，用于表示更新 TextView 显示内容的动作；然后新增了一个 Handler 对象，用于实现消息的发送和接收操作。首先，重写父类的

handleMessage()方法，在该方法中对具体的 Message 进行处理，如果发现 Message 的 what 字段的值为 UPDATE，就将 TextView 显示的内容修改；然后，在 Button 的单击事件中，创建了一个 Message 对象，并将它的 what 字段赋值为 UPDATE，然后调用 Handler 的 sendMessage()方法将这条 Message 发送出去。

编写完毕，重新运行程序，单击【按钮】按钮，则 TextView 上会显示字符串“Welcome to Android Thread!”，如图 10-2 所示。

图 10-2　UI 界面成功更新

上述就是 Android 消息处理机制的基本使用方法，使用这种机制，可以解决在子线程中不能更新 UI 的问题，下面对这一机制进行详细讲解。

10.2　消息处理机制详解

Android 异步消息处理机制主要由四个部分组成：Message，MessageQueue，Looper 和 Handler。其中，Handler 是消息处理机制的核心。在上一小节的示例中，我们就使用 Message 和 Handler 实现了更新 UI 界面的功能。下面将对消息处理机制的四个组成部分逐一进行详细讲解。

10.2.1　Message

在整个消息处理机制中，消息(Message)又叫任务(Task)，因为它封装了任务携带的信息和处理该任务的 Handler。Message 可以在内部携带少量的信息，通常用于在不同线程之间交换数据。

使用 Message 时需要注意以下四点：

(1) Message 虽然也可以通过 new 创建，但通常会使用 Message.obtain()或 Handler.obtainMessage()方法来从消息池中获得空消息对象，以节省资源。

(2) 如果一个 Message 只需要携带简单的 int 型数据，应优先使用 arg1 属性和 arg2 属性来传递数据，这样比使用 Bundle 对象节省内存。

(3) 尽可能使用 Message.what 来标识信息，以便能用不同的方式处理 Message。

(4) 如果需要从子线程(工作线程)返回很多数据信息，可以借助 Bundle 对象将这些数据集中到一起，然后存放到 obj 属性中，再返回到主线程。

10.2.2 MessageQueue

MessageQueue 即消息队列，用于存放所有通过 Handler 发送的消息。这部分消息会一直存放在 MessageQueue 中，等待被处理。每个线程只会有一个 MessageQueue 对象。

MessageQueue 是在消息的传输过程中保存消息的容器，并保证消息的传递，如果发送消息时接收者不可用，MessageQueue 会一直保留此消息，直到可以成功传递。

与 MessageQueue 相关的操作有两种：入队列操作和出队列操作。

10.2.3 Looper

Looper 的字面意思是“循环器”“轮询器”， 它是线程中 MessageQueue 的管家，被设计用来使一个普通的线程变成 Looper 线程。

所谓 Looper 线程就是循环工作的线程。在程序开发中(尤其是 GUI 开发中)，经常需要一个线程不断循环，每当 Looper 对象发现 MessageQueue 中存在一个新任务则执行新任务，将它发出，并且传递到 Handler 的 handleMessage()方法中，执行完继续等待下一个任务，这就是 Looper 线程。

1. Looper 线程创建

使用 Looper 类创建 Looper 线程非常简单，代码如下：

```
Public class LooperThread extends Thread {
    @Override
    public void run() {
    // 1.将当前线程初始化为Looper线程
    Looper.prepare();
    public void handleMessage(Message msg) {
            super.handleMessage(msg);
            switch (msg.what) {
            case 0:
                    break;
            default:
```

```
                    break;
                }
}
        // 2.开始循环处理消息队列
        Looper.loop();
    }
        }
```

使用以上代码，即可将普通线程转变为 Looper 线程。下面对代码中出现的两个重要方法进行讲解。

(1) Looper.prepare()方法。

默认情况下，一个 Android 线程是不存在消息循环的，需要调用 Looper.prepare()方法在主线程中创建一个消息循环。

Looper 预处理操作如图 10-3 所示：线程中有一个 Looper 对象，在它的内部维护着一个消息队列 MessageQueue。注意：一个线程中只能有一个 Looper 对象。

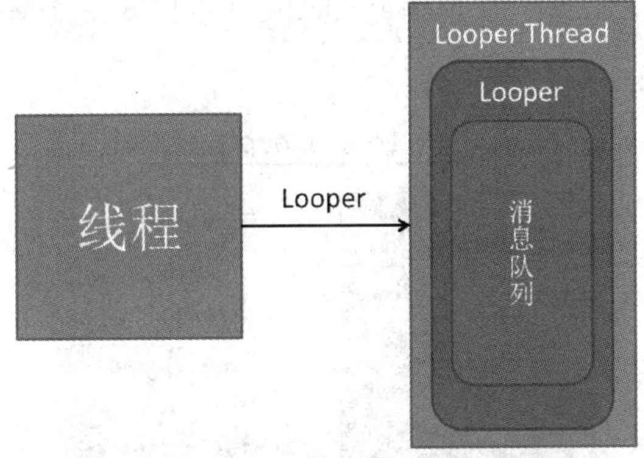

图 10-3　使用 Looper 线程维护消息队列

Looper.prepare()方法的源码如下：

```
Public class Looper{
        // 每个线程中的 Looper 对象其实是同一个 ThreadLocal，即线程本地存储(TLS)对象
        private static final ThreadLocal sThreadLocal = new ThreadLocal();
        // Looper 内的消息队列
        final MessageQueue mQueue;
        // 当前线程
        Thread mThread;

        // 每个 Looper 对象中有消息队列和其所属的线程
        private Looper() {
         mQueue = new MessageQueue();
         mRun = true;
        mThread = Thread.currentThread();
        }
```

```
    // 调用该方法会在调用线程的 TLS 中创建 Looper 对象
    public static finalvoid prepare() {
    if (sThreadLocal.get() != null) {
        // 试图在有 Looper 的线程中再次创建 Looper 将抛出异常
        thrownew RuntimeException("Only one Looper may be created per thread");
    }
    sThreadLocal.set(new Looper());
  }
}
```

通过源码来看，Looper.prepare()方法的工作方式一目了然，其核心就是将 Looper 对象定义为 ThreadLocal，即将一个唯一的 Looper 对象添加到 ThreadLocal 中。

(2) Looper.loop()。

当调用 Looper.prepare()方法创建消息循环后，需要调用 Looper.loop()方法使 Looper 线程开始工作，不断从自己维护的 MessageQueue 中取出队头的 Message 并进行处理，如图 10-4 所示。

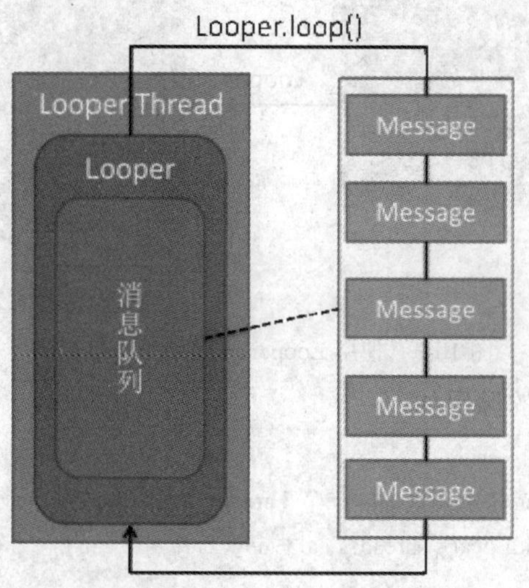

图 10-4　Looper 线程工作流程

Looper.loop()方法的源码如下：

```
Public static finalvoid loop() {
    Looper me = myLooper();  //得到当前线程 Looper
    MessageQueue queue = me.mQueue;  //得到当前 Looper 的 MQ
    Binder.clearCallingIdentity();
     final long ident = Binder.clearCallingIdentity();
     // 开始循环
    while (true) {
        Message msg
```

```
        Message msg = queue.next(); // 取出 message
      if (msg !=null) {
            if (msg.target ==null) {
            return;
            }
        }
      // 日志. . .
      if (me.mLogging!=null)
      me.mLogging.println(">>>>> Dispatching to "+ msg.target +""+ msg.callback +": "+ msg.what);
      // 非常重要！将真正的处理工作交给 Message 的 target，即后面要讲的 Handler
            msg.target.dispatchMessage(msg);

      if (me.mLogging!=null) me.mLogging.println("<<<<< Finished to"+ msg.target +""+ msg.callback);
      final long newIdent = Binder.clearCallingIdentity();
      if (ident != newIdent) {
                  Log.wtf("Looper", "Thread identity changed from 0x"
+ Long.toHexString(ident) +" to 0x"
+ Long.toHexString(newIdent) +" while dispatching to "
+ msg.target.getClass().getName() +""
+ msg.callback +" what="+ msg.what);
                  }
          }
      // 回收 message 资源
            msg.recycle();
        }
    }
}
```

2. Looper 常用方法

除 prepare()和 loop()方法以外，，Looper 类还提供了一些比较有用的其他方法：

(1) Looper.myLooper()，用于获取当前线程的 Looper 对象，代码如下：

```
publicstaticfinal Looper myLooper() {
      // 在任意线程调用 Looper.myLooper()返回的都是那个线程的 Looper
      return (Looper)sThreadLocal.get();
}
```

(2) getThread()，用于获取 Looper 对象的所属线程，代码如下：

```
public Thread getThread() {
      return mThread;
}
```

(3) quit()方法，用于结束 Looper 循环，代码如下：

```
public void quit() {
    // 创建一个空的 Message，它的 target 为 NULL，表示结束循环消息
    Message msg = Message.obtain();
    // 发出消息
    mQueue.enqueueMessage(msg, 0);
}
```

综上所述，可以将 Looper 的使用要领总结为以下几点：

(1) 每个线程有且最多只能有一个 Looper 对象，它是一个 ThreadLocal。

(2) Looper 内部有一个 MessageQueue，loop()方法调用后线程开始不断从该 MessageQueue 中取出消息(Message)执行。

(3) Looper 类能使一个线程变成 Looper 线程。

10.2.4 Handler

Handler 顾名思义就是处理者的意思，俗称“异步处理大师”，主要用于发送和处理消息。一个线程可以有多个 Handler，但是只能有一个 Looper。Handler 处理消息的原理如下：

(1) Handler 负责进行消息的发送(sendMessage()方法)和接收(handleMessage()方法)，内部会与 Looper 关联。

(2) Looper 负责传送消息，其内部包含了 MessageQueue，负责从 MessageQueue 取出消息，然后交给 Handler 处理。

(3) MessageQueue 是一个消息队列，负责存储 Handler 发来的消息，Looper 会循环地从 MessageQueue 中读取消息。

1. Handler 工作机制

Handler 的工作机制如下：

(1) Handler 可以在任意的 Looper 线程中发送消息，这些消息会被添加到该线程关联的 MessageQueue 上，如图 10-5 所示。

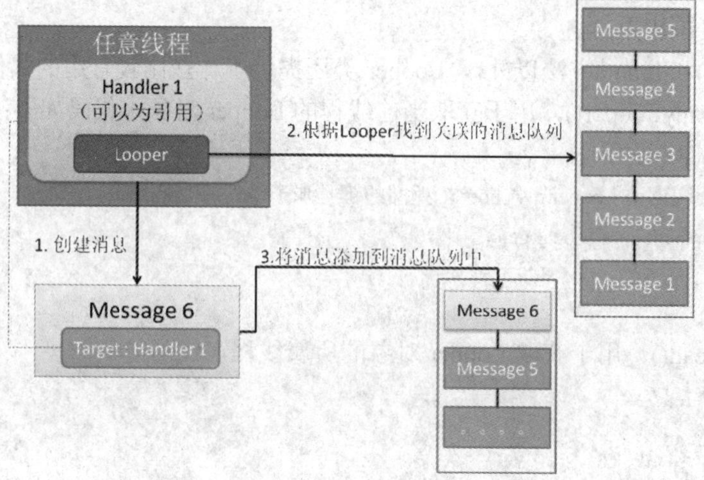

图 10-5 使用 Handler 发送消息

(2) Handler 在它关联的 Looper 线程中处理消息。由于 Android 的主线程也是一个 Looper 线程，所以，其中创建的 Handler 将默认关联主线程的 MessageQueue。因此，使用 Handler 可以解决 Android 不能在非主线程中更新 UI 的问题，如图 10-6 所示。

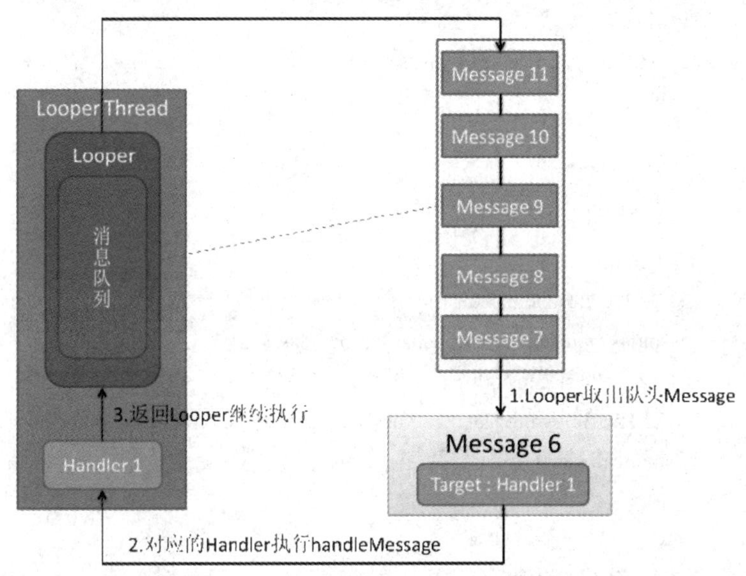

图 10-6　使用 Handler 处理消息

综上所述，Handler 的基本使用方法就是在 Activity 中创建 Handler 并将其传递给 WorkerThread(工作线程，即子线程)，WorkerThread 执行完处理任务后，再使用 Handler 发送消息给 Activity，通知主线程更新 UI，如图 10-7 所示。

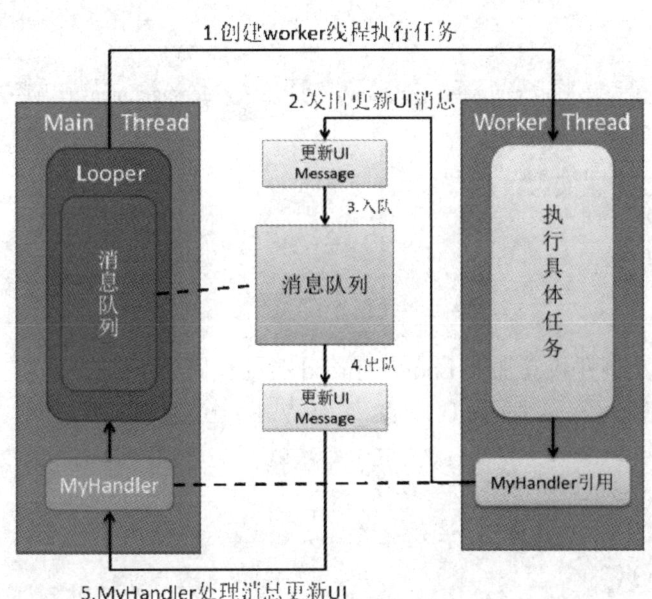

图 10-7　使用 Handler 更新 UI

2．Handler 创建流程

Handler 实例创建时会关联一个 Looper，默认会关联当前线程的 Looper。

Handler 的源码如下：

```
public class Handler {
    final MessageQueue mQueue;  // 关联的 MessageQueu final Looper mLooper;  // 关联的 Looper
    final Callback mCallback;
    …
    // 其他属性省略
    public Handler() {
    if (FIND_POTENTIAL_LEAKS) {
            final Class<?extends Handler> klass = getClass();
            if ((klass.isAnonymousClass() || klass.isMemberClass() ||
    klass.isLocalClass()) &&(klass.getModifiers() & Modifier.STATIC) ==0){
            Log.w(TAG, "The following Handler class should be static or leaks might occur: "+
            klass.getCanonicalName());
        }
    }
    // 默认将关联当前线程的 Looper
    mLooper = Looper.myLooper();
    if (mLooper ==null) {
            thrownew RuntimeException(
            "Can't create handler inside thread that has not called Looper.prepare()");
            }
    }
    // 注意：直接把关联 looper 的 MessageQueu 作为自己的 MessageQueu，因此它的消息将发送到关
联 Looper 的 MessageQueu 上
    mQueue = mLooper.mQueue;
    mCallback =null;
  }
// 其他方法
}
```

创建 Handler 实例并将其加入 LooperThread 类的代码如下：

```
public class LooperThread extends Thread {
    private Handler handler1;
    private Handler handler2;
    @Override
    public void run() {
    // 将当前线程初始化为 Looper 线程
    Looper.prepare();
    // 实例化两个 Handler
```

```
        handler1 =new Handler();
        handler2 =new Handler();

        // 开始循环处理消息队列
        Looper.loop();
    }
}
```

3. Handler 常用方法

Handler 主要用于发送和处理消息，两种操作都有各自对应的常用方法。

Handler 发送消息的常用方法如下：

(1) post(Runnable)。

(2) postAtTime(Runnable,long)。

(3) postDelayed(Runnable,long)。

(4) sendEmptyMessage(int)。

(5) sendMessage(Message)。

(6) sendMessageAtTime(Message,long)。

(7) sendMessageDelayed(Message, long)。

使用这些方法，可以向 MessageQueue 发送消息。如果只看这些方法的名称，可能会觉得 Handler 能发两种消息，一种是 Runnable 对象，一种是 Message 对象；但实际上，使用 post()等方法发出的 Runnable 对象最后都被封装成了 Message 对象。

Handler 发送的消息有如下特点：

(1) 需使用 Message.target 指明发送消息的 Handler 对象，这确保了 Looper 执行到该消息时能找到处理它的 Handler，即 loop() 方法中的关键代码 msg.target.dispatchMessage(msg)。

(2) 使用 post()方法发出的消息，其 callback(回调)结果为 Runnable 对象。

使用 Handler 发送消息的示例代码如下：

```
// 此方法用于向关联的 MessageQueue 发送 Runnable 对象，它的 run()方法将在 Handler 关联的 Looper
线程中执行
public final boolean post(Runnable r){
        // 注意 getPostMessage(r)将 Runnable 封装成 Message
        return   sendMessageDelayed(getPostMessage(r), 0);
    }

private final Message getPostMessage(Runnable r) {
        Message m = Message.obtain();   //得到空的 Message
        m.callback = r;   //将 Runnable 设为 Message 的 callback
        return m;
}
```

```
public boolean sendMessageAtTime(Message msg, long uptimeMillis){
            boolean sent =false;
      MessageQueue queue = mQueue;
            if (queue !=null) {
            msg.target =this;  // Message 的 target 必须设为该 Handler
            sent = queue.enqueueMessage(msg, uptimeMillis);
      }else {
            RuntimeException e =new RuntimeException(
this+" sendMessageAtTime() called with no mQueue");
            Log.w("Looper", e.getMessage(), e);
      }
      return sent;
}
```

Handler 消息的处理是通过核心方法 dispatchMessage(Message msg) 与方法 handleMessage(Messagemsg)完成的，示例代码如下：

```
//由Looper调用dispatchMessage()方法处理消息
public void dispatchMessage(Message msg) {
      if (msg.callback !=null) {
            // 如果Message设置了callback，则使用post()发送一个Runnable对象
         handleCallback(msg);
      } else {
            // 如果 Handler 本身设置了 callback，则执行 callback
            if (mCallback !=null) {
            /* 这种方法允许让 Activity 等来实现 Handler.Callback 接口，避免了自己编写
      Handler 重写 HandleMessage 方法。 */
            if (mCallback.handleMessage(msg)) {
                  return;
            }
         }
      }
            // 如果 Message 没有 callback，则调用 Handler 的构造方法 handleMessage
      handleMessage(msg);
      }
            // 处理 Runnable 消息
      private final void handleCallback(Message message) {
            message.callback.run();  //直接调用 run()方法!
      }
            // 由子类实现的方法
      Public void handleMessage(Message msg) {
   }
```

```
}
```

上述代码中，handleMessage(Message msg)和 Runnable 对象的 run()方法由开发者实现(实现具体逻辑)。开发者可以通过查看 API 文档来了解 Handler 的内部工作机制，这正是 Handler API 设计的精妙之处。

4．Handler 应用示例

下面通过一个示例来演示 Handler 的应用。

【示例 10.2】 在 Android Studio 中新建项目 Ch10_10D2_Handler，使用 Handler 实现网络图片的加载。

首先新建一个网络请求工具类 HttpUtil，代码如下：

```
public class HttpUtil {
        public static byte[] download(String path, String method)
                        throws ClientProtocolException, IOException {
                HttpClient client = new DefaultHttpClient();
                HttpUriRequest request = null;
                if ("get".equalsIgnoreCase(method)) {
                        request = new HttpGet(path);
                } else if ("post".equalsIgnoreCase(method)) {
                        request = new HttpPost(path);
                }
                // 响应结果
                HttpResponse response = client.execute(request);
                if (response.getStatusLine().getStatusCode() == 200) {
                        // 网络实体
                        HttpEntity entity = response.getEntity();
                        return EntityUtils.toByteArray(entity);
                }
                return null;
        }
}
```

然后编写 MainActivity，代码如下：

```
public class MainActivity extends AppCompatActivity {
    protected static final int SUCCESS_LOAD = 1;
    private ImageView ivLogo;
    private Handler handler = new Handler() {

public void handleMessage(android.os.Message msg) {
            switch (msg.what) {
                case SUCCESS_LOAD:
                    Bitmap bp = (Bitmap) msg.obj;
```

```
                    ivLogo.setImageBitmap(bp);
                    break;
                default:
                    break;
            }
        };
    };
    @Override
    protected void onCreate(Bundle savedInstanceState) {
        super.onCreate(savedInstanceState);
        setContentView(R.layout.activity_main);
        ivLogo = (ImageView) findViewById(R.id.iv_logo);
    }
    // 进行下载操作
    public void download(View v) {
        // 如果直接在主线程中进行下载操作,会报出以下异常:
        // NetWorkRunMainThread
        // 必须开启线程!
        new Thread() {
            public void run() {
                try {
                    // 得到了图片数据
                    byte[] data = HttpUtil.download(getResources().getString(R.string.img_url), "get");
                    Bitmap bitmap = BitmapFactory.decodeByteArray(data, 0,
                            data.length);
                    // 发消息
                    Message msg = Message.obtain();
                    msg.what = SUCCESS_LOAD;
                    msg.obj = bitmap;
                    handler.sendMessage(msg);
                } catch (Resources.NotFoundException | IOException e) {
                    e.printStackTrace();
                }


            };
        }.start();
    }
}
```

在模拟器上运行应用程序，单击【下载】按钮，结果如图 10-8 所示。

图 10-8　使用 Handler 下载网络图片

本 章 小 结

通过本章的学习，读者应当了解：

✧ Android 中开启子线程的方式有三种：继承方式、Runnable 接口方式和匿名内部类方式。

✧ Android 异步消息处理机制主要由四个部分组成：Message，MessageQueue，Looper 和 Handler。

✧ 将普通线程升级为 Looper 线程的方法有两种：Looper.prepare()和 Looper.loop()。

✧ Handler 的基本使用方法：在 Activity 中创建 Handler 并将其传递给 WorkerThread(工作线程，即子线程)，WorkerThread 执行完处理任务后，再使用 Handler 发送消息给 Activity，通知主线程更新 UI。

本 章 练 习

1. 在 Android 异步消息处理机制中被称为异步处理大师的是____。

A. Message

B. MessageQueue

C. Looper

D. Handler

2. 子线程的开启方式：______、______和______。
3. 简述使用 Handler 处理消息的机制和步骤。
4. 编写代码，开启子线程，获取网络图片。
5. 编写代码，分别使用三种方式开启子线程，并总结各种方式的优缺点。

第 11 章 Android 特色开发

本章目标

- 了解传感器的使用
- 掌握地图定位的原理及使用
- 掌握为 ActionBar 添加 Tabs 的方法

11.1 传感器

现代移动电话已不仅仅是简单的通信设备，随着麦克风、摄像头、加速计、指南针、温度计以及亮度探测器等功能的引入，智能电话已经变成了辅助传感设备，从而使用户能够了解更多的信息。

11.1.1 传感器简介

传感器是一种检测装置，能够探测和感受外界的信号、物理条件(如光、热、湿度)或化学组成(如烟雾)，并将探知的信息按照一定规律变换成电信号或其他所需形式的信息输出，以满足对信息的传输、处理、存储、显示、记录和控制等要求。传感器是实现自动检测和自动控制的首要环节。

Android 操作系统(以下简称 Android)中内置了很多传感器，比如一个非常实用的加速感应器(微型陀螺仪)，它能够支持重力感应、方向判断等功能，在部分游戏或软件中可以自动识别屏幕的横屏、竖屏方向，随即改变屏幕的显示布局。

Android 将设备的传感器抽象为 Sensor 类，该类描述了硬件传感器的属性，包括其类型、名称、制造商以及精确度和范围的详细信息。Sensor 类包含了一个常量集合，用于描述当前 Sensor 对象所表示的硬件传感器类型，这些常量具有 Sensor.TYPE_<TYPE>的形式。Android 的传感器类型如表 11-1 所示。

表 11-1 Android 的传感器类型

类型常量	功能描述
Sensor.TYPE_ACCELEROMETER	加速传感器，可沿着三个坐标轴返回当前的加速度
Sensor.TYPE_GYROSCOPE	陀螺仪传感器，在三个坐标轴上以角度为单位返回当前设备方向
Sensor.TYPE_LIGHT	亮度传感器，以 lux(勒克斯)为单位描述环境光的强度，用于动态控制屏幕亮度
Sensor.TYPE_MAGNETIC_FIELD	地磁传感器，可沿着三个坐标轴确定当前的磁场
Sensor.TYPE_ORIENTATION	方向传感器，可返回设备的方向
Sensor.TYPE_PRESSURE	压力传感器，以 kilopascals(千帕斯卡)为单位描述当前设备上所施加的压力
Sensor.TYPE_PROXIMITY	近程传感器，以米为单位指示设备与目标对象之间的距离，所选择的目标对象取决于探测器硬件支持的距离。邻近距离传感器的一个典型的用法是在用户的耳朵承受该设备时进行检测并自动调整屏幕的亮度或者初始化一个语音命令
Sensor.TYPE_TEMPERATURE	温度传感器，以摄氏度为单位返回温度的值，所返回的温度可以是周围房间温度、设备电池温度或者远程传感器温度

要在 Android 中使用传感器，需要使用传感器管理器类 SensorManager 和传感器事件监听类 SensorEventListener。

1. SensorManager

SensorManager 是传感器的综合管理类，可以对传感器的种类、采样率、精准度等进行管理，其常用的方法如表 11-2 所示。

表 11-2　SensorManager 类的常用方法

方　法	功 能 描 述
getSensorList()	获取指定传感器类型的所有可用的传感器列表
registerListener()	注册一个传感器监听器
unregisterListener()	注销一个传感器监听器
getDefaultSensor()	获取默认的传感器对象
getInclination()	获取地磁传感器倾斜角的弧度值
getOrientation()	获取设备旋转的方向

SensorManager 中为采样率的接收频度定义了相应的常量，这些常量按照降序排列如表 11-3 所示。

表 11-3　SensorManager 中的接收频度常量

常　量	功 能 描 述
SENSOR_DELAY_FASTEST	最快更新速率
SENSOR_DELAY_GAME	选择一个适合在游戏中使用的更新速率
SENSOR_DELAY_NORMAL	默认更新速率
SENSOR_DELAY_UI	指定一个适合 UI 功能更新的速率

SensorManager 中还为传感器精确度的反馈值定义了相应的常量，如表 11-4 所示。

表 11-4　SensorManager 中的精确度反馈值

常　量	功 能 描 述
SENSOR_STATUS_ACCURACY_LOW	传感器的精确度很低且需要校准
SENSOR_STATUS_ACCURACY_MEDIUM	传感器的数据具有平均精确度，校准可能会改善阅读效果
SENSOR_STATUS_ACCURACY_HIGH	传感器使用的是最高精确度
SENSOR_STATUS_UNRELIABLE	传感器数据不可靠，需要校准传感器且当前不能读取数据

可以通过 getSystemService()方法获取一个 SensorManager 对象，代码如下：

```
SensorManager sm=(SensorManager)getSystemService(SENSOR_SERVICE);
```

调用 SensorManager 对象的 getSensorList()方法，可以获得指定传感器类型的所有可用的传感器列表，代码如下：

```
List<Sensor> sensors=sm.getSensorList(Sensor.TYPE_ORIENTATION);
```

在 getSensorList()方法中传入参数 Sensor.TYPE_ALL，可以查找所有可用的传感器，代码如下：

```
List<Sensor> sensors=sm.getSensorList(Sensor.TYPE_ALL);
```

2. SensorEventListener

SensorEventListener 是传感器的监听接口类，包括以下两个方法：

(1) onSensorChanged(SensorEvent event)方法：用于监控传感器的值。该方法的参数是一个 SensorEvent 对象，该对象具有 sensor(传感器对象)、accuracy(精确度)、values(新

值)、timestamp(事件)4 个描述传感器事件的属性。

(2) onAccuracyChanged(Sensor sensor,int accuracy)方法：用于响应传感器精准度的变化。该方法具有两个参数：第 1 个参数表示传感器对象的名称；第 2 个参数表示传感器精准度的反馈值。

11.1.2 传感器应用

应用程序与传感器交互必须注册一个监听器，以监测一个或多个传感器相关的活动。可以使用 registerListener()方法注册一个传感器监听器，使用 unregisterListener()方法注销一个传感器监听器(在活动 Activity 的 onResume()方法中注册传感器监听器，在 onPause()方法中注销其传感器监听器，以确保仅在活动可用时使用传感器，是一种实用的好方法)，代码如下：

```
//注册一个传感器监听器
sm.registerListener(this,sensor,SensorManager.SENSOR_DELAY_NORMAL);
//注销传感器监听器
sm.unregisterListener(this);
```

registerListener()方法中包括 3 个参数：第 1 个参数是接收信号的 Listener 实例；第 2 个参数是传感器对象的名称；第 3 个参数是采样率的接收频度。

【示例 11.1】 在 Android Studio 中新建项目 Ch11_11D1_Sensor，演示传感器的应用。

在 SensorActivity 中编写代码如下：

```
public class SensorActivity extends Activity implements SensorEventListener {
    // 声明一个传感器管理器 SensorManager
    private SensorManager sm;
    private boolean regFlag;
    @Override
    public void onCreate(Bundle savedInstanceState) {
        super.onCreate(savedInstanceState);
        setContentView(R.layout.main);
        // 获取 SensorManger 实例
        sm = (SensorManager) this.getSystemService(SENSOR_SERVICE);
        regFlag=false;
    }
    @Override
    protected void onResume() {
        super.onResume();
        // 获取方向传感器列表
        List<Sensor> sensors = sm.getSensorList(Sensor.TYPE_ORIENTATION);
        if (sensors.size() > 0) {
            // 获取传感器对象
```

```
                Sensor sensor = sensors.get(0);
                // 注册传感器监听器
                regFlag=sm.registerListener(this,
                        sensor,SensorManager.SENSOR_DELAY_FASTEST);
            }
        }
        @Override
        protected void onPause() {
            //如果注册了传感器监听器，则使用 unregisterListener()方法注销
            if(regFlag){
                sm.unregisterListener(this);
                regFlag=false;
            }
            super.onPause();
        }
        @Override
        public void onAccuracyChanged(Sensor sensor, int accuracy) {
            // 处理精准度变化
        }
        @Override
        public void onSensorChanged(SensorEvent event) {
            // 处理传感器值的改变
            if(event.sensor.getType()==Sensor.TYPE_ORIENTATION){
                //获取数据
                float x=event.values[SensorManager.DATA_X];
                float y=event.values[SensorManager.DATA_Y];
                float z=event.values[SensorManager.DATA_Z];
                //由于模拟器上无法测试效果，因此将获得的数据打印如下
                Log.i("Sensor", "("+x+","+y+","+z+")");
                ......//省略已有代码
            }
        }
    }
```

上述代码执行了以下操作：① 使用 SensorActivity 类实现了一个 SensorEventListener 接口，该接口本身就是一个监听器实例。② 在 Activity 的 onResume()方法中获取方向传感器列表，并注册方向传感器的监听器；在 Activity 的 onPause()方法中注销传感器监听对象。③ 通过 onSensorChanged()方法获取传感器的数值，以进行所需操作。由于模拟器上无法进行传感器模拟，因此看不到结果数据。

11.2 地图与定位

传统 GPS 定位具有启动时间长、室内可能无法定位等弊端，为了使用户得到更好的体验，实际开发中应尽量避免使用传统 GPS 定位方法。相比之下，百度地图 SDK 综合利用 GPS、基站、Wi-Fi 等多种定位方式混合定位，适用于室内外多种定位场景，具有定位精度高、覆盖率广、网络定位请求流量小、定位速度快等特性。因此，本节将介绍如何使用百度地图 SDK 实现定位功能。

11.2.1 百度地图 SDK 介绍

百度地图 Android SDK 是一套基于 Android 2.1 及以上版本设备的应用程序接口，适用于 Android 系统移动设备的地图应用。通过调用地图 SDK 接口，可以轻松访问百度地图服务和数据，构建功能丰富、交互性强的地图类应用程序，且其提供的所有服务都是免费的，接口无使用次数限制。

百度地图 Android SDK 提供的服务主要有：

(1) 地图：提供地图(2D、3D)的展示和缩放、平移、旋转、改变视角等操作。

(2) 地理编码：提供在地理坐标和地址之间相互转换的功能。

(3) 线路规划：支持公交信息查询、公交换乘查询、驾车线路规划和步行路径检索。

(4) 覆盖物：提供多种地图覆盖物(自定义标注、几何图形、文字绘制、地形图图层、热力图图层等)，满足开发者的各种需求。

(5) 定位：采用多种定位模式，使用定位 SDK 获取位置信息，使用地图 SDK 的位置图层进行位置展示。

(6) 导航：支持调用百度地图导航和 Web 导航，以满足用户对导航功能的需求。

除此之外，百度地图 SDK 还提供了 POI 检索、LBS 云检索、离线地图等其他功能。

本节仅介绍使用百度地图 SDK 开发定位功能的方法，关于其他功能，可以登录百度地图开放平台(http://developer.baidu.com/map/index.php)获取开发文档与支持。

11.2.2 使用百度地图 SDK 开发定位功能

为了使开发者能够快速地将百度地图集成到所开发的应用当中，官方提供了示例代码供开发者学习使用。此示例代码登录开放平台即可进行下载，并可直接导入 Eclipse 或 Android Studio 中。

在正式使用百度地图 SDK 之前，还需要登录百度地图开放平台申请开发使用的密钥(ak)。该密钥与注册的百度账户相关联，因此必须先有百度账户才能获得密钥，且该密钥与创建的项目包名等参数有关。开发密钥申请地址为 http://lbsyun.baidu.com/apiconsole/key，具体申请步骤请参考官方文档。

【示例 11.2】 在 Android Studio 中新建项目 Ch11_11D2_BaiduMap，演示使用百度地图 SDK 实现地图定位功能。

(1) 在项目中新建 libs 文件夹，将开发包里的文件 BaiduLBS_Android.jar 拷贝到该文件夹下，并在 main 文件夹下新建 jniLibs 文件夹，将官方 Demo 中的.so 文件拷贝到该目

录下，拷贝完成后的项目文件目录如图 11-1 所示。

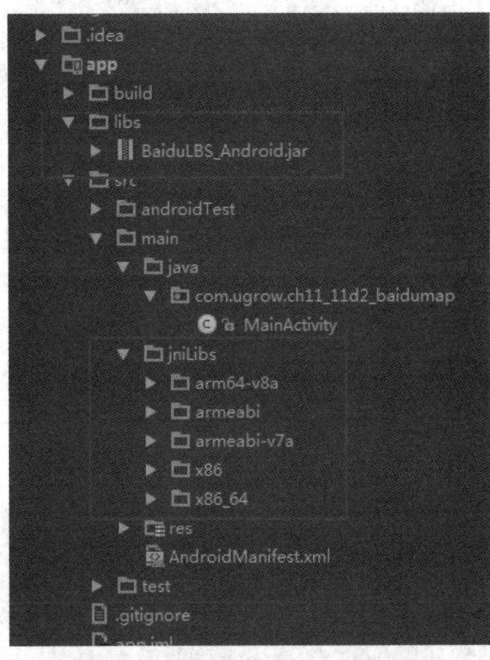

图 11-1　将百度定位 jar 包加入工作空间

(2) 在文件 BaiduLBS_Android.jar 上单击鼠标右键，在弹出的菜单中选择【Add As Library】命令，将其添加到工程的 Libraries 中，如图 11-2 所示。

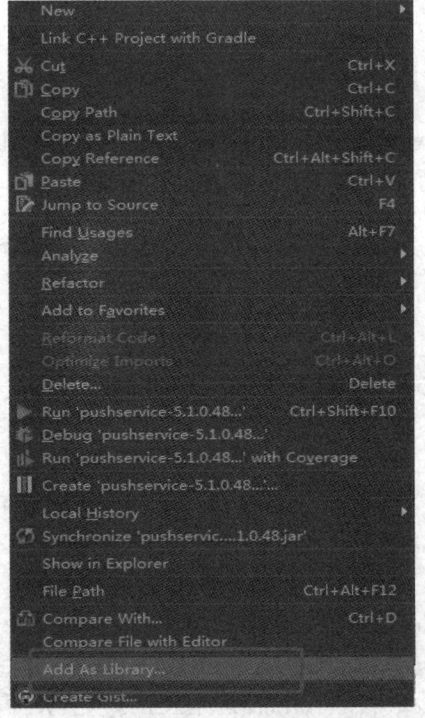

图 11-2　将百度定位 jar 包添加到项目中

(3) 在 AndroidManifest 的 application 中添加开发密钥，代码如下：

```
<application
        android:allowBackup="true"
        android:icon="@mipmap/ic_launcher"
        android:label="@string/app_name"
        android:roundIcon="@mipmap/ic_launcher_round"
        android:supportsRtl="true"
        android:theme="@style/AppTheme">

        <meta-data
            android:name="com.baidu.lbsapi.API_KEY"
            android:value="gG2jrzbv9Oto9W3UYvQlq8ThrVE32OLE" />
        <service
            android:name="com.baidu.location.f"
            android:enabled="true"
            android:process=":remote"/>
        <activity android:name=".MainActivity">
            <intent-filter>
                <action android:name="android.intent.action.MAIN" />
                <category android:name="android.intent.category.LAUNCHER" />
            </intent-filter>
        </activity>
    </application>
</manifest>
```

上述代码中的 android:value 参数的值即是已经申请的密钥，由于每个密钥仅能用于 1 个应用的验证，因此在实际开发中，需由个人自行申请。

(4) 为项目添加所需权限，代码如下：

```
    <!—以下权限用于访问网络，网络定位需要上网-->
    <uses-permission android:name="android.permission.INTERNET" />
    <!--以下权限用于进行网络定位-->
    <uses-permission android:name="android.permission.ACCESS_COARSE_LOCATION"></uses-
permission>
    <!—以下权限用于访问 GPS 定位-->
    <uses-permission android:name="android.permission.ACCESS_FINE_LOCATION"></uses-permission>
    <uses-permission android:name="android.permission.ACCESS_GPS"/>
    <!—以下权限用于访问 WIFI 网络信息，WIFI 信息可以用来进行网络定位-->
    <uses-permission android:name="android.permission.ACCESS_WIFI_STATE"></uses-permission>
    <!—以下权限用于支持提供运营商信息相关的接口，以获取运营商信息-->
    <uses-permission android:name="android.permission.ACCESS_NETWORK_STATE"></uses-permission>
    <!—以下权限用于获取 WIFI 的获取权限，WIFI 信息可以用来进行网络定位-->
```

```
    <uses-permission android:name="android.permission.CHANGE_WIFI_STATE"></uses-permission>
    <!—以下权限用于读取手机当前的状态-->
    <uses-permission android:name="android.permission.READ_PHONE_STATE"></uses-permission>
    <!—以下权限用于写入扩展存储、向扩展卡写入数据、供用户写入离线定位数据-->
    <uses-permission android:name="android.permission.WRITE_EXTERNAL_STORAGE"></uses-
permission>
    <!—以下权限用于读取 SD 卡，供用户写入离线定位数据-->
    <uses-permission android:name="android.permission.MOUNT_UNMOUNT_FILESYSTEMS"></uses-
permission>
```

(5) 在布局文件 activity_main.xml 中添加地图控件，代码如下：

```
<com.baidu.mapapi.map.MapView
        android:id="@+id/bdMapView"
        android:layout_width="fill_parent"
        android:layout_height="fill_parent"
        android:clickable="true" />
```

(6) 在应用程序创建时需要初始化 SDK 引用的 Context 全局变量。在 MainActivity 中编写代码如下：

```
public class MainActivity extends AppCompatActivity {

    // 定位相关
    LocationClient mLocClient;
    public MyLocationListenner myListener = new MyLocationListenner();
    private Double lastX = 0.0;
    private int mCurrentDirection = 0;
    private double mCurrentLat = 0.0;
    private double mCurrentLon = 0.0;
    private float mCurrentAccracy;
    boolean isFirstLoc = true; // 是否首次定位
    private MyLocationData locData;
    TextureMapView mMapView = null;
    private BaiduMap mBaiduMap;

    private static final int BAIDU_READ_PHONE_STATE = 100;

    @Override
    protected void onCreate(Bundle savedInstanceState) {
        super.onCreate(savedInstanceState);
        //在使用 SDK 各组件之前初始化 Context 信息，并传入参数 ApplicationContext
        //注意：该方法要在 setContentView()方法之前实现
```

```
        SDKInitializer.initialize(getApplicationContext());
        setContentView(R.layout.activity_main);

        // 初始化地图
        mMapView = (TextureMapView) findViewById(R.id.bmapView);
        mBaiduMap = mMapView.getMap();
        // 开启定位图层
        mBaiduMap.setMyLocationEnabled(true);
        // 初始化定位
        mLocClient = new LocationClient(this);
        mLocClient.registerLocationListener(myListener);
        LocationClientOption option = new LocationClientOption();
        option.setOpenGps(true); // 打开 Gps
        option.setCoorType("bd09ll"); // 设置坐标类型
        option.setScanSpan(1000);
        mLocClient.setLocOption(option);
        mLocClient.start();
    }

    /**
     * 定位 SDK 监听函数, 需实现 BDLocationListener 中的方法
     */
    public class MyLocationListenner implements BDLocationListener {

        public void onReceiveLocation(BDLocation location) {
            int locType = location.getLocType();
            Log.d("TAG", locType + "");

            if (location == null || mMapView == null) {
                return;
            }
            mCurrentLat = location.getLatitude();
            mCurrentLon = location.getLongitude();
            mCurrentAccracy = location.getRadius();
            locData = new MyLocationData.Builder().accuracy(location.getRadius())//120.432792,36.123604

                    .direction(100).latitude(location.getLatitude())
                    .longitude(location.getLongitude()).build();
            mBaiduMap.setMyLocationData(locData);
```

```
            if (isFirstLoc) {
                isFirstLoc = false;
                LatLng ll = new LatLng(location.getLatitude(),
                        location.getLongitude());
                MapStatus.Builder builder = new MapStatus.Builder();
                builder.target(ll).zoom(18.0f);
                mBaiduMap.animateMapStatus(MapStatusUpdateFactory.newMapStatus(builder.build()));

            }
        }

        public void onConnectHotSpotMessage(String var1, int var2) {
        }
    }


    @Override
    protected void onPause() {
        mMapView.onPause();
        super.onPause();
    }
    @Override
    protected void onResume() {
        mMapView.onResume();
        super.onResume();
    }
    @Override
    protected void onDestroy() {
        // 退出时销毁定位
        mLocClient.stop();
        // 关闭定位图层
        mBaiduMap.setMyLocationEnabled(false);
        mMapView.onDestroy();
        mMapView = null;
        super.onDestroy();
    }
}
```

运行项目，效果如图 11-3 所示。

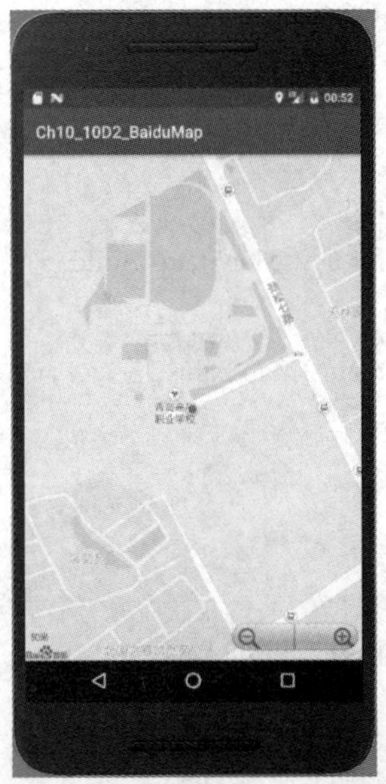

图 11-3　测试百度地图定位

注 意

在测试百度地图时，模拟器可能无法显示地图，请使用真机进行测试。

直接运行此项目会无法显示地图，这是因为申请密钥时填写的安全码构成方式为“数字签名+包名”，数字签名可能不同，因此还是建议自行申请密钥。

11.3　ActionBar 扩展功能

目前各种各样的 Tabs(选项卡)应用非常广泛，它最大的用途是可实现多个页面来回切换，使用起来非常便捷。而 ActionBar 提供了一种更加智能、更加统一的 Tabs，可以自动匹配屏幕大小，因此，ActionBar 也是官方推荐的一种展示方式。

本节主要介绍怎样为 ActionBar 添加 Tabs，鉴于 ActionBar 的基本使用方法已在第 3 章介绍过，因此这里将直接在 3.7 节的基础上进行操作。

【示例 11.3】 在 Android Studio 中新建项目 Ch11_11D3_ActionBar_Tabs，演示如何为 ActionBar 添加 Tabs。

(1) ActionBar 中的每个 Tab 都对应一个 Fragment，因此，首先需要创建 3 个 Fragment。

新建 a_fragment.xml，编写代码如下：

```
<RelativeLayout xmlns:android="http://schemas.android.com/apk/res/android"
    xmlns:tools="http://schemas.android.com/tools"
    android:layout_width="match_parent"
    android:layout_height="match_parent"
    android:background="#f00" >
<TextView
        android:layout_width="wrap_content"
```

```
        android:layout_height="wrap_content"
        android:layout_centerInParent="true"
        android:text="Fragment A"
        android:textColor="#fff"
        android:textSize="20sp"
        android:textStyle="bold" />
</RelativeLayout>
```

使用相同方法，创建 b_fragment.xml 和 c_fragment.xml。

新建 AFragment.java，编写代码如下：

```
public class AFragment extends Fragment {
    @Override
    public View onCreateView(LayoutInflater inflater, ViewGroup container,
                Bundle savedInstanceState) {
        return inflater.inflate(R.layout.a_fragment,container, false);
    }
}
```

使用相同方法，创建 BFragment.java 和 CFragment.java。

(2) 创建 TabListener，在 MainActivity.java 中添加内部类 MyTabListener，代码如下：

```
public class MainActivity extends Activity {

    @Override
    protected void onCreate(Bundle savedInstanceState) {
        super.onCreate(savedInstanceState);
        setContentView(R.layout.activity_main);
        ......//省略已有代码
    }
    ......//省略已有代码
    protected class MyTabListener implements TabListener {
        private Fragment fragment;
        private Class<?> fragmentCls;
        private String tag;
        private Activity act;
        public MyTabListener(Activity act, String tag,
                Class<?> fragmentCls) {
                super();
                this.act = act;
                this.tag = tag;
                this.fragmentCls = fragmentCls;
        }
```

```
            //Tab 被选中时调用
            @Override
            public void onTabSelected(Tab arg0, FragmentTransaction ft) {
                if (fragment == null) {
                    fragment = Fragment.instantiate(act,
                        fragmentCls.getName());
                    ft.add(android.R.id.content, fragment, tag);
                } else {
                    //附加 Fragment
                    ft.attach(fragment);
                }
            }
            //Tab 取消选中时调用
            @Override
            public void onTabUnselected(Tab arg0, FragmentTransaction ft) {
                if (fragment != null) {
                    //释放资源
                    ft.detach(fragment);
                }
            }

            //Tab 被重新选中时调用
            @Override
            public void onTabReselected(Tab arg0, FragmentTransaction ft) { }
        }
}
```

(3) 在 MainActivity.java 的 onCreate()方法中添加 Tab，主要代码如下：

```
public class MainActivity extends Activity {
        ......//省略已有代码

        @Override
        protected void onCreate(Bundle savedInstanceState) {
            super.onCreate(savedInstanceState);
            setContentView(R.layout.activity_main);
            // 获取应用的 ActionBar
            actionBar = getSupportActionBar();
            actionBar.setDisplayHomeAsUpEnabled(true);

            // 显示 ActionBar Tabs
```

```
        actionBar.setNavigationMode(ActionBar.NAVIGATION_MODE_TABS);
        // 向 Tabs 中添加 Fragment 并进行监听
        Tab tab = actionBar.newTab().setText("A Fragment")
                .setTabListener(new MyTabListener(this, "a",
                        AFragment.class));
        actionBar.addTab(tab);
        tab = actionBar.newTab().setText("B Fragment")
                .setTabListener(new MyTabListener(this, "b",
                        BFragment.class));
        actionBar.addTab(tab);
        tab = actionBar.newTab().setText("C Fragment")
                .setTabListener(new MyTabListener(this, "c",
                        CFragment.class));
        actionBar.addTab(tab);
    ......//省略已有代码
    }
    ......//省略已有代码
}
```

运行项目，效果如图 11-4 所示。

图 11-4　为 ActionBar 添加 Tabs

本 章 小 结

通过本章的学习，读者应当了解：

✧ Android 将设备的传感器抽象为 Sensor 类，该类描述了硬件传感器的属性，包括其类型、名称、制造商以及精确度和范围的详细信息。

- ✧ SensorManager 是传感器的综合管理类，可以对传感器的种类、采样率、精准度等进行管理。
- ✧ 可以使用百度地图 SDK 实现基本的地图定位功能。
- ✧ 可以将 Fragment 添加到 ActionBar 的 Tabs 中，从而通过 Tabs 在每个 Fragment 之间切换。

本 章 练 习

1. 下列不属于传感器应用的是______。
   A. 测量体温
   B. 测量海拔高度
   C. 自动横竖屏变换
   D. 自动检测是否有 Wi-Fi 连接
2. 下列对百度地图 SDK 描述错误的是______。
   A. 百度地图 SDK 属于第三方开发工具
   B. 如果不需要显示地图，只需要单纯定位，则只需将 locSDK_3.1.jar 和 liblocSDK3.so 加入到工程中即可
   C. 开发百度地图 SDK 所使用的密钥在不同应用之间是通用的
   D. 如果手机设备没有开启 GPS，只要能连接到网络，百度地图 SDK 就可以实现定位
3. AndroidManifest.xml 文件中用于修改主题样式的属性名称是______________。
4. 编写代码，实现使用 Tabs 切换页面的功能。

实践篇

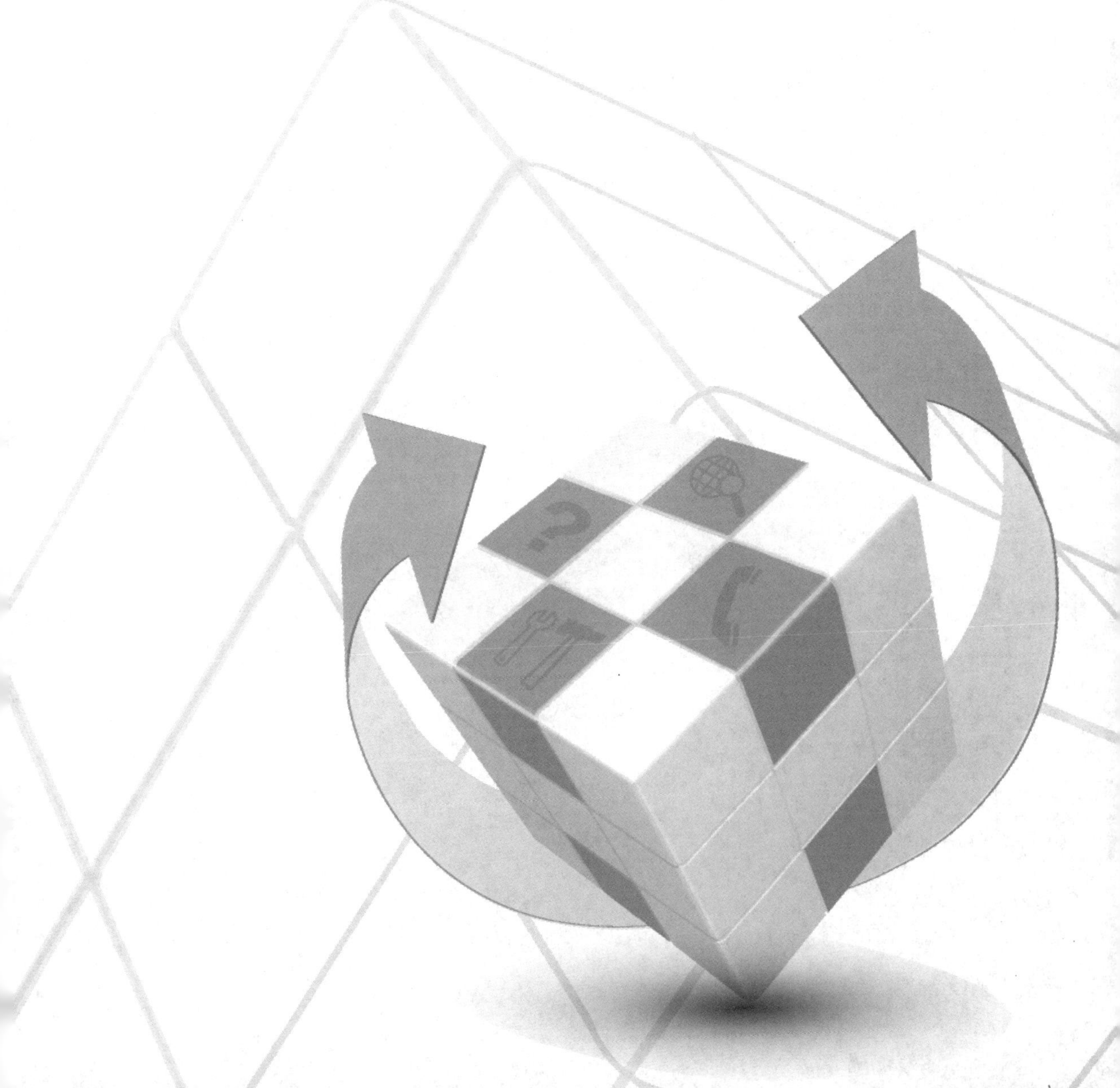

实践 1　Android 概述

实践指导

实践 1.1　开发环境搭建

在 Windows 下进行 Android 集成开发环境的下载安装及配置。

【分析】

(1) Windows 系统下搭建 Android 开发环境，所需软件如表 S1-1 所示。

表 S1-1　Android 开发环境所需软件

软件名称	版　本	说　明
JDK	JDK8.0	进行 Java 开发时所必须的开发包
Android Studio	Android Studio 2.3.2	Google 官方开发工具
Android SDK	Android SDK 4.0	Android 开发工具包

(2) Android SDK 是整个 Android 平台的核心，提供 Android 应用开发所需的 API。

(3) Android Studio 由 Google 推出，专为 Android "量身定做"，是 Google 大力支持的一款基于 IntelliJ IDEA 改造的 IDE。具有速度更快、UI 更加漂亮、更加智能、整合 Gradle 构建工具、强大的 UI 编辑器等优点。

【参考解决方案】

1. 安装 JDK

JDK 的官方网址是 http://www.oracle.com，下载 jdk-8u25-windows-i586.exe 并运行，然后按照默认步骤安装即可，如图 S1-1 所示。

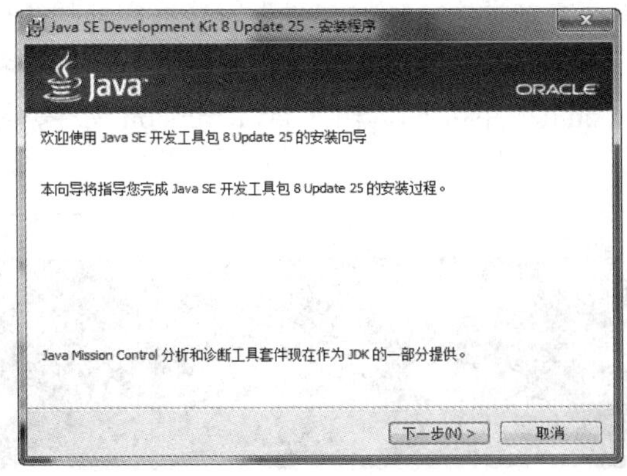

图 S1-1　安装 JDK

2. 配置 Java 环境变量

(1) 右键单击【我的电脑】图标，在弹出菜单中选择【属性】命令，在弹出的【系统】窗口中选择【高级系统设置】命令，然后在弹出的【系统属性】对话框中选择【高级】选项卡，单击其中的【环境变量】按钮，打开【环境变量】对话框。

(2) 单击【系统变量】对话框中的【新建】按钮，在弹出的【新建系统变量】对话框中将新建变量的【变量名】设置为“JAVA_HOME”，【变量值】设置为“C:\Program Files\Java\jdk1.8.0_25”(此路径是 JDK 的安装根目录)，然后单击【确定】按钮，如图 S1-2 所示。

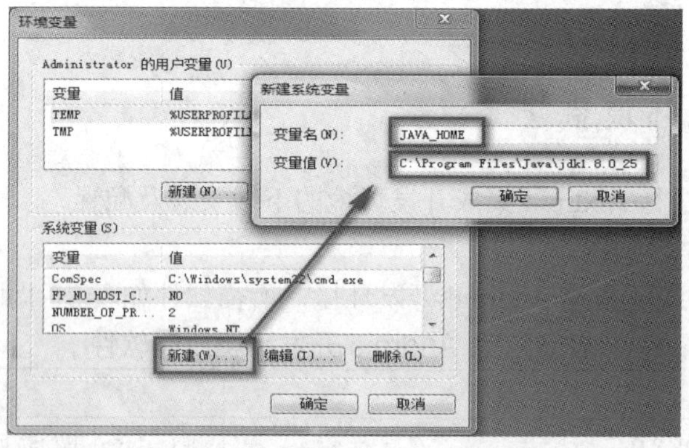

图 S1-2 创建环境变量 JAVA_HOME

(3) 使用同样的方法，继续创建 CLASSPATH 变量，将变量值设置为“.;%JAVA_HOME%\ lib\dt.jar;%JAVA_HOME%\lib\tools.jar”(Java 类、包的路径)，如图 S1-3 所示。

(4) 在【环境变量】对话框中选择【系统变量】列表中的【Path】项，然后单击【编辑】按钮，在弹出的【编辑系统变量】对话框中，在其【变量值】内添加 JDK 的 bin 路径，然后单击【确定】按钮，如图 S1-4 所示。

图 S1-3 创建环境变量 CLASSPATH

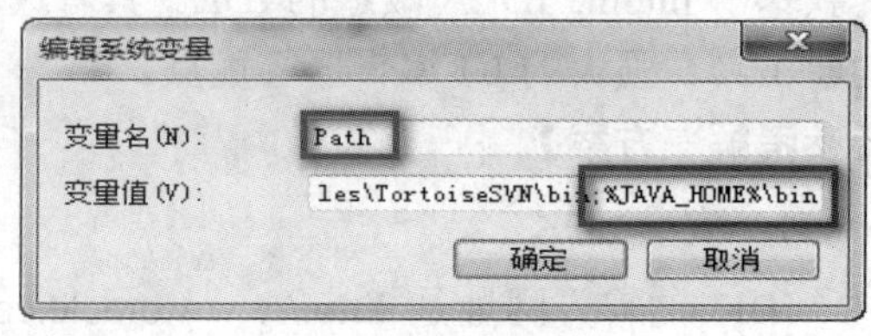

图 S1-4 编辑环境变量 Path

(5) 配置完成后，不要忘记单击【环境变量】对话框底部的【保存】按钮。

(6) 下面我们验证一下 JDK 是否配置成功：打开【命令提示符】对话框(在【开始】菜单下方搜索框中输入“cmd”并回车)，输入 java -version 命令，回车查看，如图 S1-5 所示，即表示配置成功。

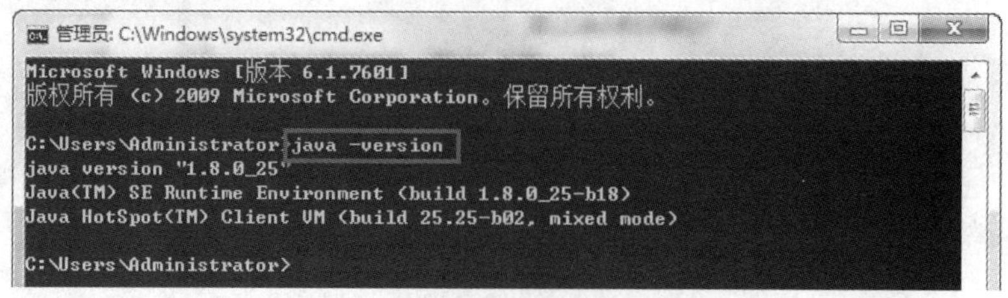

图 S1-5 测试已创建的环境变量

注　意　编辑 Path 变量时，需使用“;”将附加的路径“%JAVA_HOME%/bin”与前面的路径隔开。

3．下载 Android Studio 并解压

在 Android Studio 中文官方网站下载 Android Studio 2.3.2 压缩包，并将其解压到硬盘根目录下的 Android Studio 文件夹中。打开 Android Studio 文件夹，给 bin 文件夹中的 studio.exe 创建桌面快捷方式，双击该图标运行程序，启动界面如图 S1-6 所示。

图 S1-6　Android Studio 启动页面

启动界面过后，将显示欢迎界面【Welcome to Android Studio】，单击其中的【Start a new Android Studio project】命令，新建一个工程，如图 S1-7 所示。

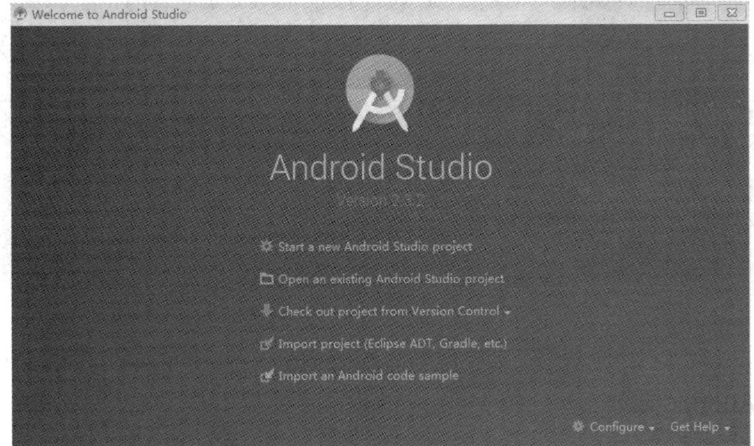

图 S1-7　Android Studio 欢迎界面

在接下来出现的界面中，设置新建项目的名称、公司域名与项目工作空间，如图 S1-8 所示。设置完毕，单击【Next】按钮，就进入了 Android Studio 的主界面。

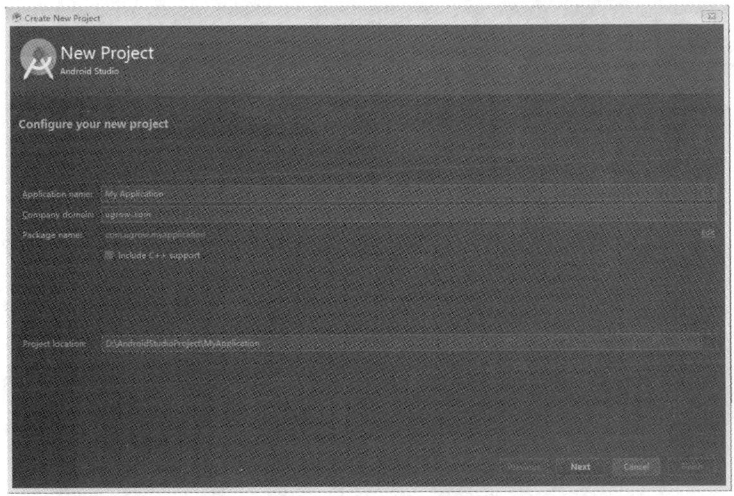

图 S1-8　设置新建项目的基本信息

4．配置 SDK 安装目录

Android Studio 启动后，需要配置 SDK 的安装目录：选择 Android Studio 导航栏中的【File】/【Settings】，在弹出界面【Setting】的左侧选择【Android SDK】项，在右侧选择 Android SDK 的安装目录，此时下面的列表中将显示所有已经安装和未安装的 SDK，选择完毕，单击【OK】按钮，如图 S1-9 所示。

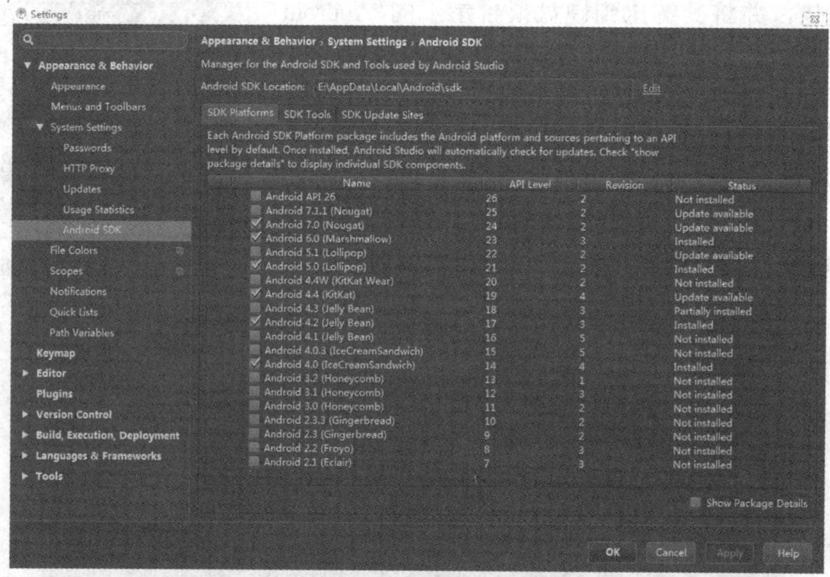

图 S1-9　配置 SDK 安装目录

至此，Android 开发环境已经安装完毕并配置成功，随时可以创建 Android 项目。

5．管理 Android SDK

如果想下载更多的 Android SDK 版本，只需单击 Android Studio 工具栏中的【 】按钮，即可打开【Android SDK Manager】对话框，可在其中选择所需版本进行下载，如图 S1-10 所示。

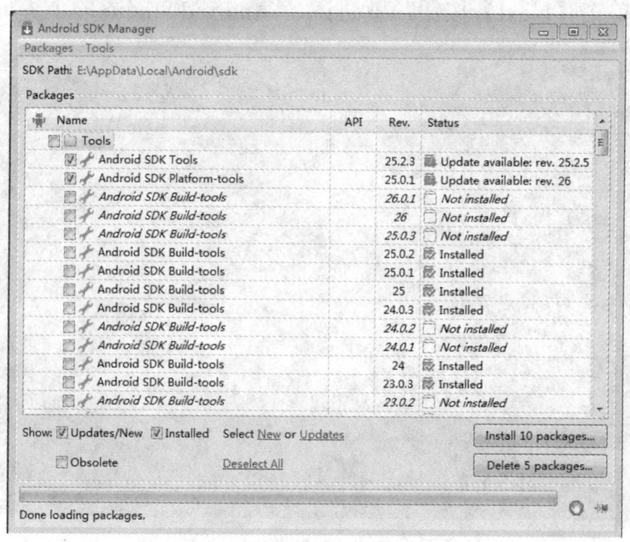

图 S1-10　管理 Android SDK

实践 1.2　创建 AVD(Android 模拟器)

配置运行 Android 应用程序所需的 AVD 虚拟设备。

【分析】

(1) 在没有真实 Android 手机的情况下，可以创建 AVD(Android 模拟器)用来模拟不同设备上可用的软件版本和硬件规范。这样可以针对各种硬件平台测试应用程序，而无需购买多款手机。

(2) 创建 AVD 时，可以配置支持的 SDK 版本、SD 卡容量、屏幕分辨率，以及一些额外的硬件设置，例如触摸屏、键盘、跟踪球、GPS、摄像头等。

(3) Android SDK 不包含任何预构建的虚拟设备，所以在模拟器内运行 Android 应用程序时需要至少创建一个 AVD 设备。

【参考解决方案】

创建一个新的 AVD 的具体步骤如下：

(1) 单击 Android Studio 工具栏中的【 】按钮，打开 AVD 管理界面，单击界面中的【Create Virtual Device...】按钮，创建一个 AVD，如图 S1-11 所示。

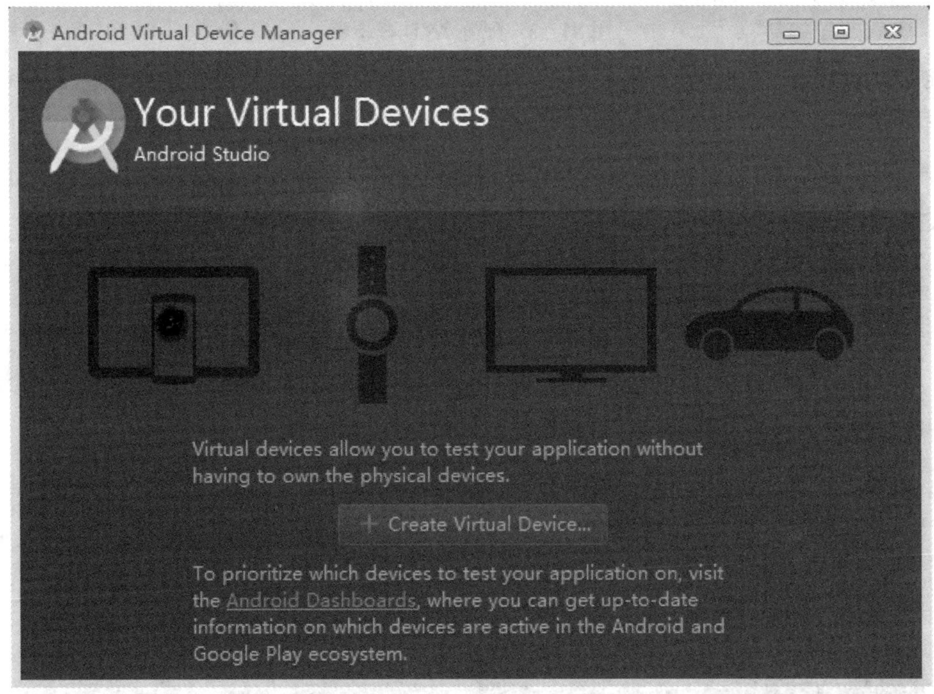

图 S1-11　AVD 模拟器管理界面

(2) 在出现的配置界面中，将新建 AVD 的名称【AVD Name】设置为“Nexus”，模拟手机设备设置为“Nexus6p”，系统版本号设置为“Android 7.0”，并设置其他一些参数，配置完毕，单击【Finish】按钮，返回图 S1-11 的管理界面，如图 S1-12 所示。

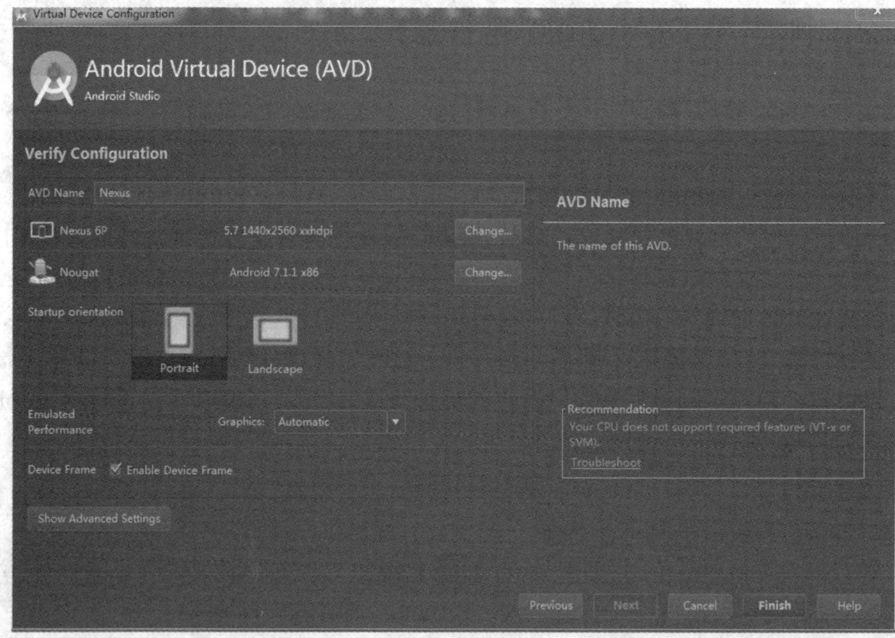

图 S1-12　配置 AVD 模拟器

此时，管理界面中会显示新创建的 AVD，单击【Actions】列表中的【▶】按钮，如图 S1-13 所示。

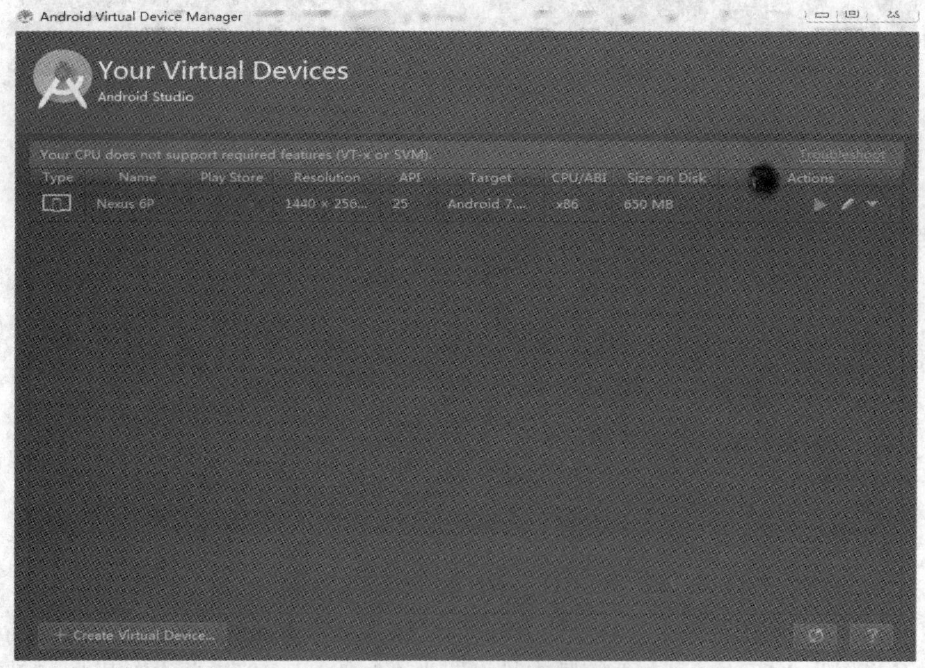

图 S1-13　运行 AVD

等待一段时间，就会看到模拟器的运行效果，如图 S1-14 所示。

图 S1-14　AVD 模拟器运行效果

实践 1.3　DDMS

使用 DDMS 对 Android 应用程序进行调试监控。

【分析】

(1) DDMS(Dalvik Debug Monitor Service，Dalvik 调试监视服务)是 Android SDK 的一个服务程序，支持模拟器和实际终端，可以提供诸如测试设备截屏、查看进程/线程及其堆信息、日志、模拟电话呼叫、模拟短信、虚拟地理坐标等功能。

(2) DDMS 是一个可视化的调试监控工具，是开发 Android 应用程序必不可少的集成工具。

【参考解决方案】

启动 Android Studio，单击 Android Studio 导航栏中的【Tools】/【Android】/【Android Device Monitor】，即可打开 DDMS 界面。

DDMS 界面中常用的窗口如下：

(1) Devices 窗口显示当前可用的设备列表，顶部的相机按钮可实现截图功能，如图 S1-15 所示。

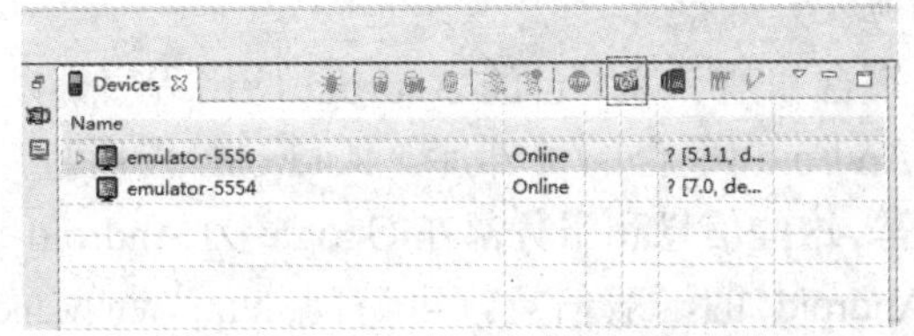

图 S1-15　设备列表窗口 Devices

(2) File Explorer 文件浏览器窗口显示当前选中设备的文件，右上角的四个按钮分别标识导出文件、导入文件、删除文件、添加文件夹，如图 S1-16 所示。

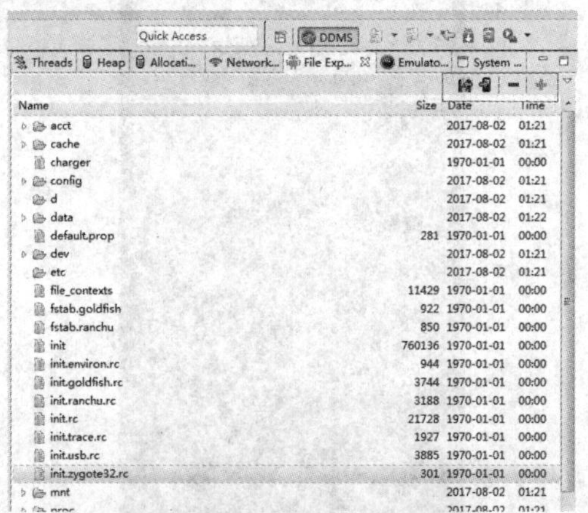

图 S1-16　文件浏览器窗口 File Explorer

(3) Emulator Control 窗口用于使用模拟器进行虚拟操作，如打电话、发短信、经纬度等，如图 S1-17 所示。

(4) LogCat 窗口是日志查看窗口，可用于查看调试信息和错误信息等，如图 S1-18 所示。

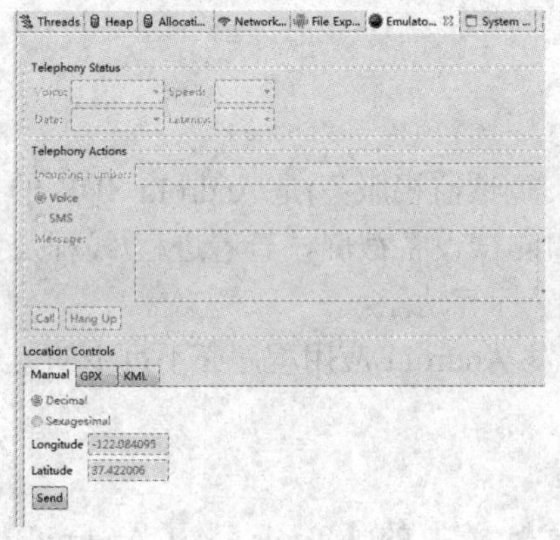

图 S1-17　虚拟操作窗口 Emulator Control

图 S1-18　日志查看窗口 LogCat

知识拓展

1. Android 网络资源

Google 为 Android 平台和基于该平台的 Android 应用程序开发提供了大量的信息和有用的服务，例如扩展 Android 平台的外部库、Android 应用程序、托管服务和 API、Android 开发人员竞赛等。这些信息和服务都在 Google 为 Android 设置的官方网站中。

如果要查找关于 Android 的一般信息，可访问 http://www.android.com；如果对用于 Android 设备的应用程序感兴趣，可访问 Android 开发人员网站 http://developer.android.com。

除了 Google 提供的 Android 官方网站之外，还有许多 Android 爱好者和相关组织建立了一些技术网站和论坛，方便 Android 开发者通过网络进行学习和技术交流，其中较有代表性的网络社区如下：

(1) eoe Android 社区：http://www.eoeandroid.com。

(2) APKBUS Android 开发者社区：http://www.apkbus.com。

(3) Android 手机资讯网：http://android.hk.cn/。

(4) 91 手机娱乐门户：http://android.sj.91.com/。

(5) 安奇网：http://www.apkcn.com/。

2. Android 与 Java ME 的区别

Android 和 Java ME 都是基于 Java 语言并针对移动设备开发的，它们之间的区别在于：

(1) Android 是一个完整的移动设备操作系统平台，由 Linux 操作系统、中间件、C 类库和核心应用程序组成。

(2) Java ME 只是 Java 的微型版本，是针对移动设备开发应用程序的开发包，必须有底层操作系统的支持，如 Symbian、Win CE 等。

Android 的优势在于开发性和开源，是一个优秀的移动设备操作系统；而 Java ME 是一个移动设备软件开发包，跨平台是其主要特点。

拓展练习

编写一个简单的 Android 应用程序，在 DDMS 日志中输出字符串“您好，欢迎来到 Android 世界！”。

实践 2　活动(Activity)

实践指导

实践 2.1　点餐系统功能结构分析

本书实践篇的后续章节将实现一个饭店使用的点餐系统，以此作为 Android 开发的实践贯穿案例。

【分析】

传统的餐饮行业中，顾客点餐是通过服务员人工完成的。顾客进入饭店后，服务员根据顾客人数选择合适的桌位，顾客就座后开始点餐，然后服务员将点好的菜单送交前台或厨师。在较大的饭店中，点餐的过程将耗费较长时间，而使用无线点餐系统可以解决此问题。

无线点餐系统分为以下两部分：

(1) Android 客户端程序：供饭店服务员使用，用于为顾客点餐。

(2) Web 服务器端程序：供 Android 客户端通过网络访问，并包含用于存储数据的数据库。

注 意

关于服务器端程序的内容将在实践 9 中讲解。

【参考解决方案】

(1) Android 终端通过无线路由器访问 Web 服务器，整个点餐系统的物理架构如图 S2-1 所示。

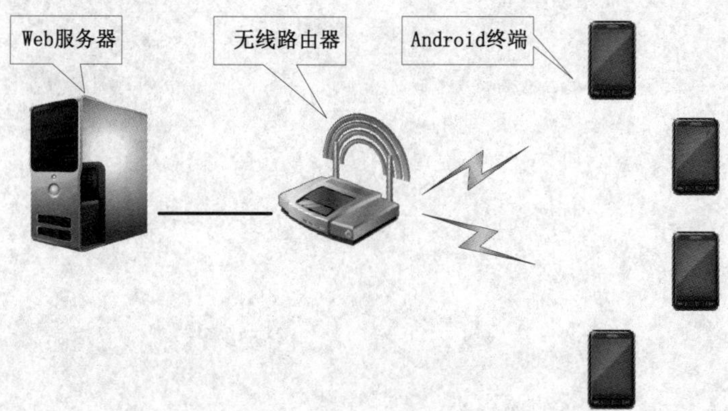

图 S2-1　点餐系统物理架构

(2) 点餐系统的功能结构如图 S2-2 所示。

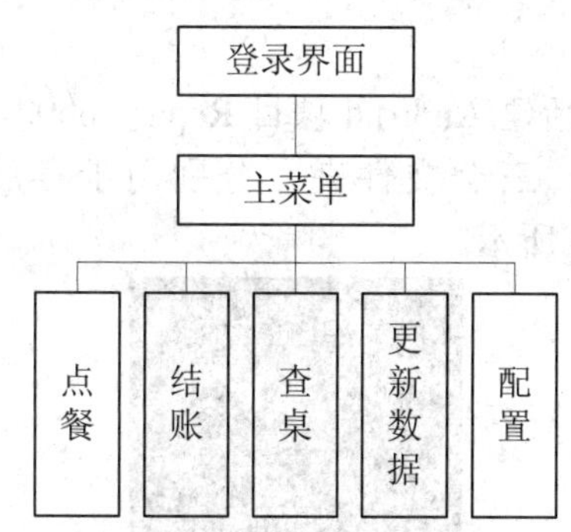

图 S2-2　点餐系统功能结构

由图 S2-2 可知，点餐系统需具备以下主要功能组件：

- ✧ 登录界面：用户输入正确的编号和密码后进入主菜单界面。
- ✧ 主菜单：主菜单页面显示点餐、结账、查桌、更新数据、配置五项功能图标。
- ✧ 点餐：用户可以录入桌号、顾客人数、备注信息，还可以选择需要的餐品及数量，确定所选餐品后，可以下单。
- ✧ 结账：用户录入订单编号，系统显示对应订单的详细信息，用户确认后，可以单击结账完成订单。
- ✧ 查桌：用户录入需要的座位数，系统显示符合条件的桌位。
- ✧ 更新数据：用户单击更新数据后，在系统数据库中查找餐品等数据，并下载到 Android 客户端作为本地缓存，以提高运行效率。
- ✧ 配置：用户可以对系统的一些选项进行配置和修改。

点餐系统 Android 客户端程序需要的 Activity 类及其功能如表 S2-1 所示。

表 S2-1　Android 客户端的 Activity 类及其功能

Activity 类	功　能
BasicActivity	所有 Activity 的父类，提供一些通用的功能
LoginActivity	登录界面
MainActivity	主菜单界面
OrderActivity	点餐界面
PayActivity	结账界面
TableActivity	查桌界面
ConfigActivity	配置界面

实践 2.2　创建点餐系统项目

在 Android Studio 中创建点餐系统的 Android 项目，建立包结构，并准备图标文件。

【分析】

(1) 项目按照模块化组织，使用不同的 Java 包存放实体类、业务类和界面类。

(2) 为了界面的美观，需要准备一些小图标，在按钮、对话框等处可以使用这些图标。

【参考解决方案】

(1) 在 Android Studio 中新建 Android 项目 Repast_ph02，并在 com.ugrow.repast_ph02 包下新建 entity、service、ui 三个文件夹，分别用于存放实体类、业务类和界面类(Activity)，项目结构如图 S2-3 所示。

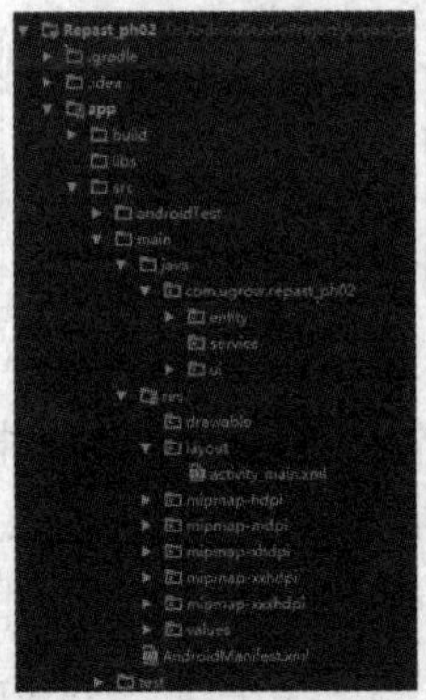

图 S2-3　点餐系统项目结构

(2) 在 res/mipmap-xhdpi 文件夹下，添加如表 S2-2 所示的图标资源。

表 S2-2　客户端需要的图标资源

icon.png		icon1.png		icon2.png	
icon3.png		icon4.png		icon5.png	
key.png		back.png		coffee.png	
save.png		search.png		money.png	
info.png		warning.png		not.png	

实践 2.3　创建点餐系统实体类

编写点餐系统需要的实体类。

【分析】

(1) 点餐系统需要用户、餐品、餐品类型、餐桌、订单、订单明细六个实体类。

(2) 餐品和餐品类型具有多对一的关系，系统需要根据餐品类型查找对应的餐品，因此餐品类中需要一个 typeId 属性来表示对应的餐品类型 ID。

(3) 订单和订单明细具有一对多的关系，系统需要根据订单查找对应的明细，因此订单类中需要一个集合属性来保存对应的订单明细。

【参考解决方案】

在 com.ugrow.repast_ph02.entity 包下创建以下实体类，并编写相应代码：

(1) User.java。

```
public class User {
        private int id; // 主键
        private String code; // 编号
        private String password; // 密码
        private String name; // 姓名
        ...Get()、Set()方法略
}
```

(2) Food.java。

```
public class Food {
        private int id; // 主键
        private String code; // 编码
        private int typeId; // 类型
        private String name; // 名称
        private int price; // 价格
        private String description; // 说明
        ...Get()、Set()方法略
}
```

(3) FoodType.java。

```
public class FoodType {
        private int id; // 主键
        private String name; // 名称
        ...Get()、Set()方法略
}
```

(4) Table.java。

```
public class Table {
        private int id; // 主键
        private String code; // 编号
        private int seats; // 座位数
        private int customers; // 当前就餐人数
        private String description; // 说明
        ...Get()、Set()方法略
}
```

(5) OrderDetail.java。

```
public class OrderDetail {
        private int id; // 主键
        private int orderId; // 订单 ID
        private int foodId; // 食品 ID
        private int num = 1; // 订购数量
        private String description; // 说明
```

```
        ...Get()、Set()方法略
}
```

(6) Order.java。

```
public class Order {
        private int id; // 主键
        private String code; // 编号
        private int tableId; // 餐桌 ID
        private int waiterId; // 服务员 ID
        private Date orderTime; // 下单时间
        private int customers; // 顾客数量
        private int status; // -1 取消，0 未结算，1 已结算
        private String description; // 说明
        private List<OrderDetail> orderDetails; // 订单明细
        ...Get()、Set()方法略
}
```

知识拓展

1．使用颜色和尺寸资源

在 Android 应用程序中，可能会使用各种颜色和尺寸，例如控件的背景色、文字的颜色、文字的大小等。为了使界面风格统一，通常会将重复使用的颜色和尺寸定义在特定的资源文件中，在代码或其他的资源文件中就可以通过资源的名称引用这些颜色和尺寸，从而避免了代码重复，也便于统一修改。

新建 Android 项目 Ph02ex，在 res/values 文件夹下新建 colors.xml 文件，编辑内容如下：

```
<?xml version="1.0" encoding="utf-8"?>
<resources>
        <color name="someBgColor">#004477</color>
        <color name="someTextColor">#AAFFDD</color>
        <color name="someTextBgColor">#000000</color>
</resources>
```

上述文件中定义了名为“someBgColor”和“someTextColor”的两个颜色资源，使用#RGB 方式指定了其颜色值，在其他资源文件中可以引用这两个颜色资源。例如，可以将主界面布局文件 res/layout/activity_main.xml 的内容修改如下：

```
<?xml version="1.0" encoding="utf-8"?>
<LinearLayout xmlns:android="http://schemas.android.com/ apk/res/android"
        android:orientation="vertical"
        android:layout_width="match_parent"
        android:layout_height="match_parent"
        android:background="@color/someBgColor">
```

```
        <TextView
            android:id="@+id/myText"
            android:layout_width="match_parent"
            android:layout_height="wrap_content"
            android:text="Hello world A!"
            android:textColor="@color/someTextColor"
            android:background="@color/someTextBgColor"/>
</LinearLayout>
```

上述布局文件通过引用“@color/someBgColor”指定了主界面背景色，通过引用“@color/someTextColor”指定了 TextView 控件的文字颜色，通过引用“@color/someTextBgColor”指定了 TextView 控件背景色。

运行项目，效果如图 S2-4 所示。

图 S2-4　设置界面控件颜色

除了在其他资源文件中引用颜色外，也可以在代码中使用。例如，可将主界面 Activity 的代码进行以下修改：

```
public class MainActivity extends AppCompatActivity {
    @Override
    public void onCreate(Bundle savedInstanceState) {
        super.onCreate(savedInstanceState);
        setContentView(R.layout.activity_main);
        // 设置窗口背景色
        getWindow().setBackgroundDrawableResource(R.color.someBgColor);
        // 获取 TextView 控件
        TextView tv = (TextView) findViewById(R.id.myText);
        // 设置 TextView 文字颜色和背景颜色
        tv.setTextColor(getResources().getColor(R.color.someTextColor));
        tv.setBackgroundColor(getResources()
                .getColor(R.color.someTextBgColor));
```

```
    }
}
```

上述代码通过 Window 对象的 setBackgroundDrawableResource()方法设置了主界面背景色，然后通过 ID 获取 TextView 控件，并调用 setTextColor()方法设置该控件的颜色。运行项目，其效果与图 S2-4 的相同。

需要特别指出，不要像以下语句这样调用 setTextColor()方法：

```
tv.setTextColor(R.color.someTextColor);
```

因为 setTextColor()方法是接收一个整数来表示颜色，而 R.color.someTextColor 是系统根据颜色名称自动生成的一个整数，其代表颜色的名称而不是颜色的值。正确方法是使用 getResources()方法获取项目资源对象 Resources，并在 Resources 中通过调用 getColor()方法根据资源名称来获取颜色资源，即使用下列方式：

```
tv.setBackgroundColor(getResources().getColor(R.color.someTextBgColor));
```

也可以将尺寸定义在资源文件中，例如在 res/values 目录下新建 dimens.xml 文件，编辑内容如下：

```
<?xml version="1.0" encoding="utf-8"?>
<resources>
    <dimen name="someHeight">200dp</dimen>
    <dimen name="someFontSize">60sp</dimen>
</resources>
```

上述文件定义了 someHeight 和 someFontSize 两个尺寸，在其他资源文件和代码中都可以引用这两个尺寸。例如，可以将 activity_main.xml 内容修改如下：

```
<?xml version="1.0" encoding="utf-8"?>
<LinearLayout xmlns:android="http://schemas.android.com/apk/res/android"
    android:orientation="vertical"
    android:layout_width="match_parent"
    android:layout_height="matc_parent"
    android:background="@color/someBgColor">
    <TextView
        android:id="@+id/myText"
        android:layout_width="matc_parent"
        android:layout_height="wrap_content"
        android:height="@dimen/someHeight"
        android:text="Hello world A!"
        android:textSize="@dimen/someFontSize"
        android:textColor="@color/someTextColor"
        android:background="@color/someTextBgColor"/>
</LinearLayout>
```

上述文件通过引用 dimens.xml 文件中定义的尺寸资源，指定了 TextView 控件的高度和文字大小。

运行项目，效果如图 S2-5 所示。

图 S2-5　设置界面控件大小

2. 使用 XML 资源

Android 项目中可以使用 XML 文件。Android 内置了开源的 XML 解析器 Pull，其使用类似于 SAX 的方式解析 XML。

XML 文件需要放置在 res/xml 目录下。例如，某个 res/xml 目录中存放有下列 XML 文件：

```
<?xml version="1.0" encoding="utf-8"?>
<corporations>
    <corporation no="1">
        <name>沃尔玛</name>
        <country>美国</country>
    </corporation>
    <corporation no="2">
        <name>壳牌</name>
        <country>荷兰</country>
    </corporation>
    <corporation no="3">
        <name>埃克森美孚</name>
        <country>美国</country>
    </corporation>
    <corporation no="4">
        <name>英国石油</name>
        <country>英国</country>
    </corporation>
    <corporation no="5">
```

```
            <name>中石油</name>
            <country>中国</country>
        </corporation>
</corporations>
```

修改主界面 Activity，使用 Android 提供的 API 解析上述 XML 文件，代码如下：

```
public class MainActivity extends AppCompatActivity {
        @Override
        public void onCreate(Bundle savedInstanceState) {
                super.onCreate(savedInstanceState);
                setContentView(R.layout.activity_main);
                TextView tv = (TextView) findViewById(R.id.myText);
                try {
                        tv.setText(readXml());
                } catch (Exception e) {
                        e.printStackTrace();
                }
        }
        String readXml() throws XmlPullParserException, IOException {
                XmlResourceParser xrp = getResources().getXml(R.xml.corporation);
                StringBuilder sb = new StringBuilder();
                int et = xrp.getEventType();
                while (et != XmlPullParser.END_DOCUMENT) {
                        String name = xrp.getName();
                        if (et == XmlPullParser.START_TAG)
                                if ("corporation".equals(name))
                                        sb.append(xrp.getAttributeValue(0)).append(" ");
                                else if ("name".equals(name)) {
                                        xrp.next();
                                        sb.append(xrp.getText()).append(" ");
                                } else if ("country".equals(name)) {
                                        xrp.next();
                                        sb.append(xrp.getText()).append("\n");
                                }
                        xrp.next();
                        et = xrp.getEventType();
                }
                return sb.toString();
        }
}
```

上述代码在 readXml()方法中使用 getResources().getXml()方法获取 XML 文件，然后调用 XmlPullParser 的相关方法解析 XML 文件内容。

运行项目，效果如图 S2-6 所示。

图 S2-6　显示解析的 XML 文件

拓展练习

练习 2.1

新建 Android 项目，定义一个颜色资源文件，并分别在其他资源文件和代码中引用该文件中定义的颜色。

练习 2.2

定义一个 XML 资源文件，在代码中读取该 XML 文件并将其内容显示在窗口中。

实践3 用 户 界 面

实践指导

实践 3.1 创建登录界面

完成点餐系统的用户登录界面搭建。

【分析】

(1) 用户登录时需要录入编号和密码。可编写用户业务类 UserService，并提供登录业务方法 login()，根据用户编号和密码查找用户，查找成功后返回 User 的实例。

(2) 用户登录成功后需要将用户信息存储，以便后续界面使用。编写 Android 提供的 Application 类的子类可完成信息存储，此类需要在 AndroidManifest.xml 文件中配置。

(3) 许多 Activity 中存在大量重复代码，如初始化、设置标题栏、显示信息提示对话框等。编写 Activity 的子类 BasicActivity，可实现这些通用的功能，且所有的 Activity 都可以继承 BasicActivity 以获取这些功能。

(4) 用户登录界面需显示用户编号输入框、密码输入框、登录按钮和取消按钮；用户录入信息错误时需弹出对话框来提示用户。

注 意

只有连接服务器端程序并查询数据库，才能确定用户是否登录成功，本实践实现的登录功能只是使用假数据模拟了登录过程，在实践 7 中将实现完整的登录功能。

【参考解决方案】

(1) 编写用户业务类 com.ugrow.repast_ph03.service.UserService，完成登录的业务方法 login()，代码如下：

```
public class UserService {
	/**
	 * 登录
	 *
	 * @param code 用户编号
	 * @param password 密码
	 * @return 用户对象
	 */
	public User login(String code, String password) {
```

```
            if ("test".equals(code) && "test".equals(password)) {
                    User u = new User();
                    u.setId(1);
                    u.setCode(code);
                    u.setPassword(password);
                    u.setName("张王李赵");
                    return u;
            }
            return null;
        }
}
```

(2) android.app.Application 类代表整个应用程序，程序启动时，系统会自动构造 Application 的实例，可以在 Activity、Service 中调用 getApplication()方法获取此实例。由于整个应用程序共用一个 Application 实例，因此可以在 Application 中保存全局变量，这需要两个步骤：

✧ 编写 Application 子类，并添加全局属性。

✧ 在 AndroidManifest.xml 文件中使用新编写的子类作为 Application。

编写 Application 的子类 com.dh.repast_ph03.App，在其中添加 User 类型的属性，用于存储登录成功的用户实例，代码如下：

```
public class App extends Application {
        /**
         * 保存当前登录者
         */
        public User user;
}
```

(3) 修改 AndroidManifest.xml 文件，配置使用 App 类，代码如下：

```
<?xml version="1.0" encoding="utf-8"?>
<manifest xmlns:android="http://schemas.android.com/apk/res/android"
    package="com.ugrow.repast_ph03">

    <application
        android:name=".App"
        android:allowBackup="true"
        android:logo="@mipmap/coffee"
        android:icon="@mipmap/icon"
        android:label="@string/app_name"
        android:roundIcon="@mipmap/ic_launcher_round"
        android:supportsRtl="true"
        android:theme="@style/AppTheme">
```

```
    </application>

</manifest>
```

上述代码在<application>元素中将 android:name 属性的值指定为“.App”，从而实现在 Activity 中调用 getApplication()方法，并返回 App 的实例；android:logo 属性指定了全局标题栏图标。

(4) 编写 com.ugrow.repast_ph03.ui.BasicActivity，作为点餐系统中 Activity 的父类，实现一些通用功能，代码如下：

```
public abstract class BasicActivity extends Activity {
    AlertDialog.Builder dialogBuilder;
    @Override
    protected void onCreate(Bundle savedInstanceState) {
        super.onCreate(savedInstanceState);
        String title = getString(R.string.app_name); // 获取应用程序标题
        App app = (App) getApplication(); // 获取 App 实例
        User user = app.user; // 获取当前登录用户实例
        if (user != null)
            title += " " + user.getCode();
        title += " " + getName();
        setTitle(title); // 设置标题栏文字
        dialogBuilder = new AlertDialog.Builder(this);
    }
    /**
     * 显示一个简单的消息对话框
     *
     * @param message 消息
     * @param iconId 图标 ID
     * @param onClickListener 单击确定按钮的事件
     */
    protected void showMessageDialog(String message, int iconId,
            DialogInterface.OnClickListener onClickListener) {
        dialogBuilder.setIcon(iconId); // 设置图标 ID
        dialogBuilder.setCancelable(false);// 单击其他区域，对话框不消失
        dialogBuilder.setTitle(message); // 设置消息
        //确定按钮事件
        dialogBuilder.setPositiveButton("确定", onClickListener);
        dialogBuilder.create().show(); // 显示对话框
    }
```

```
    /**
     * @return 界面名称
     */
    protected abstract String getName();
}
```

上述 BasicActivity 代码实现了以下功能：

✧ 设置标题栏文字，如果登录成功，标题显示为登陆者的 CODE。

✧ 提供了一个简单的对话框显示方法供子类调用。

✧ 将抽象方法 getName()用于指定界面名称。

(5) 在 res/layout 目录下，新建用户登录界面布局文件 activity_login.xml，代码如下：

```
<?xml version="1.0" encoding="utf-8"?>
<LinearLayout xmlns:android="http://schemas.android.com/apk/res/android"
    android:layout_width="match_parent"
    android:layout_height="match_parent"
    android:orientation="vertical"
    android:background="#ededed"
    android:padding="10dp" >
    <EditText
        android:id="@+id/codeEdt"
        android:layout_width="match_parent"
        android:layout_height="wrap_content"
        android:hint="请输入编号"
        android:singleLine="true"/>
    <EditText
        android:id="@+id/passwordEdt"
        android:layout_width="match_parent"
        android:layout_height="wrap_content"
        android:hint="请输入密码"
        android:inputType="textPassword"
        android:singleLine="true"/>
    <LinearLayout
        android:layout_width="match_parent"
        android:layout_height="wrap_content"
        android:paddingLeft="10dp"
        android:paddingRight="10dp" >
        <Button
            android:id="@+id/loginBtn"
            android:layout_width="wrap_content"
            android:layout_height="wrap_content"
            android:layout_weight="1"
```

```
        android:drawableLeft="@mipmap/key"
        android:text="登录" />
    <Button
        android:id="@+id/cancelBtn"
        android:layout_width="wrap_content"
        android:layout_height="wrap_content"
        android:layout_weight="1"
        android:drawableLeft="@mipmap/back"
        android:text="取消" />
  </LinearLayout>
</LinearLayout>
```

上述代码中包含两个 EditText 控件和两个 Button(按钮)控件，其中：

- ✧ 第一个 EditText 用于录入用户编号(属性 id 为“codeEdt”)。
- ✧ 第二个 EditText 用于录入密码(属性 id 为“passwordEdt”)。
- ✧ LinearLayout 里面包含登录按钮(属性 id 为“loginBtn”)和取消按钮(属性 id 为“cancelBtn”)，两个按钮各占一半的宽度，并通过 drawableLeft 属性指定了在左边显示的图标。

(6) 编写用户登录界面 com.ugrow.repast_ph03.ui.LoginActivity，代码如下：

```
public class LoginActivity extends BasicActivity{
    private EditText codeEdt; // 用户编号
    private EditText passwordEdt; // 密码
    private Button loginBtn; // 登录
    private Button cancelBtn; // 取消
    private UserService userService; // 用户业务对象
    @Override
    protected void onCreate(Bundle savedInstanceState) {
        super.onCreate(savedInstanceState);
        setContentView(R.layout.login);
        userService = new UserService();

        codeEdt = (EditText) findViewById(R.id.codeEdt);
        passwordEdt = (EditText) findViewById(R.id.passwordEdt);
        loginBtn = (Button) findViewById(R.id.loginBtn);
        cancelBtn = (Button) findViewById(R.id.cancelBtn);

        loginBtn.setOnClickListener(new OnClickListener() {
            @Override
            public void onClick(View v) {
                String code = codeEdt.getText().toString();
```

```
                String password = passwordEdt.getText().toString();
                if (code.length() == 0 || password.length() == 0) {
                    showMessageDialog("请输入编号和密码",
                            R.drawable.warning, null);
                    return;
                }
                User user = userService.login(code, password);
                if (user == null) {
                    showMessageDialog("编号或密码错误",
                            R.drawable.warning, null);
                    return;
                }
                showMessageDialog("登录成功", R.drawable.info, null);
                App app = (App) getApplication();
                app.user = user;
                // 转向主菜单界面
            }
        });
        cancelBtn.setOnClickListener(new OnClickListener() {
            @Override
            public void onClick(View v) {
                finish();
            }
        });
    }

    @Override
    protected String getName() {
        return "登录";
    }
}
```

上述代码在 onCreate()方法里初始化了各个控件，并注册了登录按钮和取消按钮的单击事件。在登录按钮的单击事件中，首先检查用户输入是否完整，然后调用 UserService 的 login()方法来验证登录是否成功，最后将登录成功的用户实例保存在 App 对象中。

(7) 在 AndroidManifest.xml 文件中对 LoginActivity 进行配置，代码如下：

```
<?xml version="1.0" encoding="utf-8"?>
<manifest xmlns:android="http://schemas.android.com/apk/res/android"
    package="com.ugrow.repast_ph03">
```

```
    <application
        android:name=".App"
        android:allowBackup="true"
        android:logo="@mipmap/coffee"
        android:icon="@mipmap/icon"
        android:label="@string/app_name"
        android:roundIcon="@mipmap/ic_launcher_round"
        android:supportsRtl="true"
        android:theme="@style/AppTheme">
        <activity android:name=".ui.MainActivity"></activity>
        <activity android:name=".ui.LoginActivity">
            <intent-filter>
                <action android:name="android.intent.action.MAIN" />

                <category android:name="android.intent.category.LAUNCHER" />
            </intent-filter>
        </activity>
    </application>

</manifest>
```

上述代码在<application>元素下添加了一个子元素<activity>，通过 name 属性将其对应的 Activity 类指定为 LoginActivity，并分别指定其子元素<intent-filter>的<action>元素和<category>元素的 name 属性值为 android.intent.action.MAIN 和 android.intent.category.LAUNCHER，以使 LoginActivity 成为应用程序的入口 Activity。

运行 Repast_ph03 项目，将在 Android 模拟器中显示登录界面，即 LoginActivity，如图 S3-1 所示。在其中输入用户编号和密码，根据输入信息的完整性和正确性，将显示不同的信息提示框。

在登录界面单击【取消】按钮，即可关闭应用项目。

图 S3-1　测试登录界面

实践 3.2　创建主菜单界面

完成点餐系统的主菜单界面搭建。

【分析】

(1) 主菜单界面中需要显示五项功能图标：点餐、结账、查桌、更新数据、配置。

(2) 可以使用 GridView 控件来显示各个功能图标，并在 Activity 中初始化各个图标及其单击事件。

(3) 需要定义 GridView 控件中每个单元格的布局，以便可以在每个功能图标下方显

示功能名称。

【参考解决方案】

(1) 编写主菜单布局文件 activity_main.xml，代码如下：

```
<?xml version="1.0" encoding="utf-8"?>
<LinearLayout xmlns:android="http://schemas.android.com/apk/res/android"
    android:layout_width="match_parent"
    android:layout_height="match_parent"
    android:background="#ededed"
    android:orientation="vertical"
    android:padding="10dp" >
    <GridView
        android:id="@+id/gdv"
        android:layout_width="match_parent"
        android:layout_height="match_parent"
        android:numColumns="3"
        android:verticalSpacing="30dp" >
    </GridView>
</LinearLayout>
```

上述代码定义了一个属性 id 为“gdv”的 GridView 控件，用于显示功能图标。GridView 通常需要结合 SimpleAdapter 来绑定数据。SimpleAdapter 构造方法则需要五个参数：

✧ Context context：上下文，通常为 Activity 当前实例。

✧ List<? extends Map<String, ?>> data：数据，每个 Map 对应视图的一项。

✧ int resource：视图项的布局资源。

✧ String[ ] from：data 中 Map 的 key。

✧ int[ ] to：视图组件的 ID，此视图用于显示 Map 的 value。

GridView 使用 SimpleAdapter 的步骤如下：

✧ 编写 GridView 中条目所需的布局文件，并确定需要绑定数据的控件的 ID。

✧ 准备数据，即填充某个 List，以便 SimpleAdapter 构造方法使用。

✧ 构造 SimpleAdapter 实例，并调用 GridView 的 setAdapter()方法来设置适配器。

(2) 编写 GridView 中每个功能图标的布局文件 main_menu_item.xml，代码如下：

```
<LinearLayout xmlns:android="http://schemas.android.com/apk/res/android"
    android:layout_width="fill_parent"
    android:layout_height="fill_parent"
    android:orientation="vertical" >
    <ImageView
        android:id="@+id/imageView"
        android:layout_width="wrap_content"
        android:layout_height="wrap_content"
```

```
        android:layout_gravity="center_horizontal" >
    </ImageView>
    <TextView
        android:id="@+id/imageTitle"
        android:layout_width="wrap_content"
        android:layout_height="wrap_content"
        android:layout_gravity="center_horizontal"
        android:textColor="#000"
        android:textSize="16sp"
        android:textStyle="bold" >
    </TextView>
</LinearLayout>
```

上述代码中，每个功能图标都包含图标和文字两部分，即一个属性 id 为“imageView”的 ImageView 和一个属性 id 为“imageTitle”的 TextView。

(3) 编写主界面 MainActivity，代码如下：

```
public class MainActivity extends BasicActivity {
        int[] icons = { R.mipmap.icon1, R. mipmap.icon2, R. mipmap.icon3,
                        R. mipmap.icon4, R. mipmap.icon5 };
        String[] iconTexts = { "点餐", "结账", "查桌", "更新数据", "设置" };
        GridView gdv;
        @Override
        public void onCreate(Bundle savedInstanceState) {
                super.onCreate(savedInstanceState);
                setContentView(R.layout.activity_main);
                gdv = (GridView) findViewById(R.id.gdv);
                List<Map<String, Object>> iconList =
                        new ArrayList<Map<String, Object>>();
                for (int i = 0, j = icons.length; i < j; i++) {
                        Map<String, Object> map = new HashMap<String, Object>();
                        map.put("imageView", icons[i]);
                        map.put("imageTitle", iconTexts[i]);
                        iconList.add(map);
                }
                gdv.setAdapter(new SimpleAdapter(this, iconList,
                                R.layout.main_menu_item, new String[] { "imageView",
                                        "imageTitle" }, new int[] { R.id.imageView,
                                        R.id.imageTitle }));
                gdv.setOnItemClickListener(
                        new AdapterView.OnItemClickListener() {
```

```
                @Override
                public void onItemClick(AdapterView<?> arg0, View arg1,int idx,long arg3) {
                    switch (idx) {
                    case 0:
                        showMessageDialog("点餐", R. mipmap.info, null);
                        break;
                    case 1:
                        showMessageDialog("结账", R. mipmap.info, null);
                        break;
                    case 2:
                        showMessageDialog("查桌", R. mipmap.info, null);
                        break;
                    case 3:
                        showMessageDialog("更新数据",
                            R. mipmap.info, null);
                        break;
                    case 4:
                        showMessageDialog("设置", R. mipmap.info, null);
                        break;
                    }
                }
            });
        }
        @Override
        protected String getName() {
            return "主菜单";
        }
}
```

上述代码中，首先定义了整型数组 icons 和字符串数组 iconTexts，用于存放每个图标项的图标 ID 和文字；然后在 onCreate()方法中初始化了一个 List<Map<String, Object>>集合，该集合用于为 SimpleAdapter 提供数据，并指定 GridView 使用 SimpleAdapter 获取数据；最后为 GridView 注册单击事件，根据编号确定单击的图标项，并显示信息提示对话框。

注 意

理论篇的第 4 章已讲解了 MainActivity，当单击图标时可显示对应的功能界面。

修改 AndroidManifest.xml 文件，将入口 Activity 指定为 MainActivity，修改方式同实践 3.1，此处不再赘述。

在 Android 模拟器中运行项目，将显示主菜单界面，单击其中的图标后，将显示对应

的信息提示对话框，如图 S3-2 所示。

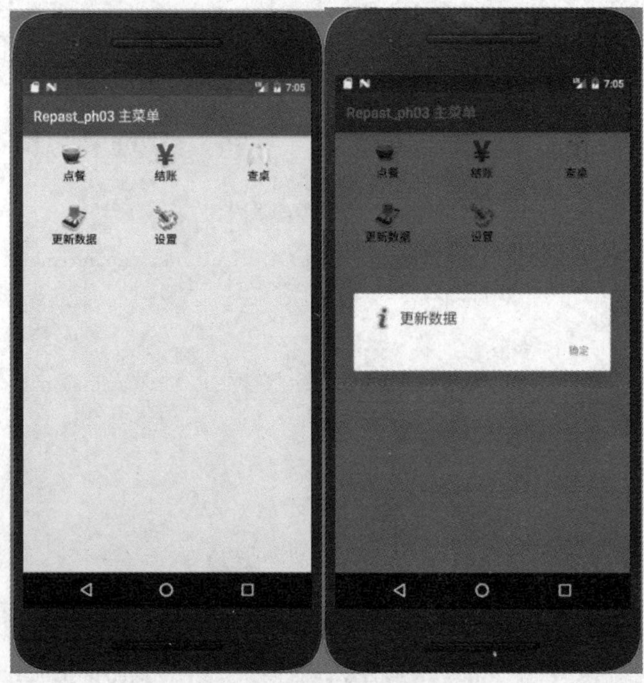

图 S3-2　完成的主菜单界面

知识拓展

1. 自动完成文本框控件

自动完成文本框控件 AutoCompleteTextView 提供了辅助输入的功能，可以对用户输入的文本进行有效的扩充提示，用户只需输入整个文本的一部分内容，系统就可即时显示出其余的内容供用户选择。

下面使用 AutoCompleteTextView 实现电话号码的自动录入功能。

首先编写布局文件 mail.xml，代码如下：

```
<?xml version="1.0" encoding="utf-8"?>
<LinearLayout xmlns:android="http://schemas.android.com/apk/res/android"
    android:orientation="vertical"
    android:layout_width="match_parent"
    android:background="#ededed"
    android:layout_height=" match__parent">
    <AutoCompleteTextView
        android:id="@+id/autoCompleteTextView"
        android:layout_width=" match__parent"
        android:layout_height="wrap_content"
        android:hint="请输入电话号码" />
```

```
</LinearLayout>
```

上述代码定义了一个 AutoCompleteTextView 控件。

然后编写 Activity，代码如下：

```
public class MainActivity extends AppCompatActivity {
        private static final String[] phonenumberStr =
                new String[] { "88888888","86668888", "7777777", "86666666",
                        "86145689", "86143429","86123889" };
        @Override
        public void onCreate(Bundle savedInstanceState) {
                super.onCreate(savedInstanceState);
                setContentView(R.layout.activity_main);
                ArrayAdapter<String> adapter = new ArrayAdapter<String>(this,
                        android.R.layout.simple_dropdown_item_1line,
                                phonenumberStr);
                AutoCompleteTextView autoCompleteTextView =
                (AutoCompleteTextView) findViewById(R.id.autoCompleteTextView);
                autoCompleteTextView.setAdapter(adapter);
        }
}
```

上述代码使用字符串数组存储了一组电话号码，并将 ArrayAdapter 作为 AutoCompleteTextView 的适配器。这样，当用户输入字符串时，AutoCompleteTextView 就会在 ArrayAdapter 中查找字符串，并显示匹配的字符串供用户选择。

运行项目，效果如图 S3-3 所示。

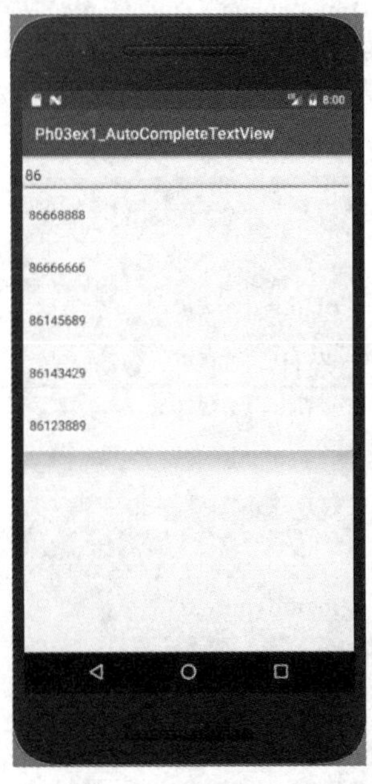

图 S3-3　自动完成文本框功能

2. 时间相关控件

Android 提供了许多与日期、时间有关的控件，其中主要的控件有：

(1) DatePicker：用于选择日期。

(2) TimePicker：用于选择时间。

(3) DigitalClock：数字时钟。

(4) AnalogClock：表状时钟。

这些控件的使用都比较简单，例如，以下布局文件就定义了这四种控件：

```
<?xml version="1.0" encoding="utf-8"?>
<RelativeLayout xmlns:android="http://schemas.android.com/apk/res/android"
    xmlns:tools="http://schemas.android.com/tools"
    android:layout_width="match_parent"
    android:layout_height="match_parent"
    android:background="#ededed"
    tools:context="com.ugrow.ph03ex2_timepicker.MainActivity">
    <ScrollView
        android:layout_width="match_parent"
        android:layout_height="wrap_content">
        <LinearLayout
            android:layout_width="match_parent"
            android:layout_height="wrap_content"
            android:orientation="vertical">
            <LinearLayout
                android:layout_width="match_parent"
                android:layout_height="wrap_content"
                android:layout_above="@+id/ll"
                android:orientation="vertical">

                <DatePicker
                    android:id="@+id/dp"
                    android:layout_width="match_parent"
                    android:layout_height="wrap_content"
                    android:calendarViewShown="false" />

                <TimePicker
                    android:id="@+id/tp"
                    android:layout_width="match_parent"
                    android:layout_height="wrap_content" />
            </LinearLayout>

            <LinearLayout
```

```
            android:id="@+id/ll"
            android:layout_width="match_parent"
            android:layout_height="wrap_content"
            android:layout_alignParentBottom="true"
            android:orientation="horizontal">

            <AnalogClock
                android:id="@+id/ac"
                android:layout_width="match_parent"
                android:layout_height="match_parent"
                android:layout_weight="1" />

            <LinearLayout
                android:layout_width="match_parent"
                android:layout_height="match_parent"
                android:layout_weight="1"
                android:orientation="vertical">

                <DigitalClock
                    android:id="@+id/dc"
                    android:layout_width="wrap_content"
                    android:layout_height="wrap_content" />

                <TextView
                    android:id="@+id/tv"
                    android:layout_width="wrap_content"
                    android:layout_height="wrap_content"
                    android:textSize="30sp" />
            </LinearLayout>
        </LinearLayout>
    </LinearLayout>
  </ScrollView>
</RelativeLayout>
```

编写 Activity，代码如下：

```
public class MainActivity extends AppCompatActivity {
    DatePicker dp;
    TimePicker tp;
    DigitalClock dc;
    AnalogClock ac;
    TextView tv;
```

```
    @Override
    public void onCreate(Bundle savedInstanceState) {
        super.onCreate(savedInstanceState);
        setContentView(R.layout.activity_main);
        dp = (DatePicker) findViewById(R.id.dp);
        tp = (TimePicker) findViewById(R.id.tp);
        dc = (DigitalClock) findViewById(R.id.dc);
        ac = (AnalogClock) findViewById(R.id.ac);
        tv = (TextView) findViewById(R.id.tv);
        Calendar c = Calendar.getInstance();
        dp.init(c.get(Calendar.YEAR), c.get(Calendar.MONTH),
            c.get(Calendar.DAY_OF_MONTH),new OnDateChangedListener(){
                    @Override
                    public void onDateChanged(DatePicker view,
                        int year,int monthOfYear, int dayOfMonth) {
                            display();
                    }
                });
        tp.setOnTimeChangedListener(new OnTimeChangedListener() {
            @Override
            public void onTimeChanged(TimePicker view, int hourOfDay, int minute){
                    display();
            }
        });
    }
    void display() {
        int y = dp.getYear();
        int m = dp.getMonth() + 1;
        int d = dp.getDayOfMonth();
        int h = tp.getCurrentHour();
        int mi = tp.getCurrentMinute();
        int s = Calendar.getInstance().get(Calendar.SECOND);
        tv.setText(y + "-" + m + "-" + d + " " + h + ":" + mi + ":" + s);
    }
}
```

上述代码为 DatePicker 控件设置了 OnDateChangedListener 监听器，为 TimePicker 控件设置了 OnTimeChangedListener 监听器，当用户修改 DatePicker 和 TimePicker 控件的值时，将调用 DatePicker 和 TimePicker 的相关方法获取日期和时间，并在 TextView 中显示修改后的结果。

运行项目，效果如图 S3-4 所示。上下划动 DatePicker 控件和 TimePicker 控件可以修改日期和时间，并在右下方的 TextView 控件中显示修改后的值，如图 S3-5 所示。

图 S3-4　时间显示控件

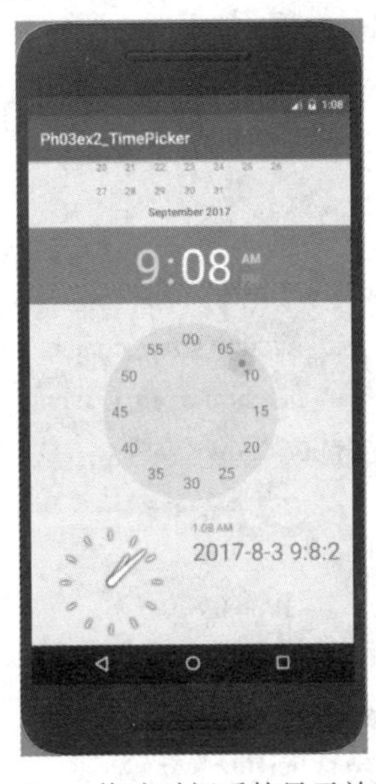

图 S3-5　修改时间后的显示效果

3. 进度条

Android 提供了进度条控件 ProgressBar，用于表示耗时操作的进度。

进度条是十分常用的控件，通常有两种表现形式：圆形进度条和条形进度条。如果操作的进度无法量化或无法预知，则通常使用圆形进度条；反之则使用条形进度条，因为条形进度条可以直观地显示当前操作的完成情况。

ProgressBar 的 getProgress()和 setProgress()方法用于获取和更改当前的进度，可以直接在其他线程中调用。如需更改 ProgressBar 的显示状态(例如使其隐藏)，不能在主线程中直接调用 ProgressBar 的相关方法，而应使用 Handler 和 Message 对象通过传递消息来实现。

下面通过一个示例来演示进度条的用法。首先编写布局文件 activity_main.xml，代码如下：

```
<?xml version="1.0" encoding="utf-8"?>
<LinearLayout xmlns:android="http://schemas.android.com/apk/res/android"
    android:layout_width="match_parent"
    android:layout_height="match_parent"
    android:background="#ededed"
    android:gravity="center_horizontal"
    android:orientation="vertical" >
    <Button
        android:id="@+id/startBtn"
        android:layout_width="match_parent"
        android:layout_height="wrap_content"
        android:text="开始" />
```

```
    <ProgressBar
        android:id="@+id/pb1"
        android:layout_width="wrap_content"
        android:layout_height="wrap_content"
        android:visibility="gone" />
    <ProgressBar
        android:id="@+id/pb2"
        style="?android:attr/progressBarStyleSmall"
        android:layout_width="wrap_content"
        android:layout_height="wrap_content"
        android:visibility="gone" />

    <ProgressBar
        android:id="@+id/pb3"
        style="?android:attr/progressBarStyleLarge"
        android:layout_width="wrap_content"
        android:layout_height="wrap_content"
        android:visibility="gone" />
    <ProgressBar
        android:id="@+id/pb4"
        style="?android:attr/progressBarStyleHorizontal"
        android:layout_width="match_parent"
        android:layout_height="wrap_content"
        android:visibility="gone" />
</LinearLayout>
```

上述代码定义了四个进度条：第一个没有指定 style 属性，采用默认样式，即中等大小的圆形进度条；第二个 style 值为“?android:attr/progressBarStyleSmall”，是小的圆形进度条；第三个 style 值为“?android:attr/progressBarStyleLarge”，是大的圆形进度条；第四个值为“?android:attr/progressBarStyleHorizontal”，是条形进度条。四个进度条的 visibility 值皆为“gone”，因此初始都是不显示的。

然后编写 Activity，代码如下：

```
public class MainActivity extends AppCompatActivity {
    ProgressBar pb1;
    ProgressBar pb2;
    ProgressBar pb3;
    ProgressBar pb4;
    Button startBtn;
    @Override
    public void onCreate(Bundle savedInstanceState) {
        super.onCreate(savedInstanceState);
```

```
setContentView(R.layout.activity_main);
pb1 = (ProgressBar) findViewById(R.id.pb1);
pb2 = (ProgressBar) findViewById(R.id.pb2);
pb3 = (ProgressBar) findViewById(R.id.pb3);
pb4 = (ProgressBar) findViewById(R.id.pb4);
startBtn = (Button) findViewById(R.id.startBtn);

final int max = pb4.getMax(); // 最大值

final Handler handler = new Handler() {
    @Override
    public void handleMessage(Message msg) {
        int p = msg.what;
        // 如果达到最大值，隐藏圆形进度条
        if (p == max) {
            pb1.setVisibility(View.GONE);
            pb2.setVisibility(View.GONE);
            pb3.setVisibility(View.GONE);
        }
        pb4.setProgress(p); // 修改条形进度条的进度
        // 修改条形进度条的第二进度
        pb4.setSecondaryProgress(p * 2);
    };
};

startBtn.setOnClickListener(new OnClickListener() {
    @Override
    public void onClick(View v) {
        // 显示四个进度条
        pb1.setVisibility(View.VISIBLE);
        pb2.setVisibility(View.VISIBLE);
        pb3.setVisibility(View.VISIBLE);
        pb4.setVisibility(View.VISIBLE);

        // 模拟耗时操作
        new Thread() {
            @Override
            public void run() {
                for (int i = 1; i <= max; i++) {
```

```
                    try {
                        Thread.sleep(50);
                        Message msg = new Message();
                        msg.what = i;
                        // 发送消息
                        handler.sendMessage(msg);
                    } catch (InterruptedException e) {
                        e.printStackTrace();
                    }
                }
            }
        }.start();
    }
});
}
}
```

上述代码进行了以下操作：① 定义了 Handler 对象，并重写 handleMessage()方法，以根据消息的内容改变四个进度条的状态；② 使用新线程在按钮的单击事件中模拟了一个耗时操作，操作过程中会定时向 Handler 对象发送消息；③ 当 Handler 对象接收到消息时，会调用其 handleMessage()方法，从而改变进度条的状态。

运行项目，初始界面不会显示进度条；单击【开始】按钮后，四个进度条出现并开始滚动；模拟操作执行完毕后，圆形进度条被隐藏，只显示已经完成的条形进度条，如图 S3-6 所示。

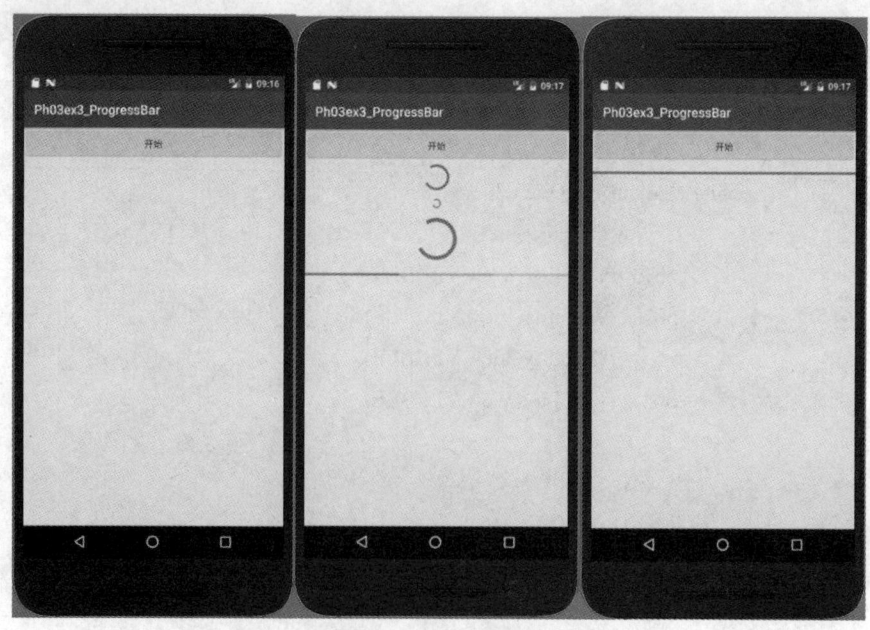

图 S3-6　测试进度条控件

4. 拖动条

拖动条控件 SeekBar 与普通进度条控件 ProgressBar 类似(SeekBar 继承于 ProgressBar)，但是 SeekBar 支持用户修改进度。

下面通过一个示例来演示 SeekBar 的用法。首先编写布局文件，代码如下：

```
<?xml version="1.0" encoding="utf-8"?>
<LinearLayout xmlns:android="http://schemas.android.com/apk/res/android"
    android:layout_width="match_parent"
     android:layout_height="match_parent"
     android:background="#ededed"
    android:gravity="center_horizontal"
    android:orientation="vertical" >
    <SeekBar
        android:id="@+id/sb"
        android:layout_width="match_parent"
        android:layout_height="wrap_content" />
    <TextView
        android:id="@+id/tv"
        android:layout_width="match_parent"
        android:layout_height="wrap_content"
        android:textSize="30sp" />
</LinearLayout>
```

上述代码定义了一个 SeekBar 和一个 TextView。

然后编写 Activity，代码如下：

```
public class MainActivity extends AppCompatActivity {
        SeekBar sb;
        TextView tv;
        boolean finished; // 是否已完成
        boolean dragging; // 是否正在拖动
        @Override
        public void onCreate(Bundle savedInstanceState) {
                super.onCreate(savedInstanceState);
                setContentView(R.layout.activity_main);
                sb = (SeekBar) findViewById(R.id.sb);
                tv = (TextView) findViewById(R.id.tv);
                final int max = sb.getMax(); // 最大值
                sb.setOnSeekBarChangeListener(new OnSeekBarChangeListener() {
                        @Override
                        public void onStopTrackingTouch(SeekBar seekBar) {
                                dragging = false; // 停止拖动
```

```
            }
            @Override
            public void onStartTrackingTouch(SeekBar seekBar) {
                dragging = true; // 开始拖动
            }
            @Override
            public void onProgressChanged(SeekBar seekBar,
                int progress,boolean fromUser) {
                    int p = sb.getProgress();
                    tv.setText("进度:" + p * 100 / max + "%");
                    if (p == max)
                        finished = true;
            }
        });
        new Thread() {
            @Override
            public void run() {
                while (!finished)
                    if (!dragging)
                        try {
                            Thread.sleep(50);
                            sb.incrementProgressBy(1);
                        } catch (InterruptedException e) {
                            e.printStackTrace();
                        }
            }
        }.start();
    }
}
```

上述代码进行了两项操作：① 开启了一个新线程，用于模拟一个耗时操作，在该线程中循环判断任务是否执行完成，当任务没有完成时，如果不再拖动进度条，则会从当前进度位置开始继续自动修改 SeekBar 的进度。② 为 SeekBar 注册了状态改变监听器 OnSeekBarChangeListener，当用户开始拖动、停止拖动、进度发生改变时会调用 OnSeekBarChangeListener 的 onStartTrackingTouch()、onStopTrackingTouch()以及 onProgressChanged()方法；在 onStartTrackingTouch()和 onStopTrackingTouch()方法中修改拖动标志的值，在 onProgressChanged()方法中更新 TextView 的显示，并在到达终点时修改完成标志，使模拟耗时操作的线程结束。

运行项目，将显示状态持续变化的 SeekBar：可以向前、向后拖动 SeekBar，拖动过程中进度会停止前进；松开拖动后，进度会从当前位置继续前进，如图 S3-7 所示。

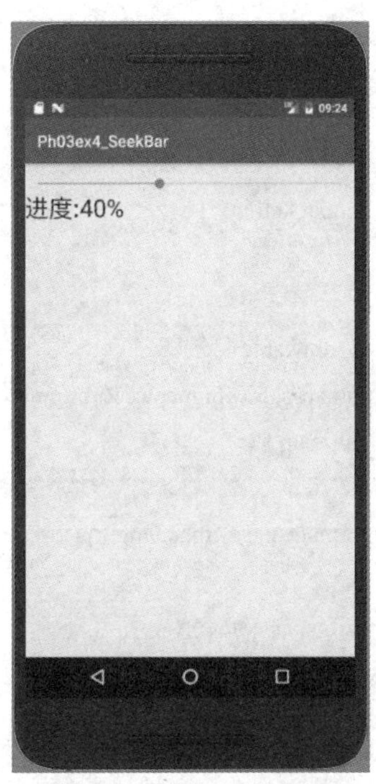

图 S3-7　测试拖动条控件

5. 图片切换效果

使用 Android 提供的 ImageSwitcher 和 Gallery 类可以方便地实现常见的图片切换效果。ImageSwitcher 通常用于显示并切换图片，而 Gallery 是存储图片的一个容器。

首先编写布局文件，代码如下：

```
<?xml version="1.0" encoding="utf-8"?>
<RelativeLayout xmlns:android="http://schemas.android.com/apk/res/android"
    android:layout_width="match_parent"
    android:layout_height="match_parent"
    android:background="#ededed">
    <ImageSwitcher
        android:id="@+id/imgSwt"
        android:layout_width="match_parent"
        android:layout_height="match_parent"
        android:paddingBottom="5dp"
        android:layout_above="@+id/glr" />
    <Gallery
        android:id="@+id/glr"
        android:layout_width="match_parent"
        android:layout_height="100dp"
        android:layout_alignParentBottom="true"
        android:gravity="center"
```

```
        android:spacing="5dp" />
</RelativeLayout>
```

上述代码定义了一个 ImageSwitcher 和一个 Gallery。

然后编写 Activity，代码如下：

```
public class MainActivity extends AppCompatActivity {
    ImageSwitcher imgSwt;
    Gallery glr;
    // 需要显示的图片资源，保存在 drawable 目录下
    int[] imgs = { R.mipmap.a, R. mipmap.b, R. mipmap.c, R. mipmap.d,
                R. mipmap.e, R. mipmap.f };
    // 缓存在 Gallery 中使用的 ImageView
    ImageView[] galleryViews = new ImageView[imgs.length];
    @Override
    public void onCreate(Bundle savedInstanceState) {
        super.onCreate(savedInstanceState);
        setContentView(R.layout.activity_main);

        imgSwt = (ImageSwitcher) findViewById(R.id.imgSwt);
        glr = (Gallery) findViewById(R.id.glr);
        // 设置 ImageSwitcher 显示图片的工厂
        imgSwt.setFactory(new ViewFactory() {
            @Override
            public View makeView() {
                ImageView iv = new ImageView(A.this);
                iv.setBackgroundColor(0xFF000000);
                iv.setScaleType(ImageView.ScaleType.FIT_CENTER);
                iv.setLayoutParams(new ImageSwitcher.LayoutParams(
                LayoutParams.FILL_PARENT,LayoutParams.FILL_PARENT));
                return iv;
            }
        });
        // 设置 ImageSwitcher 的进入动画
        imgSwt.setInAnimation(AnimationUtils.loadAnimation(this,android.R.anim.fade_in));
        // 设置 ImageSwitcher 的退出动画
        imgSwt.setOutAnimation(AnimationUtils.loadAnimation(this,android.R.anim.fade_out));
        // 初始化缓存 Gallery 使用的 ImageView
        for (int i = 0; i < galleryViews.length; i++) {
            ImageView iv = new ImageView(A.this);
            iv.setImageResource(imgs[i]);
```

```
            iv.setAdjustViewBounds(true);
            iv.setLayoutParams(new Gallery.LayoutParams(
                    LayoutParams.WRAP_CONTENT,LayoutParams.WRAP_CONTENT));
            galleryViews[i] = iv;
        }
        // 设置 Gallery 的适配器
        glr.setAdapter(new BaseAdapter() {
            @Override
            public int getCount() {
                return imgs.length;
            }
            @Override
            public Object getItem(int position) {
                return null;
            }
            @Override
            public long getItemId(int position) {
                return 0;
            }
            @Override
            public View getView(int position, View convertView,
                    ViewGroup parent){
                        return galleryViews[position];
            }
        });
        // 编辑 Gallery 被选中的事件
        glr.setOnItemSelectedListener(new OnItemSelectedListener() {
            @Override
            public void onItemSelected(AdapterView<?> arg0, View arg1,int arg2, long arg3) {
                imgSwt.setImageResource(imgs[arg2]);
            }
            @Override
            public void onNothingSelected(AdapterView<?> arg0) {
                // TODO
            }
        });
    }
}
```

上述代码对 ImageSwitcher 和 Gallery 对象执行了以下操作：

(1) ImageSwitcher：设置显示图片的工厂、进入动画和退出动画。

(2) Gallery：设置适配器，其中需要重写 getView()方法；设置选中图片的事件，并更新 ImageSwitcher 显示的图片。

运行项目，将显示第一张图片，单击界面底部的小图片，界面上部将显示对应的大图，如图 S3-8 所示。

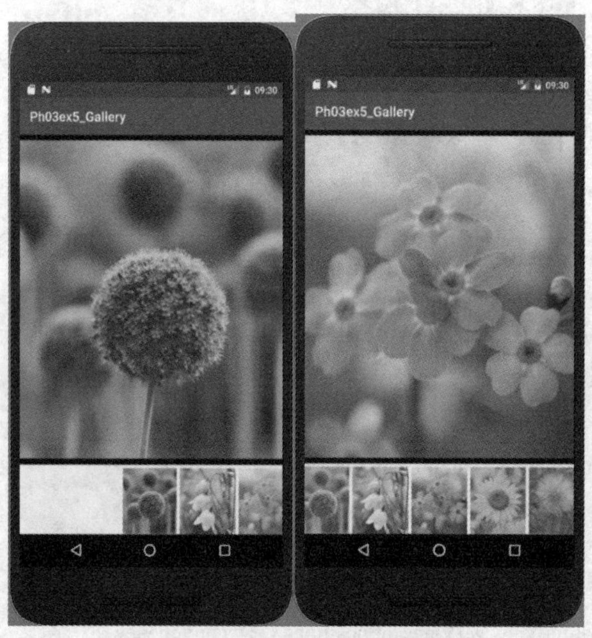

图 S3-8　测试图片切换效果

6．进度对话框

进度对话框实际上是一个包含进度条的对话框，其使用方式与进度条非常类似，也有两种表现形式：圆形进度条对话框与条形进度条对话框。

下面通过一个示例来演示进度对话框的用法。首先编写布局文件，代码如下：

```
<?xml version="1.0" encoding="utf-8"?>
<LinearLayout xmlns:android="http://schemas.android.com/apk/res/android"
    android:layout_width="match_parent"
     android:layout_height="match_parent"
     android:background="#ededed"
    android:orientation="vertical" >
    <Button
        android:id="@+id/startBtn1"
        android:layout_width="match_parent"
        android:layout_height="wrap_content"
        android:text="圆形进度条对话框" />
    <Button
```

```
        android:id="@+id/startBtn2"
        android:layout_width="match_parent"
        android:layout_height="wrap_content"
        android:text="条形进度条对话框" />
</LinearLayout>
```

上述代码声明了【圆形进度条对话框】和【条形进度条对话框】两个按钮。

然后编写 Activity，代码如下：

```
public class MainActivity extends AppCompatActivity {
    Button startBtn1;
    Button startBtn2;
    ProgressDialog pd1;
    ProgressDialog pd2;
    @Override
    public void onCreate(Bundle savedInstanceState) {
        super.onCreate(savedInstanceState);
        setContentView(R.layout.activity_main);
        startBtn1 = (Button) findViewById(R.id.startBtn1);
        startBtn2 = (Button) findViewById(R.id.startBtn2);
        pd1 = new ProgressDialog(this);
        pd2 = new ProgressDialog(this);
        pd1.setMessage("请稍候......");
        pd2.setMessage("请稍候......");
        //设置进度条为条形
        pd2.setProgressStyle(ProgressDialog.STYLE_HORIZONTAL);
        final int max = 100;// 进度最大值
        final Handler handler = new Handler() {
            @Override
            public void handleMessage(Message msg) {
                int p = msg.what;
                // 如果达到最大值，隐藏进度条对话框
                if (p == max) {
                    pd1.hide();
                    pd2.hide();
                }
                // 修改条形进度条对话框的进度
                pd2.setProgress(p);
            };
        };
        startBtn1.setOnClickListener(new OnClickListener() {
```

```
            @Override
            public void onClick(View v) {
                pd1.show();
                new Thread() {
                    @Override
                    public void run() {
                        for (int i = 1; i <= max; i++) {
                            try {Thread.sleep(50);
                                Message msg = new Message();
                                msg.what = i;
                                // 发送消息
                                handler.sendMessage(msg);
                            } catch (InterruptedException e) {
                                e.printStackTrace();
                            }
                        }
                    }
                }.start();
            }
        });
        startBtn2.setOnClickListener(new OnClickListener() {
            @Override
            public void onClick(View v) {
                pd2.show();
                new Thread() {
                    @Override
                    public void run() {
                        for (int i = 1; i <= max; i++) {
                            try {
                                Thread.sleep(50);
                                Message msg = new Message();
                                msg.what = i;
                                // 发送消息
                                handler.sendMessage(msg);
                            } catch (InterruptedException e) {
                                e.printStackTrace();
                            }
                        }
                    }
```

```
                    }.start();
                }
            });
    }
}
```

上述代码中定义了两个进度条对话框 pd1 和 pd2，其中 pd2 通过调用 setProgressStyle(ProgressDialog.STYLE_HORIZONTAL)方法来显示条形进度条，pd1 则采用默认的圆形进度条。

与普通进度条类似，上述代码在按钮单击事件中模拟了一个耗时操作，并在 Handler 对象的 handleMessage()方法中，根据消息的内容改变 pd1 和 pd2 的状态。

运行项目，分别单击两个按钮，将显示两种风格的进度对话框，如图 S3-9 所示。

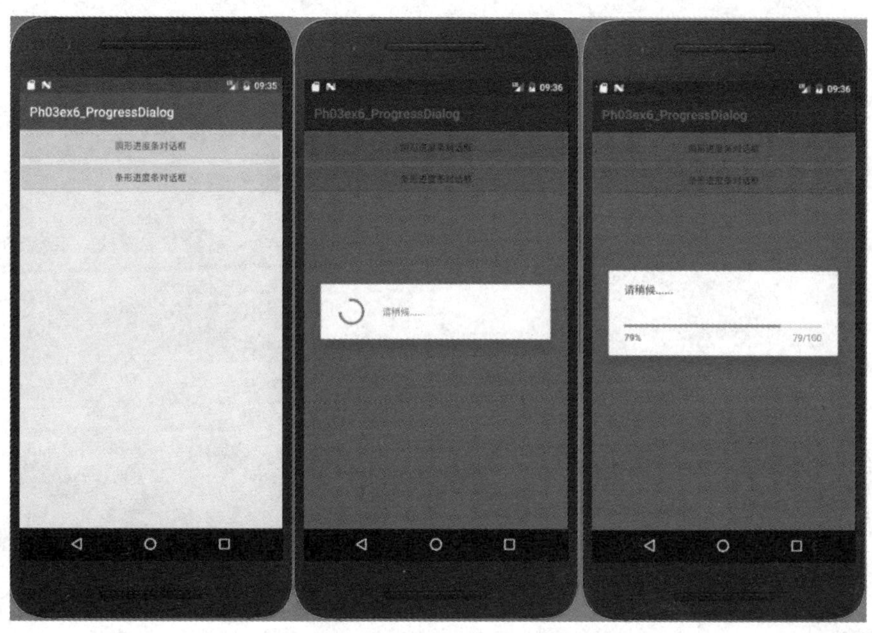

图 S3-9 测试进度对话框

拓展练习

练习 3.1

修改知识拓展中进度条的示例代码，使进度条到达最大值后再逐渐缩小，缩小到最小值后再逐渐变大，如此循环执行。

练习 3.2

修改知识拓展中拖动条的示例代码，添加以下四个按钮，并实现对应的功能。

(1) 暂停：使进度暂停。

(2) 执行：继续执行。

(3) 加速：执行速度变为原来的 2 倍。

(4) 减速：执行速度变为原来的一半。

练习 3.3

修改知识拓展中图片切换效果的示例代码，使底部 Gallery 中的图片自动缓慢滚动显示，上部大图也相应地自动切换。

实践 4　意图(Intent)

实践指导

实践 4.1　完善登录功能

修改 LoginActivity，使登录成功后转到主菜单界面。

【分析】

(1) 实现登录功能需要在多个 Activity 之间多次跳转，因此要在 BasicActivity 中添加 showActivity()方法，负责显示 Activity，BasicActivity 的子类可以直接调用此方法。

(2) 在 LoginActivity 中验证登录成功后，调用 showActivity()方法来显示主菜单界面 MainActivity。

【参考解决方案】

(1) 修改 BasicActivity，在其中添加 showActivity()方法，代码如下：

```
public abstract class BasicActivity extends Activity {
	...... // 省略其他代码
	/**
	 * 显示 Activity
	 *
	 * @param context 上下文
	 * @param contextClass 需要显示的界面 Class
	 */
	protected void showActivity(Context context,Class<? extends Context> contextClass) {
		Intent intent = new Intent(context, contextClass);
		startActivity(intent);
	}
}
```

上述代码通过 Context 和 Class<? extends Context>类型的对象构造了 Intent，并调用 startActivity()方法来显示对应的 Context。

(2) 修改 LoginActivity，在登录信息验证成功后，调用 showActivity()方法显示主菜单界面 MainActivity，代码如下：

```
public class LoginActivity extends BasicActivity {
    ...... // 省略其他代码
    @Override
    protected void onCreate(Bundle savedInstanceState) {
        ...... // 省略其他代码
        loginBtn.setOnClickListener(new OnClickListener() {
            @Override
            public void onClick(View v) {
                ...... // 省略其他代码
                // showMessageDialog("登录成功",R.drawable.info,null);
                App app = (App) getApplication();
                app.user = user;
                // TODO 转向主菜单界面
                showActivity(LoginActivity.this, MainActivity.class);
                finish();//打开主界面后，关闭登录界面
            }
        });
    }
}
```

(3) 在 AndroidManifest.xml 文件中配置 LoginActivity 和 MainActivity，代码如下：

```
<?xml version="1.0" encoding="utf-8"?>
<manifest xmlns:android="http://schemas.android.com/apk/res/android"
    package="com.ugrow.repast_ph04">

    <application
        android:name=".App"
        android:allowBackup="true"
        android:icon="@mipmap/icon"
        android:label="@string/app_name"
        android:logo="@mipmap/coffee"
        android:roundIcon="@mipmap/ic_launcher_round"
        android:supportsRtl="true"
        android:theme="@style/AppTheme">
        <activity android:name=".ui.MainActivity">

        </activity>
        <activity android:name=".ui.LoginActivity">
            <intent-filter>
                <action android:name="android.intent.action.MAIN" />
```

```
            <category android:name="android.intent.category.LAUNCHER" />
        </intent-filter>
    </activity>

</application>

</manifest>
```

运行 Repast_ph04 项目，登录成功后，将显示主菜单界面。

实践 4.2　点餐功能

实现点餐功能。

【分析】

(1) 点餐界面可以选择桌号，并录入顾客数量和备注信息。

(2) 点餐界面可以选择多个餐品，并录入每个餐品的数量和备注信息。为方便录入，应提供按照餐品类型选择餐品的功能。

(3) 信息录入完毕后，在点餐界面确定下单。

(4) 选择桌号时需要调用餐桌业务类 TableService 的 getTables()方法来获取所有的餐桌信息。

(5) 选择餐品时需要调用餐品业务类 FoodService 的 getFoodTypes()方法来获取所有的餐品类型，调用 getFoodsByTypeId()方法可获取某种类型的所有餐品。

(6) 下单时需要调用订单业务类 OrderService 的 addOrder()方法来添加订单。

(7) 点餐界面下部需要显示已点餐品列表，选择餐品时需要使用对话框供用户进行选择，因此点餐功能的实现需要三个布局文件：

✧ order.xml：点餐界面的总体布局文件。

✧ ordered.xml：点餐界面下部已点餐品列表的布局文件。

✧ food.xml：选择餐品对话框的布局文件。

(8) 需要使用 Spinner 显示餐桌和餐品，可以重写 Table 和 Food 类的 toString()方法，使 Spinner 能够正确显示餐桌和餐品中合适的属性。

注意　上述业务操作需要查询数据库或连接服务器获取数据，本实践中将使用假设数据模拟，在介绍数据库和网络操作后，将在后续实践中完善此功能。

【参考解决方案】

(1) 编写餐桌业务类 TableService，代码如下：

```
public class TableService {
    static List<Table> tables = new ArrayList<Table>();
    // 餐桌假数据
    static {
```

```
		for (int i = 1; i <= 20; i++) {
			Table t = new Table();
			t.setId(i);
			t.setCode("TABLE" + i);
			t.setSeats(i % 5 * 2 + 2);
			t.setCustomers(i % 3 == 0 ? t.getSeats() : 0);
			t.setDescription(i % 4 == 0 ? "靠窗" : "");
			tables.add(t);
		}
	}
	/**
	 * 获取所有餐桌。本地查询，查询结果中的当前就餐人数不可用。
	 *目前使用假数据
	 *
	 * @return 餐桌
	 */
	public List<Table> getTables() {
		return tables;
	}
}
```

上述代码使用 TableService 的 getTables()方法返回所有的餐桌数据。

(2) 编写餐品业务类 FoodService，代码如下：

```
public class FoodService {
	static List<FoodType> types = new ArrayList<FoodType>();
	static List<Food> foods = new ArrayList<Food>();
	// 餐品类型假数据
	static {
		FoodType t = new FoodType();
		t.setId(0);
		t.setName("全部");
		types.add(t);

		t = new FoodType();
		t.setId(1);
		t.setName("热菜");
		types.add(t);

		t = new FoodType();
		t.setId(2);
		t.setName("凉菜");
```

```
        types.add(t);

        t = new FoodType();
        t.setId(3);
        t.setName("烧烤");
        types.add(t);

        t = new FoodType();
        t.setId(4);
        t.setName("酒水");
        types.add(t);

        t = new FoodType();
        t.setId(5);
        t.setName("主食");
        types.add(t);
    }
    // 餐品假数据
    static {
        for (int i = 1; i <= 40; i++) {
            f.setId(i);
            f.setCode("FOOD" + i);
            f.setTypeId(i % 5 + 1);
            f.setName("餐品" + i);
            f.setPrice((i % 7 + 1) * 10);
            foods.add(f);
        }
    }
    /**
     * 获取所有的餐品类型。本地查询
     *目前使用假数据
     * @return 餐品类型
     */
    public List<FoodType> getFoodTypes() {
        return types;
    }

    /**
     * 根据类型获取对应的餐品。本地查询
     *目前使用假数据
     * @param foodTypeId 餐品类型，0 表示不限定类型
```

```
     *
     * @return 餐品
     */
    public List<Food> getFoodsByTypeId(int foodTypeId) {
        if (foodTypeId == 0)
            return foods;
        List<Food> list = new ArrayList<Food>();
        for (Food f : foods)
            if (f.getTypeId() == foodTypeId)
                list.add(f);
        return list;
    }
}
```

上述代码使用 FoodService 的 getFoodTypes()方法返回所有的餐品类型，使用 getFoodsByTypeId()方法返回特定类型的所有餐品。

(3) 编写订单业务类 OrderService，代码如下：

```
public class OrderService {
    static List<Order> orders = new ArrayList<Order>();
    // 订单、订单明细假数据
    static {
        Order o = new Order();
        o.setId(1);
        o.setCode("1");
        o.setTableId(1);
        o.setWaiterId(1);
        o.setOrderTime(new Date());
        o.setCustomers(4);
        o.setStatus(1);// 已结算
        o.setOrderDetails(new ArrayList<OrderDetail>());
        for (int i = 1; i <= 5; i++) {
            OrderDetail od = new OrderDetail();
            od.setId(i);
            od.setOrderId(1);
            od.setFoodId(i);
            od.setNum(1);
            o.getOrderDetails().add(od);
        }
        orders.add(o);
```

```
            o = new Order();
            o.setId(2);
            o.setCode("2");
            o.setTableId(2);
            o.setWaiterId(2);
            o.setOrderTime(new Date());
            o.setCustomers(10);
            o.setStatus(0);// 未结算
            o.setOrderDetails(new ArrayList<OrderDetail>());
            for (int i = 6; i <= 15; i++) {
                    OrderDetail od = new OrderDetail();
                    od.setId(i);
                    od.setOrderId(2);
                    od.setFoodId(i);
                    od.setNum(2);
                    o.getOrderDetails().add(od);
            }
            orders.add(o);
    }
    /**
     * 添加订单
     *
     * @param order 订单
     */
    public void addOrder(Order order) {
            // TODO 添加订单到服务器，后续章节完成
    }
}
```

(4) 编写点餐界面的布局文件 order.xml，代码如下：

```
<?xml version="1.0" encoding="utf-8"?>
<RelativeLayout xmlns:android="http://schemas.android.com/apk/res/android"
    android:layout_width="match_parent"
    android:layout_height="match_parent"
    android:background="#ededed"
    android:padding="10dp" >
    <LinearLayout
        android:id="@+id/top_ll"
        android:layout_width="match_parent"
        android:layout_height="wrap_content"
```

```
android:orientation="vertical" >
<LinearLayout
    android:layout_width="match_parent"
    android:layout_height="wrap_content"
    android:orientation="horizontal" >
    <Spinner
        android:id="@+id/tableSpn"
        android:layout_width="wrap_content"
        android:layout_height="wrap_content"
        android:layout_weight="2"
        android:hint="桌号" />
    <EditText
        android:id="@+id/customersEdt"
        android:layout_width="wrap_content"
        android:layout_height="wrap_content"
        android:layout_weight="1"
        android:hint="顾客人数"
        android:inputType="number"
        android:singleLine="true" />
</LinearLayout>
<RelativeLayout
    android:layout_width="match_parent"
    android:layout_height="wrap_content" >
    <EditText
        android:id="@+id/descriptionEdt"
        android:layout_width="match_parent"
        android:layout_height="wrap_content"
        android:layout_toLeftOf="@+id/ll"
        android:hint="备注"
        android:maxLines="3" />
    <LinearLayout
        android:id="@+id/ll"
        android:layout_width="wrap_content"
        android:layout_height="wrap_content"
        android:layout_alignParentRight="true"
        android:orientation="horizontal" >
        <TextView
            android:layout_width="wrap_content"
            android:layout_height="wrap_content"
            android:text="合计：¥"
            android:textColor="#FF8080"
```

```
            android:textSize="20sp"
            android:textStyle="bold" />
        <TextView
            android:id="@+id/sumTxv"
            android:layout_width="wrap_content"
            android:layout_height="wrap_content"
            android:text="0"
            android:textColor="#FF8080"
            android:textSize="20sp"
            android:textStyle="bold" />
    </LinearLayout>
</RelativeLayout>
<LinearLayout
    android:layout_width="match_parent"
    android:layout_height="wrap_content"
    android:orientation="horizontal" >
    <Button
        android:id="@+id/addFoodBtn"
        android:layout_width="match_parent"
        android:layout_height="wrap_content"
        android:layout_weight="1"
        android:drawableLeft="@drawable/coffee"
        android:text="点菜" />
    <Button
        android:id="@+id/orderBtn"
        android:layout_width="match_parent"
        android:layout_height="wrap_content"
        android:layout_weight="1"
        android:drawableLeft="@drawable/save"
        android:text="下单" />
    <Button
        android:id="@+id/cancelBtn"
        android:layout_width="match_parent"
        android:layout_height="wrap_content"
        android:layout_weight="1"
        android:drawableLeft="@drawable/back"
        android:text="取消" />
</LinearLayout>
</LinearLayout>
<ListView
    android:id="@+id/orderedLtv"
```

```
        android:layout_width="match_parent"
        android:layout_height="wrap_content"
        android:layout_below="@+id/top_ll"
        android:background="@drawable/bg1"
        android:visibility="gone" />
</RelativeLayout>
```

上述代码在点餐界面的上部显示桌号 Spinner、顾客人数 EditText、备注 EditText 等需要录入的信息以及点菜按钮、下单按钮、取消按钮；在界面的底部设置了一个 ListView，用于显示已点餐品列表。

(5) 编写已点餐品列表的布局文件 ordered.xml，代码如下：

```
<?xml version="1.0" encoding="utf-8"?>
<TableLayout xmlns:android="http://schemas.android.com/apk/res/android"
    android:layout_width="match_parent"
    android:layout_height="wrap_content"
    android:stretchColumns="1" >
    <TableRow
        android:layout_width="match_parent"
        android:textColor="#80FFFF" >
        <TextView
            android:id="@+id/noTxv"
            android:layout_height="wrap_content"
            android:gravity="right"
            android:textColor="#fff"
            android:width="20dp" />
        <TextView
            android:id="@+id/nameTxv"
            android:layout_width="wrap_content"
            android:layout_height="wrap_content"
            android:gravity="left"
            android:paddingLeft="3dp"
            android:textColor="#00FFFF" />
        <TextView
            android:id="@+id/descriptionTxv"
            android:layout_width="wrap_content"
            android:layout_height="wrap_content"
            android:gravity="left"
            android:paddingLeft="3dp"
            android:textColor="#A0FF42" />
        <TextView
            android:id="@+id/numTxv"
```

```
            android:layout_height="wrap_content"
            android:gravity="right"
            android:paddingLeft="3dp"
            android:textColor="#FFFF80"
            android:textStyle="bold"
            android:width="30dp" />
        <TextView
            android:id="@+id/priceTxv"
            android:layout_height="wrap_content"
            android:gravity="right"
            android:paddingLeft="3dp"
            android:textColor="#FF8080"
            android:textStyle="bold"
            android:width="40dp" />
        <CheckBox
            android:id="@+id/checkCkb"
            android:layout_width="wrap_content"
            android:layout_height="wrap_content"
            android:checked="true"
            android:paddingRight="5dp" />
    </TableRow>
</TableLayout>
```

上述代码使用 TableLayout 布局定义了已点餐品列表中每一行的布局格式，即在每一行的最后设置一个 CheckBox，供用户取消已点的餐品。点餐界面中已点餐品列表的 ListView 里将使用该布局显示每个餐品。

(6) 编写选择餐品对话框的布局文件 order_dialog.xml，代码如下：

```
<?xml version="1.0" encoding="utf-8"?>
<LinearLayout xmlns:android="http://schemas.android.com/apk/res/android"
    android:layout_width="match_parent"
    android:layout_height="match_parent"
    android:background="#ededed"
    android:orientation="vertical"
    android:padding="10dp" >
    <TableLayout
        android:layout_width="match_parent"
        android:layout_height="match_parent"
        android:stretchColumns="1" >
        <GridView
            android:id="@+id/foodTypeGdv"
            android:layout_width="match_parent"
```

```
            android:layout_height="wrap_content"
            android:columnWidth="80dp"
            android:gravity="center"
            android:horizontalSpacing="5dp"
            android:numColumns="auto_fit"
            android:stretchMode="columnWidth"
            android:verticalSpacing="10dp" >
        </GridView>
        <Spinner
            android:id="@+id/foodSpn"
            android:layout_width="match_parent"
            android:layout_height="wrap_content" />
        <TableRow>
            <EditText
                android:id="@+id/numEdt"
                android:layout_height="wrap_content"
                android:hint="数量"
                android:inputType="number"
                android:singleLine="true"
                android:width="60dp" />
            <EditText
                android:id="@+id/descriptionEdt"
                android:layout_width="match_parent"
                android:layout_height="wrap_content"
                android:hint="备注"
                android:maxLines="3" />
        </TableRow>
    </TableLayout>
</LinearLayout>
```

上述代码使用 GridView 控件显示所有的餐品类型，使用 Spinner 控件显示对应的餐品，以及供用户录入餐品数量和备注的 EditText。

(7) 编写点餐界面类 OrderActivity，代码如下：

```
public class OrderActivity extends BasicActivity {
    private Spinner tableSpn;
    private EditText customersEdt;
    private EditText descriptionEdt;
    private TextView sumTxv;
    private Button addFoodBtn;
    private Button orderBtn;
    private Button cancelBtn;
```

```
private ListView orderedLtv;
// 已点酒菜
private List<Map<String, Object>> orderedList = new ArrayList<Map<String, Object>>();
private String[] orderedLtvKeys = new String[] { "no", "name","description", "num", "price" };
private int[] orderedLtvIds = new int[] { R.id.noTxv, R.id.nameTxv,
            R.id.descriptionTxv, R.id.numTxv, R.id.priceTxv };
// 点菜 Dialog
private Spinner foodSpn;
private EditText numEdt;
private EditText foodDescriptionTxt;
private TableService tableService;
private FoodService foodService;
private OrderService orderService;
@Override
protected void onCreate(Bundle savedInstanceState) {
    super.onCreate(savedInstanceState);
    setContentView(R.layout.order);

    tableService = new TableService();
    foodService = new FoodService();
    orderService = new OrderService();

    ...... // 使用 findViewById()方法初始化各控件

    initTableSpn();
    initOrderedLtv();

    LayoutInflater li = LayoutInflater.from(this);
    View foodView = li.inflate(R.layout.order_dialog, null);
    foodSpn = (Spinner) foodView.findViewById(R.id.foodSpn);
    numEdt = (EditText) foodView.findViewById(R.id.numEdt);
    foodDescriptionTxt = (EditText) foodView.findViewById(R.id.descriptionEdt);
    GridView foodTypeGdv = (GridView) foodView.findViewById(R.id.foodTypeGdv);
    initFoodTypeGv(foodTypeGdv);
    numEdt.setText("1");
    AlertDialog.Builder dialogBuilder = new AlertDialog.Builder(this);
    dialogBuilder.setTitle("选择酒菜");
    dialogBuilder.setIcon(R.drawable.coffee);
    dialogBuilder.setView(foodView);
    dialogBuilder.setCancelable(true);
    dialogBuilder.setPositiveButton("确定",
            new DialogInterface.OnClickListener() {
                @Override
```

```
                public void onClick(DialogInterface dialog, int which) {
                    Food food=(Food)foodSpn.getSelectedItem();
                    int num = 1;
                    String n = numEdt.getText().toString();
                    if (n.length() > 0)
                        num = Integer.parseInt(n);
                    String description =
                    foodDescriptionTxt.getText().toString();
                    Map<String, Object> line = new HashMap<String, Object>();
                    line.put("foodId", food.getId());
                    line.put("no", orderedList.size() + 1);
                    line.put("name", food.getName());
                    line.put("description", description);
                    line.put("num", num);
                    line.put("price", food.getPrice());
                    line.put("checked", true);
                    orderedList.add(line);
                    ((SimpleAdapter) orderedLtv.getAdapter())
                            .notifyDataSetChanged();
                    orderedLtv.setVisibility(View.VISIBLE);
                    // 合计
                    refreshSum(num * food.getPrice());
                    foodDescriptionTxt.setText(null);
                }
            });
dialogBuilder.setNegativeButton("取消", null);

final AlertDialog dialog = dialogBuilder.create();
addFoodBtn.setOnClickListener(new OnClickListener() {
    @Override
    public void onClick(View v) {
        numEdt.setText("1");
        dialog.show();
    }
});

orderBtn.setOnClickListener(new OnClickListener() {
    @Override
    public void onClick(View v) {
        int customers = 0;
```

```
                String c = customersEdt.getText().toString();
                if (c.length() > 0)
                    customers = Integer.parseInt(c);
                if (customers <= 0) {
                    showMessageDialog("请输入顾客数量", R.drawable.warning, null);
                    return;
                }
                int sum=Integer.parseInt(sumTxv.getText().toString());
                if (sum <= 0) {
                    showMessageDialog("尚未选择任何餐品", R.drawable.warning, null);
                    return;
                }

                Table table = (Table) tableSpn.getSelectedItem();

                Order order = new Order();
                User waiter = ((App) getApplication()).user;
                order.setTableId(table.getId());
                order.setWaiterId(waiter.getId());
                order.setCustomers(Integer.parseInt(customersEdt.getText().toString()));
                order.setDescription(descriptionEdt.getText().toString());
                order.setOrderDetails(new ArrayList<OrderDetail>());

                for (Map<String, Object> line : orderedList) {
                    boolean checked = (Boolean) line.get("checked");
                    if (!checked)
                        continue;
                    OrderDetail od = new OrderDetail();
                    od.setFoodId((Integer) line.get("foodId"));
                    od.setNum((Integer) line.get("num"));
                    od.setDescription((String) line.get("description"));
                    order.getOrderDetails().add(od);
                }
                orderService.addOrder(order);
                finish();
            }
        });

        cancelBtn.setOnClickListener(new OnClickListener() {
            @Override
            public void onClick(View v) {
```

```
                    OrderActivity.this.finish();
                }
            });
        }
        privatevoid initTableSpn() {
            List<Table> tables = tableService.getTables();
            tableSpn.setAdapter(new ArrayAdapter<Table>(this,
                    android.R.layout.simple_spinner_item, tables));
        }
        privatevoid initFoodSpn(int foodTypeId) {
            List<Food> foods = foodService.getFoodsByTypeId(foodTypeId);
            foodSpn.setAdapter(new ArrayAdapter<Food>(this,
                    android.R.layout.simple_spinner_item, foods));
        }
        privatevoid initOrderedLtv() {
            SimpleAdapter sa = new SimpleAdapter(this, orderedList,
                    R.layout.ordered, orderedLtvKeys, orderedLtvIds) {
                // 覆盖此方法，以使 Checkbox 可修改合计
                @Override
                public View getView(int position, View convertView,
                ViewGroup parent) {
                    final int lineNo = position;
                    View view = super.getView(position, convertView, parent);
                    // 为 Checkbox 添加事件，修改选中状态时计算合计
                    CheckBox checkCkb = (CheckBox) view.findViewById(R.id.checkCkb);
                    checkCkb.setOnCheckedChangeListener(
                    new CheckBox.OnCheckedChangeListener() {
                        @Override
                        public void onCheckedChanged(CompoundButton
                            buttonView,boolean isChecked) {
                            @SuppressWarnings("unchecked")
                            Map<String, Object> line = (Map<String,
                                Object>) getItem(lineNo);
                            int num = (Integer) line.get("num");
                            int price = (Integer) line.get("price");
                            if (isChecked)
                                refreshSum(num * price);
                            else
                                refreshSum(-num * price);
```

```
                    line.put("checked", isChecked);
                }
            });
            return view;
        }
    };
    orderedLtv.setAdapter(sa);
}
privatevoid initFoodTypeGv(GridView foodTypeGdv) {
    final List<FoodType> foodTypes = foodService.getFoodTypes();
    final List<RadioButton> rbs = new ArrayList<RadioButton>();
    for (FoodType ft : foodTypes) {
        RadioButton rb = new RadioButton(OrderActivity.this);
        // rb.setText(ft.name);
        // rb.setTag(ft.id);
        rb.setText(ft.getName());
        rb.setTag(ft.getId());
        rb.setOnCheckedChangeListener(
        new RadioButton.OnCheckedChangeListener() {
            @Override
            public void onCheckedChanged(CompoundButton
            buttonView,boolean isChecked) {
                if (!isChecked)
                    return;
                for (RadioButton b : rbs) {
                    if (b != buttonView)
                        b.setChecked(false);
                }
                int foodTypeId = Integer.parseInt(buttonView.getTag().toString());
                initFoodSpn(foodTypeId);
            }
        });
        rbs.add(rb);
    }
    // 默认选中全部类型
    rbs.get(0).setChecked(true);
    initFoodSpn(0);
    foodTypeGdv.setAdapter(new BaseAdapter() {
        @Override
```

```
                    public int getCount() {
                        return foodTypes.size();
                    }
                    @Override
                    public Object getItem(int position) {
                        return null;
                    }
                    @Override
                    public long getItemId(int position) {
                        return 0;
                    }
                    @Override
                    public View getView(int position, View convertView, ViewGroup parent) {
                        return rbs.get(position);
                    }
            });
        }
        privatevoid refreshSum(int newAmount) {
            int sum = Integer.parseInt(sumTxv.getText().toString());
            sum += newAmount;
            sumTxv.setText(sum + "");
        }
        @Override
        protected String getName() {
            return "点餐";
        }
}
```

(8) 修改 Table 和 Food 实体，在其中添加 toString()方法，代码如下：

```
public class Table {
        private int id;
        private String code; // 编号
        private int seats; // 座位数
        private int customers; // 当前就餐人数
        private String description; // 说明
        ...Get()、Set()方法略
        @Override
        public String toString() {
            return code;
        }
```

```
}
public class Food {
        private int id;
        private String code; // 编码
        private int typeId; // 类型
        private String name; // 名称
        private int price; // 价格
        private String description; // 说明
        ...Get()、Set()方法略
        @Override
        public String toString() {
                return code + " " + name + " ¥" + price;
        }
}
```

(9) 修改主菜单界面 MainActivity，使在单击点餐图标时显示 OrderActivity，代码如下：

```
public class MainActivity extends BasicActivity {
        ...... // 省略其他代码
        @Override
        public void onCreate(Bundle savedInstanceState) {
                super.onCreate(savedInstanceState);
                ...... // 省略其他代码
                gdv.setOnItemClickListener(
                new AdapterView.OnItemClickListener() {
                        @Override
                        public void onItemClick(AdapterView<?> arg0, View arg1,int idx,long arg3) {
                        switch (idx) {
                                case 0:
                                //showMessageDialog("点餐", R.drawable.info,null);
                                showActivity(MainActivity.this,OrderActivity.class);
                                break;
                                ...... // 省略其他代码
                                }
                        }
                });
        }
}
```

(10) 在 AndroidManifest.xml 文件中配置 OrderActivity，代码如下：

```
<activity android:name=".ui.OrderActivity"/>
```

运行 Repast_ph04 项目，登录后，在主菜单界面单击点餐图标，将显示点餐界面，并

在餐桌列表中显示所有的餐桌编号，如图 S4-1 所示。

单击【点菜】按钮，将显示餐品选择对话框【选择酒菜】，并在其中显示所有的餐品类型，选择某个餐品类型后，餐品列表将自动显示此类型对应的餐品，如图 S4-2 所示。

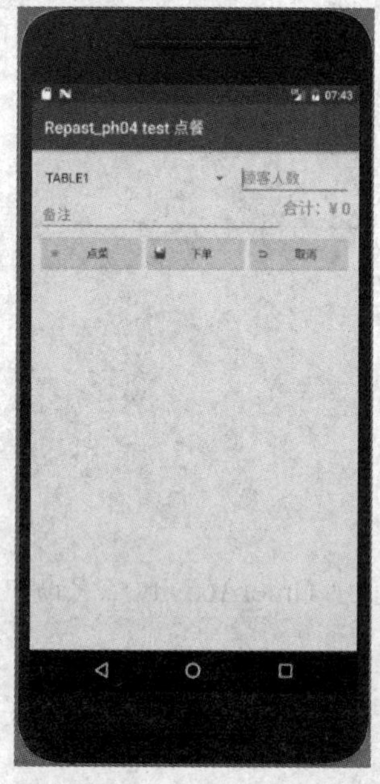

图 S4-1　点餐主界面

图 S4-2　点餐功能-选择餐品

选择某个餐品，录入数量和备注信息后，单击【确定】按钮，将返回点餐界面，并显示已点的餐品，单击已点餐品列表行末的 CheckBox，可取消或再次选中某个餐品，并自动计算总金额，多次操作后的点餐界面如图 S4-3 所示。点餐完毕后，单击【下单】按钮，此时如果未录入顾客人数，会显示信息提示对话框；录入顾客数量后，再次单击【下单】按钮，即可完成下单操作。

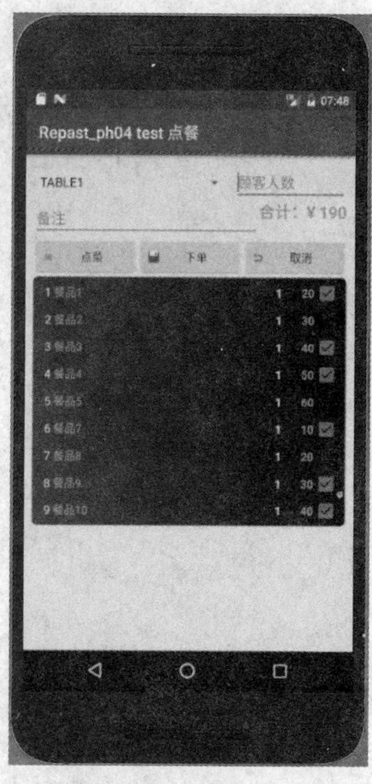

图 S4-3　点餐功能-下单

实践 4.3　结账功能

实现结账功能。

【分析】

(1) 结账界面需提供根据订单编号查询订单详细信息的功能，显示的信息包括订单本身信息以及对应的多条订单明细信息。用户确认信息无误后，单击结账按钮，完成结账操作。

(2) 调用订单业务类 OrderService 的 getOrder()方法完成根据订单编号查找订单的功能，调用 pay()方法来实现结账功能。

(3) 显示订单明细信息时需要显示对应的餐品名称、价格等，订单明细实体 OrderDetail 中只保存了对应餐品的 ID，因此需要调用餐品业务类 FoodService 的 getFood()方法来完成根据 ID 查找餐品的功能。

(4) 结账界面底部需要显示订单对应的明细信息，可重新使用点餐界面中已点餐品列表的布局文件 ordered.xml。

(5) 订单明细信息使用 ListView 显示，使用 SimpleAdapter 填充数据，要求数据类型为 List<Map<String, Object>>，而 List<OrderDetail>类型的数据无法直接使用，需要构造一个数据容器 OrderDto，其中包括三项数据：

- ✧ 订单 order。
- ✧ 订单的合计金额 sum。
- ✧ List<Map<String, Object>>类型的订单明细信息 orderedList。

(6) OrderService 的 getOrder()方法需要返回 OrderDto 类。

查找订单和结账都需要连接服务器完成，本实践中将使用假设数据模拟，后续实践在介绍数据库和网络操作后，后续实践将完善此功能。

【参考解决方案】

(1) 修改餐品业务类 FoodService，在其中添加 getFood()方法，代码如下：

```
public class FoodService {
	...... // 省略其他代码
	/**
	 * 根据 ID 获取餐品。本地查询
	 *
	 * @param foodId
	 *            餐品 ID
	 * @return 餐品
	 */
	public Food getFood(int foodId) {
		// TODO 查询本地数据库获取餐品
		for (Food f : foods)
			if (f.getId()== foodId)
				return f;
		return null;
	}
}
```

(2) 修改订单业务类 OrderService，在其中添加 getOrder()和 pay()方法，并添加静态内部类 OrderDto 作为数据容器，代码如下：

```
public class OrderService {
	...... // 省略其他代码
	private FoodService foodService = new FoodService();
```

```
/**
 * 根据 CODE 获取订单
 *
 * @param code 订单 Code
 * @return 订单
 */
public OrderDto getOrder(String code) {
        OrderDto dto = new OrderDto();
        for (Order order : orders)
                if (order.getCode().equals(code)) {
                        dto.order = order;
                        break;
                }
        if (dto.order == null)
                return null;
        for (OrderDetail od : dto.order.getOrderDetails()) {
                Food food = foodService.getFood(od.getFoodId());
                Map<String, Object> line = new HashMap<String, Object>();
                line.put("no", dto.orderedList.size() + 1);
                line.put("name", food.getName());
                line.put("description", od.getDescription());
                line.put("num", od.getNum());
                line.put("price", food.getPrice());
                dto.orderedList.add(line);
                dto.sum += od.getNum()* food.getPrice();
        }
        return dto;
}

/**
 * 结账
 *
 * @param ordered 对应订单 ID
 */
public void pay(int orderId) {
        // TODO 连接服务器完成结账
}

/**
```

```
     * Order 数据传输对象
     */
    public static class OrderDto {
        /**
         * 订单
         */
        public Order order;

        /**
         * 订单合计金额
         */
        public int sum;

        /**
         * 明细记录。key 为 no、name、description、num、price
         */
        public List<Map<String, Object>> orderedList = new ArrayList<Map<String, Object>>();
    }
}
```

上述代码中，定义了数据传输对象 OrderDto，它用于封装订单及明细信息，在 getOrder()方法中，查询订单、明细并转换为 OrderDto 对象返回。在 getOrder()方法中调用 FoodService 的 getFood()方法根据餐品 ID 查询对应的餐品信息。

(3) 编写结账界面布局文件 pay.xml，代码如下：

```
<?xml version="1.0" encoding="utf-8"?>
<LinearLayout xmlns:android="http://schemas.android.com/apk/res/android"
    android:layout_width="match_parent"
    android:layout_height="match_parent"
    android:background="#ededed"
    android:orientation="vertical"
    android:padding="10dp" >
    <EditText
        android:id="@+id/orderCodeEdt"
        android:layout_width="match_parent"
        android:layout_height="wrap_content"
        android:hint="请输入订单编号"
        android:inputType="number"
        android:singleLine="true" />
    <LinearLayout
        android:layout_width="match_parent"
```

```
    android:layout_height="wrap_content" >
    <Button
        android:id="@+id/queryBtn"
        android:layout_width="match_parent"
        android:layout_height="wrap_content"
        android:layout_weight="1"
        android:drawableLeft="@drawable/search"
        android:text="查询" />
    <Button
        android:id="@+id/payBtn"
        android:layout_width="match_parent"
        android:layout_height="wrap_content"
        android:layout_weight="1"
        android:drawableLeft="@mipmap/money"
        android:text="结账" />
    <Button
        android:id="@+id/cancelBtn"
        android:layout_width="match_parent"
        android:layout_height="wrap_content"
        android:layout_weight="1"
        android:drawableLeft="@mipmap/back"
        android:text="取消" />
</LinearLayout>
<TextView
    android:id="@+id/sumTxv"
    android:layout_width="match_parent"
    android:layout_height="wrap_content"
    android:gravity="right"
    android:textColor="#FF8080"
    android:textSize="20sp"
    android:textStyle="bold" />
<LinearLayout
    android:layout_width="match_parent"
    android:layout_height="wrap_content" >
    <TextView
        android:id="@+id/orderCodeTxv"
        android:layout_width="match_parent"
        android:layout_height="wrap_content"
        android:layout_weight="1"
        android:textColor="#D5BFFF"
```

```
        android:textStyle="bold" />
    <TextView
        android:id="@+id/tableCodeTxv"
        android:layout_width="match_parent"
        android:layout_height="wrap_content"
        android:layout_weight="1"
        android:textColor="#D5BFFF"
        android:textStyle="bold" />
</LinearLayout>
<LinearLayout
    android:layout_width="match_parent"
    android:layout_height="wrap_content" >
    <TextView
        android:id="@+id/waiterCodeTxv"
        android:layout_width="match_parent"
        android:layout_height="wrap_content"
        android:layout_weight="1"
        android:textColor="#D5BFFF"
        android:textStyle="bold" />
    <TextView
        android:id="@+id/customersTxv"
        android:layout_width="match_parent"
        android:layout_height="wrap_content"
        android:layout_weight="1"
        android:textColor="#D5BFFF"
        android:textStyle="bold" />
</LinearLayout>
<TextView
    android:id="@+id/orderTimeTxv"
    android:layout_width="match_parent"
    android:layout_height="wrap_content"
    android:textColor="#D5BFFF"
    android:textStyle="bold" />
<TextView
    android:id="@+id/descriptionTxv"
    android:layout_width="match_parent"
    android:layout_height="wrap_content"
    android:textColor="#8080FF"
    android:textStyle="bold" />
<ListView
```

```
        android:id="@+id/orderedLtv"
        android:layout_width="wrap_content"
        android:layout_height="wrap_content"
        android:background="@drawable/bg1"
        android:visibility="gone" />
</LinearLayout>
```

上述代码在结账界面的上部显示了订单编号以及查询、结账、取消按钮，底部显示查询出的订单信息，使用 ListView 控件显示订单明细信息。

(4) 编写结账界面 PayActivity，代码如下：

```
public class PayActivity extends BasicActivity {
    private static final SimpleDateFormat sdf = new SimpleDateFormat("yyyy-MM-dd HH:mm");
    private EditText orderCodeEdt;
    private Button queryBtn;
    private Button payBtn;
    private Button cancelBtn;
    private TextView sumTxv;
    private ListView orderedLtv;

    private TextView orderCodeTxv;
    private TextView tableCodeTxv;
    private TextView waiterCodeTxv;
    private TextView orderTimeTxv;
    private TextView customersTxv;
    private TextView descriptionTxv;

    // 已点酒菜
    private List<Map<String, Object>> orderedList =
            new ArrayList<Map<String, Object>>();
    private String[] orderedLtvKeys = new String[] { "no", "name", "description","num", "price" };
    private int[] orderedLtvIds = new int[] { R.id.noTxv, R.id.nameTxv,
                R.id.descriptionTxv, R.id.numTxv, R.id.priceTxv };

    private int orderId;
    private OrderService orderService;
    @Override
    protected void onCreate(Bundle savedInstanceState) {
        super.onCreate(savedInstanceState);
        setContentView(R.layout.pay);
        orderService = new OrderService();
        ...... // 使用 findViewById()方法初始化各控件
```

```
initOrderedLtv();
queryBtn.setOnClickListener(new OnClickListener() {
        @Override
        public void onClick(View v) {
                PayActivity.this.orderId = 0;
                String orderCode = orderCodeEdt.getText().toString();
                clearDisplay();
                if (orderCode.length() == 0) {
                        showMessageDialog("请输入订单编号", R.drawable.warning, null);
                        return;
                }
                OrderDto dto = null;
                dto = orderService.getOrder(orderCode);
                if (dto == null) {
                        showMessageDialog("未查找到订单：" + orderCode,
                                        R.drawable.warning, null);
                        return;
                }
                orderCodeTxv.setText("订单编号：" + dto.order.getCode());
                tableCodeTxv.setText("    餐桌 ID：" + dto.order.getTableId()+ "");
                waiterCodeTxv.setText("服务员 ID：" + dto.order.getWaiterId()+ "");
                orderTimeTxv.setText("        时间：" + sdf.format(dto.order.getOrderTime()));
                customersTxv.setText("顾客数量：" + dto.order.getCustomers());
                descriptionTxv.setText(dto.order.getDescription()==
                        null ? "": ("        备注：" + dto.order.getDescription()));
                orderedList.addAll(dto.orderedList);
                SimpleAdapter sa = (SimpleAdapter) orderedLtv.getAdapter();
                sa.notifyDataSetChanged();
                orderedLtv.setVisibility(View.VISIBLE);
                if (dto.order.getStatus()== 1) {
                        sumTxv.setText("此订单已结算，合计：¥" + dto.sum);
                        PayActivity.this.orderId = -1;
                } else {
                        sumTxv.setText("合计：¥" + dto.sum);
                        PayActivity.this.orderId = dto.order.getId();
                }
        }
});

payBtn.setOnClickListener(new OnClickListener() {
```

```
                @Override
                public void onClick(View v) {
                    if (orderId == 0) {
                        showMessageDialog("请选择订单", R.drawable.warning, null);
                        return;
                    }
                    if (orderId == -1) {
                        showMessageDialog("此订单已结算",
                                R.drawable.warning, null);
                        return;
                    }
                    orderService.pay(orderId);
                    orderId = -1;
                    showMessageDialog("结账完成", R.drawable.info,
                            new DialogInterface.OnClickListener() {
                                @Override
                                public void onClick(DialogInterface
                                        dialog,int which) {
                                    finish();
                                }
                            });
                }
            });
            cancelBtn.setOnClickListener(new OnClickListener() {
                @Override
                public void onClick(View v) {
                    PayActivity.this.finish();
                }
            });
        }

        // 清空显示的数据
        void clearDisplay() {
            orderCodeEdt.setText(null);
            orderCodeTxv.setText(null);
            tableCodeTxv.setText(null);
            waiterCodeTxv.setText(null);
            orderTimeTxv.setText(null);
            customersTxv.setText(null);
            descriptionTxv.setText(null);
```

```
            sumTxv.setText(null);
            orderedList.clear();
            SimpleAdapter sa = (SimpleAdapter) orderedLtv.getAdapter();
            sa.notifyDataSetChanged();
            orderedLtv.setVisibility(View.GONE);
        }

        void initOrderedLtv() {
            SimpleAdapter sa = new SimpleAdapter(this, orderedList,
                        R.layout.ordered, orderedLtvKeys, orderedLtvIds) {
                @Override
                public View getView(int position, View convertView, ViewGroup parent){
                    View view = super.getView(position, convertView, parent);
                    CheckBox checkCkb = (CheckBox) view.findViewById(R.id.checkCkb);
                    checkCkb.setVisibility(View.GONE); // 不显示 CheckBox
                    return view;
                }
            };
            orderedLtv.setAdapter(sa);
        }

        @Override
        protected int getLayoutId() {
            return R.layout.pay;
        }

        @Override
        protected String getName() {
            return "结账";
        }
}
```

(5) 修改 MainActivity，使在单击结账图标时显示 PayActivity，代码如下：

```
public class MainActivity extends BasicActivity {
        ...... // 省略其他代码
        @Override
        public void onCreate(Bundle savedInstanceState) {
            super.onCreate(savedInstanceState);
            ...... // 省略其他代码
            gdv.setOnItemClickListener(new AdapterView.OnItemClickListener() {
                @Override
```

```
                public void onItemClick(AdapterView<?> arg0, View arg1, int idx,long arg3) {
                    switch (idx) {
                    ...... // 省略其他代码
                    case 1:
                    //showMessageDialog("结账", R.drawable.info, null);
                    showActivity(MainActivity.this, PayActivity.class);
                    break;
                    ...... // 省略其他代码
                    }
                }
            });
    }
}
```

(6) 在 AndroidManifest.xml 文件中配置 PayActivity，代码如下：

```
<activity android:name=".ui.PayActivity" />
```

运行 Repast_ph04 项目，登录后在主菜单界面单击【结账】按钮，将显示结账界面。输入订单编号后，单击【查询】按钮，将显示对应的订单信息，如图 S4-4 所示。

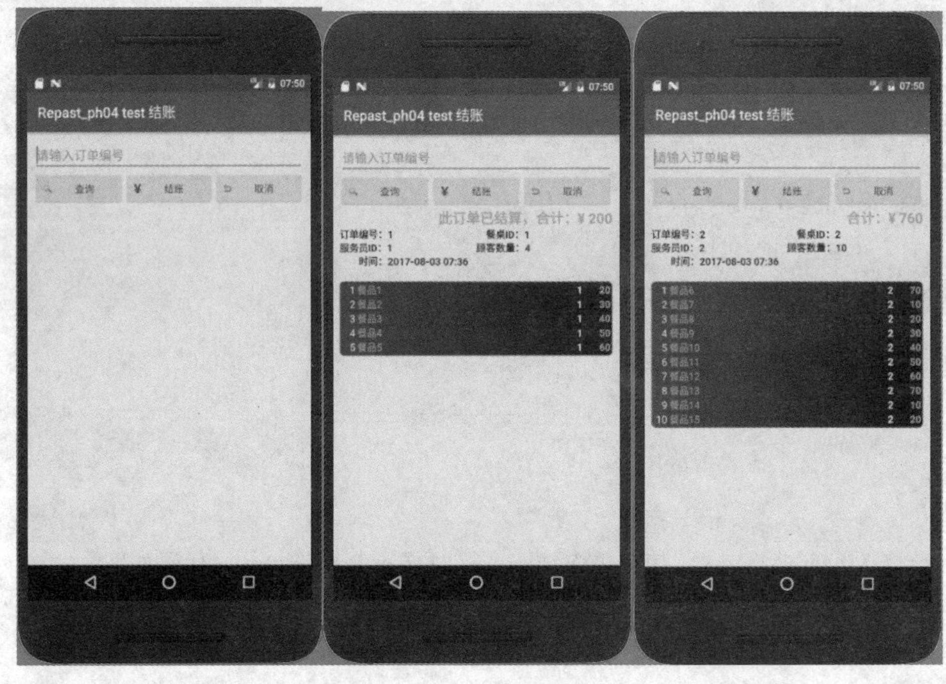

图 S4-4　测试结账功能

如果未输入订单编号就单击【查询】按钮，或查询到的订单已结算，均会显示提示对话框；而确认无误后，单击【结账】按钮，同样会显示对话框，提示用户结账已完成，如图 S4-5 所示。

图 S4-5 结账操作的各种提示

知识拓展

在程序中直接调用 Intent 可以访问 Android 内置的许多系统功能，其中主要功能的相关代码如下：

(1) 发送短信。

```
Uri uri = Uri.parse("smsto:12345678");
Intent intent = new Intent(Intent.ACTION_VIEW, uri);
intent.putExtra("sms_body", "短信内容");
```

(2) 发送彩信。

```
Uri pic = Uri.parse("file:///sdcard/some.jpg");
Intent intent = new Intent(Intent.ACTION_SEND);
intent.putExtra("sms_body", "短信内容");
intent.putExtra(Intent.EXTRA_STREAM, pic);
intent.setType("image/jpeg");
```

(3) 打开拨打电话界面。

```
Uri uri = Uri.parse("tel:13812345678");
Intent intent = new Intent(Intent.ACTION_DIAL, uri);
```

(4) 直接拨打电话。

```
// 需要添加权限 android.permission.CALL_PHONE
```

```
Uri uri = Uri.parse("tel:13812345678");
Intent intent = new Intent(Intent.ACTION_CALL, uri);
```

(5) 发送邮件。

```
Uri uri = Uri.parse("mailto:aaa@bbb.com");
Intent intent = new Intent(Intent.ACTION_SENDTO, uri);
```

(6) 播放多媒体。

```
Uri uri = Uri.parse("file:///sdcard/some.mp3");
Intent intent = new Intent(Intent.ACTION_VIEW, uri);
intent.setType("audio/mp3");
```

(7) 打开网站。

```
Uri uri = Uri.parse("http://www.dong-he.cn");
Intent intent = new Intent(Intent.ACTION_VIEW, uri);
```

拓展练习

为本章知识拓展中的每一项功能编写 Activity 并添加相应按钮，实现单击按钮时执行相应功能。

实践 5　广播(Broadcast)

实践指导

实践　完善点餐功能

修改 OrderActivity，实现单击已点餐品列表行末的 CheckBox 时自动计算总金额的功能。

【分析】

(1) 多个餐品使用 ListView 显示时，ListView 中每一个 item 的末尾都有一个 CheckBox 复选框，选中或取消该选框，即可实现总金额的实时更改。

(2) 使用 Broadcast 可实现此功能：当改变复选框的状态时，使用 sendBroadcast()方法发送一条广播，同时使用 Intent 传递数据。

(3) 广播发送之后，在 OrderActivity 中新建一个 MyReceiver，让其继承 Broadcast Receiver，重写 onReceive()方法接收数据。

(4) 使用 BroadcastReceiver 时需要注册，可以使用 registerReceiver()方法进行动态注册。

【参考解决方案】

修改 OrderActivity，代码如下：

```
public class OrderActivity extends BasicActivity {

    ……//省略其他代码
    @Override
    protected void onCreate(Bundle savedInstanceState) {
            super.onCreate(savedInstanceState);
            setContentView(R.layout.order);
            ……//省略其他代码
    /**
      * 动态注册广播
      */
      initBroadCastReceiver();
    AlertDialog.Builder dialogBuilder = new AlertDialog.Builder(this);
      dialogBuilder.setTitle("选择酒菜");
      dialogBuilder.setIcon(R.mipmap.coffee);
      dialogBuilder.setView(foodView);
      dialogBuilder.setCancelable(true);
      dialogBuilder.setPositiveButton("确定",
```

```
                new DialogInterface.OnClickListener() {
                    @Override
                    public void onClick(DialogInterface dialog, int which) {
                        Food food = (Food) foodSpn.getSelectedItem();
                        int num = 1;
                        String n = numEdt.getText().toString();
                        if (n.length() > 0)
                            num = Integer.parseInt(n);
                        String description = foodDescriptionTxt.getText()
                                .toString();
                        Map<String, Object> line = new HashMap<String, Object>();
                        line.put("foodId", food.getId());
                        line.put("no", orderedList.size() + 1);
                        line.put("name", food.getName());
                        line.put("description", description);
                        line.put("num", num);
                        line.put("price", food.getPrice());
                        line.put("checked", true);
                        orderedList.add(line);
                        ((SimpleAdapter) orderedLtv.getAdapter())
                                .notifyDataSetChanged();
                        orderedLtv.setVisibility(View.VISIBLE);
                        // 发送广播
                        Intent intent = new Intent("filter");
                        intent.putExtra("num",num);
                        intent.putExtra("price",food.getPrice());
                        sendBroadcast(intent);
                        foodDescriptionTxt.setText(null);
                    }
                });
        dialogBuilder.setNegativeButton("取消", null);
    }

    /**
     * 注册广播接收者
     */
    private void initBroadCastReceiver() {
        myReceiver = new MyReceiver();
        IntentFilter filter = new IntentFilter("filter");
        registerReceiver(myReceiver,filter);
    }
```

```
class MyReceiver extends BroadcastReceiver {
    @Override
    public void onReceive(Context context, Intent intent) {
        int num = intent.getIntExtra("num", 0);
        int price = intent.getIntExtra("price", 0);
        int newAmount = num * price;
            int sum = Integer.parseInt(sumTxv.getText().toString());
            sum += newAmount;
            sumTxv.setText(sum + "");
    }
}

    @Override
protected void onDestroy() {
    super.onDestroy();
    if (myReceiver != null){
        unregisterReceiver(myReceiver);
    }
}
  ……//省略其他代码
  private void initOrderedLtv() {
    SimpleAdapter sa = new SimpleAdapter(this, orderedList,
            R.layout.ordered, orderedLtvKeys, orderedLtvIds) {
        // 覆盖此方法，以使 Checkbox 可修改合计金额
        @Override
        public View getView(int position, View convertView, ViewGroup parent) {
            final int lineNo = position;
            View view = super.getView(position, convertView, parent);
            // 为 CheckBox 添加事件，修改选中状态时计算合计金额
            CheckBox checkCkb = (CheckBox) view.findViewById(R.id.checkCkb);
            checkCkb.setOnCheckedChangeListener(new CheckBox.OnCheckedChangeListener() {
                @Override
                public void onCheckedChanged(CompoundButton buttonView,
                                                    boolean isChecked) {
                    @SuppressWarnings("unchecked")
                    Map<String, Object> line = (Map<String, Object>) getItem(lineNo);
                    int num = (Integer) line.get("num");
                    int price = (Integer) line.get("price");
                    if (isChecked) {
                        //发送广播
```

```
                            Intent intent = new Intent("filter");
                            intent.putExtra("num",num);
                            intent.putExtra("price",price);
                            sendBroadcast(intent);
                        } else {
                            //发送广播
                            Intent intent = new Intent("filter");
                            intent.putExtra("num",-num);
                            intent.putExtra("price",price);
                            sendBroadcast(intent);
                        }
                        line.put("checked", isChecked);
                    }
                });
                return view;
            }
        };
        orderedLtv.setAdapter(sa);
    }
}
```

上述代码在 onCreate()方法中使用 registerReceiver()方法动态注册 BroadcastReceiver，其中两个参数分别为 MyReceiver 和 IntentFilter；注册完成后，使用 sendBroadcast()方法发送广播，用 Intent 把数据传递给 BroadcastReceiver。注意：最后要在 onDestroy()方法中把 BroadcastReceiver 注销。

编写完毕，在 Android Studio 中运行项目 Repast_ph05，单击选择或取消选择已点餐品列表行末的 CheckBox，就会自动计算总金额，如图 S5-1 所示。

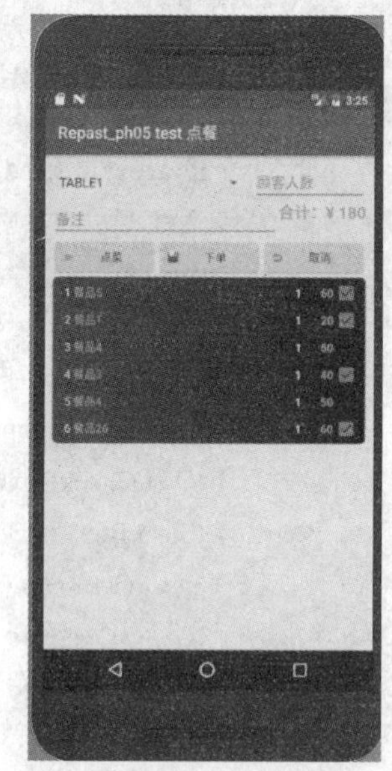

图 S5-1　测试点餐金额计算功能

知识拓展

在 Android 应用程序中，当退出 App 时，应用程序会弹出一些提示，如“再按一次退出程序”，或者“确定退出程序？”。一方面，这种提示优化了人机交互，使程序 UI 更加美观和人性化；另一方面，由于移动应用场景中用户的注意力较分散，容易出现误操作行为，此功能也可降低用户误操作退出的概率。本知识拓展将简单介绍 Android 实现此功能的两种方式。

1. 单击两次退出程序

在 Android Studio 中新建项目，实现“再按一次退出程序”功能。

编写 MainActivity，代码如下：

```
public class MainActivity extends AppCompatActivity {

    private long exitTime = 0;

    @Override
    protected void onCreate(Bundle savedInstanceState) {
        super.onCreate(savedInstanceState);
        setContentView(R.layout.activity_main);
    }
    @Override
    public boolean onKeyDown(int keyCode, KeyEvent event) {
        if (keyCode == KeyEvent.KEYCODE_BACK){
            if (System.currentTimeMillis() - exitTime > 2000){
                Toast.makeText(MainActivity.this,"再按一次退出程序",Toast.LENGTH_SHORT).show();
                exitTime = System.currentTimeMillis();
            }else {
                System.exit(0);
            }
            return true;
        }
        return super.onKeyDown(keyCode, event);
    }
}
```

上述代码中通过重写退出页面的 Activity 中的 onKeyDown()方法来实现此功能：首先，判断用户是否点击了手机系统的返回键；其次，通过 System. currentTimeMillis()方法获取系统时间(注意：该方法的返回值类型为 long 类型，其开始时间是从 1970 年 1 月 1 月开始，截止到点击时的时间)。

编写完毕，在 Android Studio 中运行程序，单击一次手机系统返回键，界面显示字符串“再按一次退出程序”，如图 S5-2 所示；当连续单击两次返回键时，才会退出应用程序。

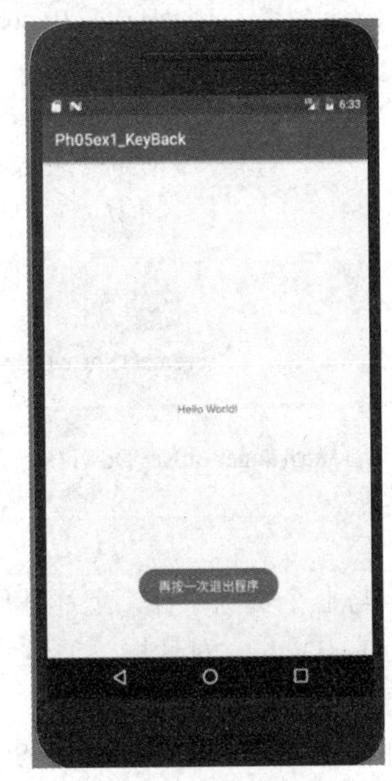

图 S5-2　再按一次退出应用程序

2．询问后退出程序

前面实现了“再按一次退出程序”的功能，而在另一些 Android 应用程序中，退出时会弹出一个询问对话框，提示“确定退出程序吗？”，单击对话框中的确认按钮，才会退出应用程序。二者的实现原理相同，只是显示的形式不同。下面在 Android Studio 中新建一个项目，实现弹出询问对话框后退出程序的功能。

编写 MainActvity，代码如下：

```
public class MainActivity extends AppCompatActivity {

    @Override
    protected void onCreate(Bundle savedInstanceState) {
        super.onCreate(savedInstanceState);
        setContentView(R.layout.activity_main);
    }
    @Override
    public boolean onKeyDown(int keyCode, KeyEvent event) {
        if (keyCode == KeyEvent.KEYCODE_BACK){
            new AlertDialog.Builder(this)
                    .setTitle("确认退出应用吗？")
                    .setIcon(R.mipmap.ic_launcher)
                    .setPositiveButton("确定", new DialogInterface.OnClickListener() {

                        @Override
                        public void onClick(DialogInterface dialog, int which) {
                            // 单击“确认”后的操作
                            finish();
                        }
                    })
                    .setNegativeButton("返回", new DialogInterface.OnClickListener() {

                        @Override
                        public void onClick(DialogInterface dialog, int which) {
                            // 单击“返回”后的操作，这里没有设置任何操作
                        }
                    })
                    .create().show();
        }
        return super.onKeyDown(keyCode, event);
    }
}
```

与上个实例一样，上述代码重写了 onKeyDown()方法，但不同之处在于：该实例使用了 AlertDialog 对话框，通过设置对话框的图片、标题、提示信息和两个按钮来实现询问后退出程序的功能。

编写完毕，在 Android Studio 中运行程序，单击手机系统的返回键，出现界面如图 S5-3 所示。单击对话框中的【返回】按钮，对话框消失；单击【确定】按钮，则退出应用

程序。

图 S5-3　询问后退出应用程序

拓展练习

参考本章知识拓展 1 和 2，编写一个应用程序，实现使用广播方式退出应用程序的功能。

实践 6　服务(Service)

实践指导

实践　更新数据功能

实现更新数据的功能。

【分析】

(1) 在点餐系统中，餐桌、餐品、餐品类型的数据较少被修改，因此可以在 Android 客户端保存这些数据，需要使用时直接从客户端查询，而不需要连接服务器，这样可以提高查询速度。

(2) 当服务器端数据发生变化时，可以通知用户更新数据，即连接服务器并下载数据到客户端，单击主菜单中的更新数据图标即可开始更新。

(3) 更新数据可能需要较长的时间，适合使用 Android 的 Service 实现。

(4) 当更新开始时，需要显示对话框来提示用户正在更新数据；更新完毕或者更新过程出现错误时，也需要使用对话框显示相应的提示信息。通过 Android 的广播功能，可使更新数据的 Service 在更新完毕或出现错误时向外发送广播，Activity 接收广播后即可做出相应的操作。

【参考解决方案】

(1) 编写 Android 服务类 com.ugrow.repast_ph06.service.UpdateDataService，代码如下：

```
public class UpdateDataService extends Service {

	public static final String OK = "OK";
	public static final String EXCEPTION = "EXCEPTION";

	@Override
	publicIBinderonBind(Intent intent) {
		return null;
	}

	@Override
```

```
    public void onCreate() {
        super.onCreate();
        new Thread() {
            @Override
            public void run() {
                try {
                    updateData();
                    sendBroadcast(new Intent(OK)); // 广播更新成功
                } catch (Exception e) {
                    e.printStackTrace();
                    // 广播更新失败
                    Intent intent = new Intent(EXCEPTION);
                    intent.putExtra(EXCEPTION, e.getMessage());
                    sendBroadcast(intent);
                }
                stopSelf();
            }
        }.start();
    }

    /**
     * 从服务器下载数据，并更新到本地数据库
     */
    public void updateData() throws InterruptedException {
        // TODO
        Thread.sleep(3000); // 模拟耗时操作
    }
}
```

上述代码中，UpdateDataService 继承了 Service，是 Android 的服务。更新数据操作需要在 updateData()方法中连接服务器下载数据并更新到本地数据库，此处使当前线程暂停 3 秒钟以模拟下载操作，并在 onCreate()方法中开启一个新线程来调用 updateData()方法，updateData()方法执行完毕后，会发送更新成功的广播信息，如果出现异常，则发送更新失败的广播信息，并在 Intent 中存储异常信息。

(2) 修改 MainActivity，实现在单击【更新数据】图标时启动 UpdateDataService 服务，代码如下：

```
public class MainActivity extends BasicActivity {

    private int[] icons = { R.mipmap.icon1, R.mipmap.icon2,
            R. mipmap.icon3,R. mipmap.icon4, R. mipmap.icon5 };
```

```
private String[] iconTexts = { "点餐", "结账", "查桌", "更新数据",
        "设置" };
private GridView gdv;

private ProgressDialog updateDialog;
private BroadcastReceiver boradcastReceiver;

@Override
public void onCreate(Bundle savedInstanceState) {
    super.onCreate(savedInstanceState);
    setContentView(R.layout.activity_main);

    gdv = (GridView) findViewById(R.id.gdv);
    List<Map<String, Object>>iconList = new ArrayList<Map<String, Object>>();
    for (int i = 0, j = icons.length; i < j; i++) {
        Map<String, Object> map = new HashMap<String, Object>();
        map.put("imageView", icons[i]);
        map.put("imageTitle", iconTexts[i]);
        iconList.add(map);
    }
    gdv.setAdapter(new SimpleAdapter(this,iconList,R.layout.menuitem,
        new String[] { "imageView", "imageTitle" },
            new int[] {R.id.imageView, R.id.imageTitle }));
gdv.setOnItemClickListener(new AdapterView.OnItemClickListener(){
        @Override
        public void onItemClick(AdapterView<?> arg0, View arg1, intidx,long arg3) {
            switch (idx) {
            ...... // 省略其他代码
            case 3:
                updateData();
                break;
            ...... // 省略其他代码                  }
        }
    });

    updateDialog = new ProgressDialog(MainActivity.this);
    updateDialog.setMessage("正在更新数据，请稍候......");
    updateDialog.setCancelable(false);
    IntentFilter f = new IntentFilter();
```

```
        f.addAction(UpdateDataService.OK);
        f.addAction(UpdateDataService.EXCEPTION);
        boradcastReceiver = new BroadcastReceiver() {
            @Override
            public void onReceive(Context context, Intent intent) {
                updateDialog.hide();
                String action = intent.getAction();
                if (UpdateDataService.OK.equals(action))
                    showMessageDialog("更新完成", R. mipmap.info, null);
                else if (UpdateDataService.EXCEPTION.equals(action))
                    showMessageDialog("更新失败："+
                    intent.getStringExtra(UpdateDataService.EXCEPTION),
                        R. mipmap.not, null);
            }
        };
        registerReceiver(boradcastReceiver, f);
    }

    void updateData() {
        AlertDialog.Builder b=new AlertDialog.Builder(MainActivity.this);
    b.setIcon(R. mipmap.warning);
    b.setTitle("更新需要较长时间，确定需要更新吗？");
    b.setPositiveButton("确定",new DialogInterface.OnClickListener(){
            @Override
            public void onClick(DialogInterface dialog, int which) {
                updateDialog.show();
                startService(new Intent(MainActivity.this, UpdateDataService.class));
            }
        });
        b.setNegativeButton("取消", null);
        b.create().show();
    }

    @Override
    protected void onDestroy() {
        super.onDestroy();
        unregisterReceiver(boradcastReceiver);
    }
    @Override
```

```
        protected String getName() {
                return "主菜单";
        }
}
```

上述代码中，首先在主菜单界面中定义了一个 ProgressDialog，用于在更新数据时显示提示信息；在单击【更新数据】图标时调用 updateData()方法，在 updateData()方法中显示 ProgressDialog，并开始执行 UpdateDataService 服务；然后在 MainActivity 的 onCreate()方法中注册了广播接收器，并指定接收 UpdateDataService.OK 和 UpdateDataService.EXCEPTION 的 Intent；在广播接收器的 onReceive()方法中先关闭 ProgressDialog，然后判断 Intent 的 action，如果是 OK，则显示更新成功对话框，否则显示更新失败对话框，并显示异常信息；最后在 MainActivity 的 onDestroy()方法中注销广播接收器。

注 意　ProgressDialog 的使用可参见实践 3 中知识拓展的第 6 部分内容。

(3) 在 AndroidManifest.xml 文件中配置 UpdateDataService，代码如下：

```
<service android:name=".service.UpdateDataService" />
```

运行项目，登录成功后，在主菜单界面单击【更新数据】图标，将显示确认对话框，单击对话框中的【确定】按钮，即开始更新数据，并显示进度对话框。更新成功后将显示对话框，提示更新完成，如图 S6-1 所示。

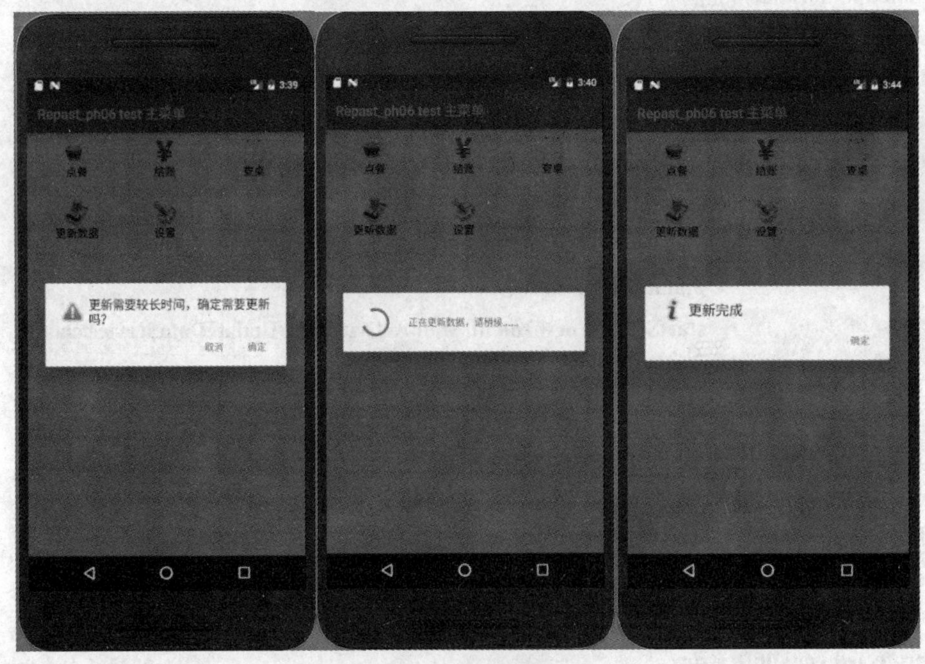

图 S6-1　测试更新数据功能

修改 UpdateDataService，在 updateData()方法中故意抛出异常，以模拟发生了错误，代码如下：

```
public class UpdateDataService extends Service {
```

```
    ......// 省略其他代码
    public void updateData() throws Exception {
        // TODO
        Thread.sleep(3000); // 模拟耗时操作
        throw new Exception("连接服务器失败");
    }
}
```

重新运行项目，执行更新数据操作，将显示对话框提示更新错误，如图 S6-2 所示。

图 S6-2　数据更新失败提示

知识拓展

Android 提供了针对多种常见多媒体格式的编码和解码 API，可以非常方便地操作音频、视频等多媒体文件。其中，使用 android.media.MediaPlayer 类可以实现音频、视频文件的播放，包括播放、暂停、停止、重复播放等功能。这些文件可以位于本地文件系统或者项目资源目录中，也可以是网络上的文件流。

1. 播放音频

下面是实现简单音频播放功能的示例。首先编写布局文件，代码如下：

```
<?xml version="1.0" encoding="utf-8"?>
<LinearLayoutxmlns:android="http://schemas.android.com/apk/res/android"
```

```
        android:layout_width="match_parent"
        android:layout_height="wrap_content">
        <Button
            android:id="@+id/startBtn"
            android:layout_width="match_parent"
            android:layout_height="wrap_content"
            android:text="开始"
            android:layout_weight="1"/>
        <Button
            android:id="@+id/pauseBtn"
            android:layout_weight="1"
            android:text="暂停"
            android:enabled="false"
            android:layout_width="match_parent"
            android:layout_height="wrap_content"/>
        <Button
            android:id="@+id/stopBtn"
            android:text="停止"
            android:enabled="false"
            android:layout_weight="1"
            android:layout_width="match_parent"
            android:layout_height="wrap_content"/>
</LinearLayout>
```

上述代码定义了【开始】、【暂停】、【停止】三个按钮。

然后编写 Activity，代码如下：

```
public class SoundPlayer extends Activity {

    MediaPlayer player;

    Button startBtn;
    Button pauseBtn;
    Button stopBtn;

    @Override
    public void onCreate(Bundle savedInstanceState) {
        super.onCreate(savedInstanceState);
        setContentView(R.layout.sound);

        // 播放 res/raw 目录下的 windowsxp 文件
```

```
player = MediaPlayer.create(SoundPlayer.this, R.raw.windowsxp);
startBtn = (Button) findViewById(R.id.startBtn);
pauseBtn = (Button) findViewById(R.id.pauseBtn);
stopBtn = (Button) findViewById(R.id.stopBtn);

// 播放完成事件
player.setOnCompletionListener(new OnCompletionListener() {
        @Override
        public void onCompletion(MediaPlayermp) {
                startBtn.setEnabled(true);
                pauseBtn.setEnabled(false);
                stopBtn.setEnabled(false);
        }
});

// 开始按钮
startBtn.setOnClickListener(new OnClickListener() {
        @Override
        public void onClick(View v) {
                player.start(); // 开始播放
                startBtn.setEnabled(false);
                pauseBtn.setEnabled(true);
                stopBtn.setEnabled(true);
        }
});

// 暂停按钮
pauseBtn.setOnClickListener(new OnClickListener() {
        @Override
        public void onClick(View v) {
                player.pause(); // 暂停播放
                startBtn.setEnabled(true);
                pauseBtn.setEnabled(false);
                stopBtn.setEnabled(true);
        }
});

// 停止按钮
stopBtn.setOnClickListener(new OnClickListener() {
        @Override
```

```
                    public void onClick(View v) {
                        player.stop(); // 停止播放
                        startBtn.setEnabled(true);
                        pauseBtn.setEnabled(false);
                        stopBtn.setEnabled(false);
                        try {
                            player.prepare(); // 为下次播放准备
                            player.seekTo(0); // 回到音频起点
                        } catch (Exception e) {
                            e.printStackTrace();
                        }
                    }
                });
            }

    @Override
    protected void onDestroy() {
        super.onDestroy();
        player.release(); // 释放资源
    }
}
```

上述代码中，首先调用 MediaPlayer.create()方法创建了 MediaPlayer 对象，并指定播放 res/raw 目录下的 windowsxp 资源文件；然后分别为【开始】、【暂停】、【停止】按钮注册了单击事件；最后为 MediaPlayer 注册了 OnCompletionListener 事件，当播放完毕时会自动调用其 onCompletion()方法。

运行项目，单击【开始】按钮，将播放 res/raw 目录下的 windowsxp.wmv 文件；单击【暂停】按钮会暂停播放；再次单击【开始】按钮，将继续播放；单击【停止】按钮，将结束播放，如图 S6-3 所示。

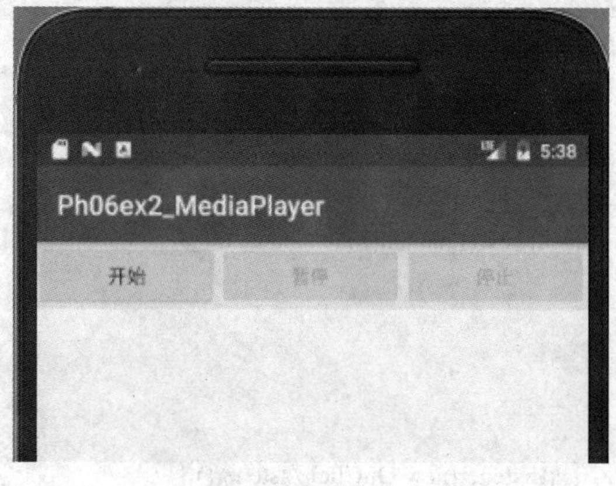

图 S6-3　测试音频播放功能

2．播放视频

使用 MediaPlayer 类也可以播放视频文件，但是比较复杂。而 Android 提供了 VideoView 控件，其内置了 MediaPlayer，使用 VideoView 结合 MediaController 类可以方便地实现视频播放。

下面是实现视频播放功能的示例。首先编写布局文件，代码如下：

```
<?xml version="1.0" encoding="utf-8"?>
<LinearLayoutxmlns:android="http://schemas.android.com/apk/res/android"
    android:layout_width="match_parent"
    android:layout_height="wrap_content"
    android:gravity="center_horizontal" >
    <VideoView
        android:id="@+id/videoView"
        android:layout_width="match_parent"
        android:layout_height="match_parent"/>
</LinearLayout>
```

上述代码定义了一个 VideoView 控件。

然后编写 Activity，代码如下：

```
public class VideoPlayer extends AppCompatActivity {
    VideoViewvideoView;
    MediaController controller;
    @Override
    public void onCreate(Bundle savedInstanceState) {
        super.onCreate(savedInstanceState);
        // 不显示标题
        requestWindowFeature(Window.FEATURE_NO_TITLE);
        // 全屏
        getWindow().setFlags(WindowManager.LayoutParams.FLAG_FULLSCREEN,
                WindowManager.LayoutParams.FLAG_FULLSCREEN);
        // 横屏
        setRequestedOrientation(
            ActivityInfo.SCREEN_ORIENTATION_LANDSCAPE);
        setContentView(R.layout.video);
        videoView = (VideoView) findViewById(R.id.videoView);
        controller = new MediaController(this);
        // 设置播放的文件
        videoView.setVideoURI(Uri.parse("android.resource://"
                + getPackageName() + "/" + R.raw.td));
        // 关联 VideoView 和 MediaController
        controller.setMediaPlayer(videoView);
```

```
            videoView.setMediaController(controller);
            // 开始播放
            videoView.start();
    }
}
```

上述代码中，首先设置了不显示应用程序标题，并分别设置了全屏以及横屏显示，这样更适合播放视频；然后指定了 VideoView 播放的视频文件，注意 VideoView 无法直接访问资源文件，而是需要以 URI 的形式指定资源；接着将 VideoView 和 MediaController 相关联；最后调用 VideoView 的 start()方法，开始播放视频。

运行项目，将显示正在播放的视频文件，如图 S6-4 所示。

图 S6-4　播放视频

单击屏幕后，将显示播放控制区，其中包括前进、后退、暂停按钮，以及一个可拖动的进度条，使用这些按钮和进度条可以调整播放进度，如图 S6-5 所示。

图 S6-5　测试播放控制区

拓展练习

结合本实践的知识拓展，编写一个综合的多媒体程序，实现播放音频和视频的功能。

实践 7　数据存储

实践指导

实践 7.1　创建数据库

为点餐系统在 Android 客户端创建 SQLite 数据库，保存餐品、餐品类型、餐桌等的数据。

【分析】

(1) 在点餐系统中，餐桌、餐品、餐品类型的数据较少被修改，因此可以在 Android 客户端使用内置的 SQLite 数据库保存这些数据，需要使用时直接从客户端查询，而无需再连接服务器，这样可以提高查询速度。用户可通过主菜单的更新数据功能连接服务器并下载数据到客户端。

(2) 编写 SQLiteOpenHelper 的子类 DbHelper，在业务类中使用 DbHelper 来完成数据库操作。

(3) 在 DbHelper 的 onCreate()方法中创建餐桌、餐品、餐品类型表，可使用以下 SQL 语句：

```
create table if not exists food(
    id integer primary key,
    code varchar(5),
    type_id integer,
    name varchar(20),
    price integer,
    description varchar(100))
    create table if not exists food_type(
        id integer primary key,
        name varchar(10)))
    create table if not exists tables(
        id integer primary key,
        code varchar(5),
        seats integer,
```

```
        description varchar(50))
```

(4) 用户更新数据时需要删除原有的数据，因此 DbHelper 中需提供删除数据库的方法 deleteDb()。

【参考解决方案】

编写 com.ugrow.repast_ph07.db.DbHelper，代码如下：

```
public class DbHelper extends SQLiteOpenHelper {
    /**
     * 数据库文件
     */
    privatestatic String dbName = "repast.db";
    privateContext context;

    /**
     * @param context 上下文
     */
    public DbHelper(Context context) {
        super(context, dbName, null, 1);
        this.context = context;
    }
    /**
     * 删除数据库
     */
    public void deleteDb() {
        context.deleteDatabase(dbName);
    }
    @Override
    public void onCreate(SQLiteDatabase db) {
        db.execSQL("create table if not exists food (id integer primary key, code varchar(5), type_id
    integer, name varchar(20), price integer, description varchar(100))");
        db.execSQL("create table if not exists food_type (id integer primary key, name varchar(10))");
        db.execSQL("create table if not exists tables (id integer primary key, code varchar(5), seats
    integer, description varchar(50))");
    }

    @Override
    public void onUpgrade(SQLiteDatabase db, int oldVersion, int newVersion) {
        //
    }
}
```

实践 7.2　数据更新功能

修改更新数据服务 UpdateDataService 的 updateData()方法，使用 DbHelper 将数据插入到本地数据库。

【分析】

(1) 更新数据时首先需要从服务器下载完整的餐桌、餐品、餐品类型数据，然后调用 DbHelper 的 getWritableDatabase()方法来获取 SQLiteDatabase 对象，通过 SQLiteDatabase 的 execSQL()方法执行插入语句，即可将数据保存到本地数据库。

(2) 更新数据时会操作餐桌、餐品、餐品类型三个表，并执行多条 INSERT 语句，因此需要使用 SQLite 的事务管理功能。

(3) 构造 DbHelper 对象时需要 Context 对象，通过 UpdateDataService 的 getApplicationContext()方法可获取当前应用的 Context 对象，将其传入 DbHelper 构造方法即可。

【参考解决方案】

修改 com.ugrow.repast.ph07.service.UpdateDataService，代码如下：

```
public class UpdateDataService extends Service {
	...... // 省略原有代码
	/**
	 * 从服务器下载数据，并更新到本地数据库
	 */
	@SuppressWarnings({ "rawtypes", "unchecked" })
	void updateData() {
		DbHelper dbHelper = new DbHelper(getApplicationContext());
		// 删除已有数据
		dbHelper.deleteDb();
		// TODO data 是从服务器获取的数据
		ArrayList[] data = downloadFromServer();
		ArrayList<Food> foods = data[0];
		ArrayList<FoodType> foodTypes = data[1];
		ArrayList<Table> tables = data[2];
		// 插入到本地数据库
		SQLiteDatabase db = dbHelper.getWritableDatabase();
		// 开始事务
		db.beginTransaction();
		String sql1 = "insert into food(id,code,type_id,name,price,
				description) values (?,?,?,?,?,?)";
		String sql2 = "insert into food_type(id,name) values (?,?)";
		String sql3 = "insert into tables(id,code,seats,description) values (?,?,?,?)";
```

```
        for (Food f : foods)
            db.execSQL(
                sql1,new Object[] { f.getId(), f.getCode(),f.getTypeId(),
                f.getName(), f.getPrice(), f.getDescription()});
        for (FoodType ft : foodTypes)
            db.execSQL(sql2, new Object[] { ft.getId(), ft.getName() });
        for (Table t : tables)
            db.execSQL(sql3,new Object[] { t.getId(), t.getCode(), t.getSeats(),t.getDescription() });

        // 提交
        db.setTransactionSuccessful();
        db.endTransaction();
        db.close();
    }
    // 模拟从服务器获取数据
    @SuppressWarnings("rawtypes")
    ArrayList[] downloadFromServer() {
        ArrayList[] data = new ArrayList[3];
        ArrayList<FoodType> types = new ArrayList<FoodType>();
        ArrayList<Food> foods = new ArrayList<Food>();
        ArrayList<Table> tables = new ArrayList<Table>();
        data[0] = foods;
        data[1] = types;
        data[2] = tables;

        // 餐品类型假数据
        FoodType ft = new FoodType();
        ft.setId(1);
        ft.setName("热菜");
        types.add(ft);
        ft = new FoodType();
        ft.setId(2);
        ft.setName("凉菜");
        types.add(ft);
        ft = new FoodType();
        ft.setId(3);
        ft.setName("烧烤");
        types.add(ft);
        ft = new FoodType();
```

```
            ft.setId(4);
            ft.setName("酒水");
            types.add(ft);
            ft = new FoodType();
            ft.setId(5);
            ft.setName("主食");
            types.add(ft);
            // 餐品假数据
            for (int i = 1; i <= 40; i++) {
                Food f = new Food();
                f.setId(i);
                f.setCode("FOOD" + i);
                f.setTypeId(i % 5 + 1);
                f.setName("餐品" + i);
                f.setPrice((i % 7 + 1) * 10);
                foods.add(f);
            }
            // 餐桌假数据
            for (int i = 1; i <= 20; i++) {
                Table t = new Table();
                t.setId(i);
                t.setCode("TABLE" + i);
                t.setSeats(i % 5 * 2 + 2);
                t.setCustomers(i % 3 == 0 ? t.getSeats() : 0);
                t.setDescription(i % 4 == 0 ? "靠窗" : "");
                tables.add(t);
            }
            return data;
    }
}
```

上述代码首先在 updateData()方法里构造了 DbHelper 的实例，并调用 deleteDb()方法删除了原有数据；然后从服务器获取餐桌、餐品、餐品类型的完整数据；接着调用 DbHelper 的 getWritableDatabase()方法获取了 SQLiteDatabase 实例，并调用 beginTransaction()方法开始事务；再遍历每个餐桌、餐品、餐品类型对象，将相应的 SQL 语句插入对应的表中；最后提交事务并关闭 SQLiteDatabase。

注 意

downloadFromServer()方法负责实现从服务器下载数据的功能，本实践使用的是模拟数据，在实践 9 中将完善此功能。

实践 7.3　操作数据库

修改餐品、餐桌业务类，将获取数据的操作改为从本地数据库中查询获取。

【分析】

(1) 业务类中需要使用 DbHelper 操作数据库，因此要在构造方法中传入 Context 对象来构造 DbHelper 实例，同时修改使用这些业务类的 Activity，在其中传入 Context 对象。

(2) 使用 SQLiteDatabase 的 query()方法可以查询数据库。

【参考解决方案】

(1) 修改 com.ugrow.repast_ph07.service.TableService，代码如下：

```
public class TableService {
	private DbHelper dbHelper;
	public TableService(Context context) {
		dbHelper = new DbHelper(context);
	}
	/**
	 * 获取所有餐桌。本地查询，查询结果中的当前就餐人数不可用
	 *
	 * @return 餐桌
	 */
	public List<Table> getTables() {
		List<Table> list = new ArrayList<Table>();
		SQLiteDatabase db = dbHelper.getWritableDatabase();
		Cursor cursor = db.query("tables",null,null,null,null,null,null);
		for (cursor.moveToFirst(); !cursor.isAfterLast();
		cursor.moveToNext()) {
			Table t = new Table();
			int id = cursor.getInt(cursor.getColumnIndex("id"));
			String code=cursor.getString(cursor.getColumnIndex("code"));
			int seats = cursor.getInt(cursor.getColumnIndex("seats"));
			String description = cursor.getString(cursor
					.getColumnIndex("description"));
			t.setId(id);
			t.setCode(code);
			t.setSeats(seats);
			t.setDescription(description);
			list.add(t);
		}
		cursor.close();
```

```
            db.close();
            return list;
        }
}
```

(2) 修改 com.ugrow.repast_ph07.service.FoodService，代码如下：

```
public class FoodService {
        private DbHelper dbHelper;
        public FoodService(Context context) {
                dbHelper = new DbHelper(context);
        }
        /**
         * 获取所有的餐品类型。本地查询
         *
         * @return 餐品类型
         */
        public List<FoodType> getFoodTypes() {
                List<FoodType> list = new ArrayList<FoodType>();
                SQLiteDatabase db = dbHelper.getWritableDatabase();
                Cursor cursor = db.query("food_type", null, null, null, null, null,null);
                for (cursor.moveToFirst(); !cursor.isAfterLast();
                cursor.moveToNext()) {
                        FoodType t = new FoodType();
                        int id = cursor.getInt(cursor.getColumnIndex("id"));
                        String name = cursor.getString(cursor.getColumnIndex("name"));
                        t.setId(id);
                        t.setName(name);
                        list.add(t);
                }
                cursor.close();
                db.close();
                return list;
        }
        /**
         * 根据类型获取对应的餐品。本地查询
         *
         * @param foodTypeId
         *   餐品类型，0 表示不限定类型
         *
         * @return 餐品
```

```
 */
public List<Food> getFoods(int foodTypeId) {
        List<Food> list = new ArrayList<Food>();
        SQLiteDatabase db = dbHelper.getWritableDatabase();
        String where = null;
        if (foodTypeId != 0)
                where = "type_id = " + foodTypeId;
        Cursor cursor=db.query("food",null,where,null,null,null,null);
        for (cursor.moveToFirst(); !cursor.isAfterLast();
        cursor.moveToNext()) {
                Food t = new Food();
                int id = cursor.getInt(cursor.getColumnIndex("id"));
                String code = cursor.getString(cursor.getColumnIndex("code"));
                int typeId = cursor.getInt(cursor.getColumnIndex("type_id"));
                String name = cursor.getString(cursor.getColumnIndex("name"));
                int price = cursor.getInt(cursor.getColumnIndex("price"));
                String description = cursor.getString(cursor.getColumnIndex("description"));
                t.setId(id);
                t.setCode(code);
                t.setTypeId(typeId);
                t.setName(name);
                t.setPrice(price);
                t.setDescription(description);
                list.add(t);
        }
        cursor.close();
        db.close();
        return list;
}

/**
 * 根据 ID 获取餐品。本地查询
 *
 * @param foodId
 * 餐品 ID
 * @return 餐品
 */
public Food getFood(int foodId) {
        Food t = null;
```

```
            SQLiteDatabase db = dbHelper.getWritableDatabase();
            Cursor cursor = db.query("food",null,"id="+foodId, null, null,
                    null, null);
            if (cursor.moveToFirst()) {
                    t = new Food();
                    int id = cursor.getInt(cursor.getColumnIndex("id"));
                    String code = cursor.getString(cursor.getColumnIndex("code"));
                    int typeId = cursor.getInt(cursor.getColumnIndex("type_id"));
                    String name = cursor.getString(cursor.getColumnIndex("name"));
                    int price = cursor.getInt(cursor.getColumnIndex("price"));
                    String description = cursor.getString(cursor.getColumnIndex("description"));
                    t.setId(id);
                    t.setCode(code);
                    t.setTypeId(typeId);
                    t.setName(name);
                    t.setPrice(price);
                    t.setDescription(description);
            }
            cursor.close();
            db.close();
            return t;
    }
}
```

(3) 修改 com.ugrow.repast_ph07.service.OrderService，代码如下：

```
public class OrderService {
    privateFoodService foodService;
    public OrderService(Context context) {
        foodService = new FoodService(context);
    }
    ...... // 省略已有代码
}
```

(4) 修改 com.ugrow.repast_ph07.ui.OrderActivity，代码如下：

```
public class OrderActivity extends BasicActivity {
    @Override
    protected void onCreate(Bundle savedInstanceState) {
        super.onCreate(savedInstanceState);

        tableService = new TableService(this);
        foodService = new FoodService(this);
```

```
        orderService = new OrderService(this);
        ...... // 省略已有代码
    }
    ...... // 省略已有代码
}
```

(5) 修改 com.ugrow.repast_ph07.ui.PayActivity，代码如下：

```
public class PayActivity extends BasicActivity {

    @Override
    protected void onCreate(Bundle savedInstanceState) {
        super.onCreate(savedInstanceState);

        orderService = new OrderService(this);
        ...... // 省略已有代码
    }
    ...... // 省略已有代码
}
```

运行 Repast_ph07 项目，登录后，首先执行更新数据操作，将数据保存到本地 SQLite 数据库中；然后执行点餐操作，其中的餐品、餐品类型、餐桌列表等信息都是从本地数据库中查询出来的。

实践 7.4　点餐系统的配置功能

实现点餐系统的配置功能。

【分析】

(1) 点餐系统需要连接服务器获取数据，服务器的地址可能发生变化，因此需要在点餐系统中加入服务器地址的配置项，使用户可以输入新的地址。

(2) 用户修改服务器地址后，需要将地址保存在 SharedPreferences(共享参数)中。程序启动时，可以从 SharedPreferences 中读取服务器地址供连接服务器使用。

(3) 需要编写 Tools 工具类并创建 SharedPreferences 存取方法；同时编写界面 ConfigActivity，用于输入配置项。

【参考解决方案】

(1) 编写 com.ugrow.repast_ph07.utils.Tools.java，代码如下：

```
public class Tools {
    private static final String CONFIG_FILE = "config"; // 文件名
    /**
     * 获取共享参数 - String
     *
     * @param ctx
```

```
 * @param key
 * @return
 */
public static String getStringPre(Context ctx, String key) {
        SharedPreferences pre = ctx.getSharedPreferences(CONFIG_FILE,
                        Activity.MODE_PRIVATE);
        return pre.getString(key, "");
}
/**
 * 获取共享参数 - int
 *
 * @param ctx
 * @param key
 * @return
 */
public static int getIntPre(Context ctx, String key) {
        SharedPreferences pre = ctx.getSharedPreferences(CONFIG_FILE,
                Activity.MODE_PRIVATE);
        return pre.getInt(key, -1);
}
/**
 * 添加共享参数 - String
 *
 * @param ctx
 * @param key
 * @param value
 * @return
 */
public static boolean putStringPre(Context ctx, String key,
String value) {
        Editor editor = getEditor(ctx);
        editor.putString(key, value);
        return editor.commit();
}
/**
 * 添加共享参数 - int
 *
 * @param ctx
 * @param key
```

```
     * @param value
     * @return
     */
    public static boolean putIntPre(Context ctx, String key, int value) {
        Editor editor = getEditor(ctx);
        editor.putInt(key, value);
        return editor.commit();
    }
    private static Editor getEditor(Context ctx) {
        SharedPreferences pre = ctx.getSharedPreferences(CONFIG_FILE,
                Activity.MODE_PRIVATE);
        return pre.edit();
    }
}
```

上述代码分别创建了 SharedPreferences 存取 String 和 int 类型数据的公共方法。在本项目中将会用到许多公共方法，这些公共方法都可以单独提取出来放到公共的工具类中，以便代码的重用。

(2) 修改 com.dh.repast_ph06.App 类，在其中添加服务器地址属性，代码如下：

```
public class App extends Application {
    /**
     * 保存当前登录者
     */
    public User user;
    /**
     * 获取服务器 url
     *
     * @return
     */
    public String getServerUrl() {
        return Tools.getStringPre(this, "serverUrl");
    }

}
```

上述代码在 App 中添加了获取服务器地址的方法 getServerUrl()，以供其他类使用。

(3) 编写布局文件 config.xml，代码如下：

```
<?xml version="1.0" encoding="utf-8"?>
<LinearLayout
    xmlns:android="http://schemas.android.com/apk/res/android"
    android:orientation="vertical"
```

```
        android:background="#ededed"
        android:layout_width="match_parent"
        android:layout_height="match_parent"
        android:padding="10dp" >
        <TextView
                android:text="服务器地址："
                android:textSize="18sp"
                android:textStyle="bold"
                android:gravity="bottom"
                android:layout_height="40dp"
                android:layout_width="wrap_content"
                android:textColor="#000"/>
        <EditText
                android:id="@+id/serverUrlEdt"
                android:hint="请输入服务器地址"
                android:layout_width="match_parent"
                android:layout_height="wrap_content"/>
        <LinearLayout
                android:layout_width="match_parent"
                android:layout_height="wrap_content">
                <Button
                        android:id="@+id/okBtn"android:text="保存"
                        android:drawableLeft="@mipmap/save"
                        android:layout_width="match_parent"
                        android:layout_height="wrap_content"
                        android:layout_weight="1"/>
                <Button
                        android:id="@+id/cancelBtn"
                        android:text="取消"
                        android:drawableLeft="@mipmap/back"
                        android:layout_width="match_parent"
                        android:layout_height="wrap_content"
                        android:layout_weight="1"/>
        </LinearLayout>
</LinearLayout>
```

(4) 编写选项配置界面类 com.ugrow.repast.ph07.ui.ConfigActivity，代码如下：

```
public class ConfigActivity extends BasicActivity {
        private EditText serverUrlEdt;
        private Button okBtn;
```

```
    private Button cancelBtn;

    @Override
    protected void onCreate(Bundle savedInstanceState) {
        super.onCreate(savedInstanceState);
        setContentView(R.layout.config);

        serverUrlEdt = (EditText) findViewById(R.id.serverUrlEdt);
        okBtn = (Button) findViewById(R.id.okBtn);
        cancelBtn = (Button) findViewById(R.id.cancelBtn);

        serverUrlEdt.setText(Tools.getStringPre(this, "serverUrl"));
        okBtn.setOnClickListener(new OnClickListener() {
            @Override
            public void onClick(View v) {
                String serverUrl = serverUrlEdt.getText().toString();
                Tools.putStringPre(ConfigActivity.this, "serverUrl", serverUrl);
                finish();

            }
        });
        cancelBtn.setOnClickListener(new OnClickListener() {
            @Override
            public void onClick(View v) {
                finish();
            }
        });
    }
    @Override
    protected String getName() {
        return "配置";
    }
}
```

上述代码在 ConfigActivity 的 onCreate()方法中调用 Tools 的 getStringPre(this, "serverUrl")方法以获取服务器地址；并在单击【保存】按钮时，调用 Tools 的 putStringPre(ConfigActivity.this, "serverUrl", serverUrl)方法保存配置信息。

(5) 修改 AndroidMenifest.xml 文件，在其中添加 ConfigActivity，代码如下：

```
<activity android:name=".ui.ConfigActivity" />
```

(6) 修改主菜单界面 MainActivity，使在单击配置图标时显示界面 ConfigActivity，代码如下：

```
public class MainActivity extends BasicActivity {
    ...... // 省略其他代码
    @Override
    public void onCreate(Bundle savedInstanceState) {
        super.onCreate(savedInstanceState);
        ...... // 省略其他代码
    gdv.setOnItemClickListener(new AdapterView.OnItemClickListener(){
            @Override
            public void onItemClick(AdapterView<?> arg0, View arg1,
            int idx,long arg3) {
                switch (idx) {
                ...... // 省略其他代码
                case 4:
                    showActivity(MainActivity.this, ConfigActivity.class);
                    break;
                }
            }
        });
    }
}
```

运行 Repast_ph07 项目，登录成功后在主菜单界面单击【设置】图标，会显示配置界面，如图 S7-1 所示；在界面中输入服务器地址后，单击【保存】按钮，该地址将保存到共享参数中，再次打开设置界面，将会显示上次保存的服务器地址，如图 S7-2 所示。

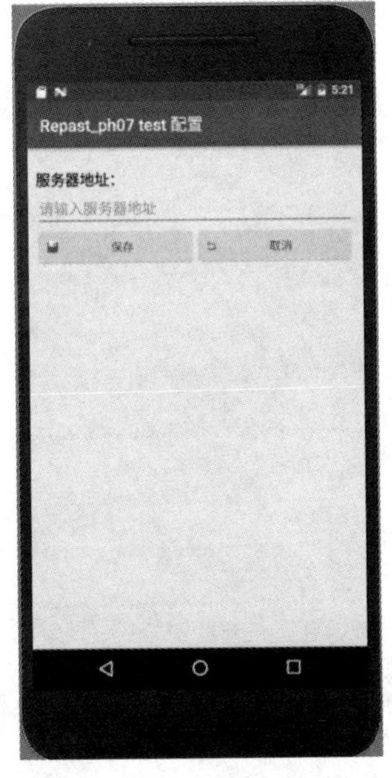

图 S7-1　服务器地址配置界面

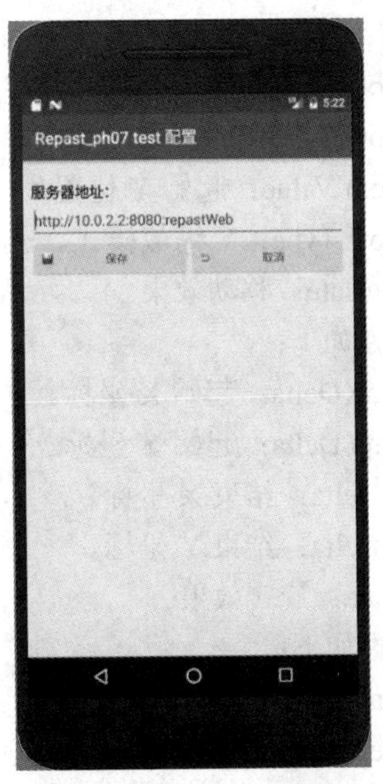

图 S7-2　显示已保存的地址

知识拓展

1. 制作动画

人眼具有“视觉暂留”的特点，当大量图片快速切换(通常认为每秒超过 24 幅)时，就会使人误认为是连续的动作，这是动画的基本原理。如果所有的画面都是手工制作的，则称为逐帧动画；如果只有某些画面是手工制作的(关键帧)，而其余画面是由程序自动生成的(中间帧)，则称为补间动画。电影是典型的逐帧动画，以每秒 24 幅的速度播放胶片上的图像，从而形成动态的影像；常见的 Flash 动画则绝大部分都是补间动画。

Android 支持实现补间和逐帧动画，并可采用在 XML 文件中配置和直接编码两种方式完成动画。本实践采用 XML 文件完成动画制作。注意 Android 的动画文件需存放于 res/anim 目录下。

针对补间动画，Android 提供了四种变换效果，分别具有不同的控制属性。Android 补间动画变换方式及常用的属性如下：

(1) Alpha：透明效果。

对应属性如下：

✧ fromAplha：起始透明度。
✧ toAlpha：结束透明度。

(2) Scale：缩放效果。

对应属性如下：

✧ fromX：起始 X 缩放比例。
✧ fromY：起始 Y 缩放比例。
✧ toX：结束 X 缩放比例。
✧ toY：结束 Y 缩放比例。
✧ pivotXValue：起始 X 位置。
✧ pivotXType：X 缩放模式。
✧ pivotYValue：起始 Y 位置。
✧ pivotYType：Y 缩放模式。

(3) Translate：移动效果。

对应属性如下：

✧ fromXDelta：起始 X 坐标。
✧ fromYDelta：起始 Y 坐标。
✧ toXDelta：结束 X 坐标。
✧ toYDelta：结束 Y 坐标。

(4) Rotate：旋转效果。

对应属性如下：

✧ fromDegrees：起始角度。
✧ toDegrees：结束角度。
✧ pivotXValue：起始 X 位置。

✧ pivotXType：X 缩放模式。

✧ pivotYValue：起始 Y 位置。

✧ pivotYType：Y 缩放模式。

Android 逐帧动画则是通过顺序显示图片以形成连续的动态效果，可结合 XML 动画文件和 AnimationDrawable 类来实现。

下面举例演示 Android 中补间动画和逐帧动画的简单应用。将需要使用的图片资源放入 res/mipmap-mdpi 目录下，如表 S7-1 所示。

表 S7-1　制作动画需要使用的图片资源

android.png		r0.png		r1.png	
r2.png		r3.png		r4.png	
r5.png		r6.png		r7.png	

其中的 android.png 是补间动画使用的图片，其余图片供逐帧动画使用。

(1) 在 res/anim 目录下添加补间动画文件，代码如下：

✧ alpha.xml。

```
<?xml version="1.0" encoding="utf-8"?>
<set xmlns:android="http://schemas.android.com/apk/res/android" >
<alpha
xmlns:android="5000"
android:fromAlpha="0.1"
android:toAlpha="1" />
</set>
```

✧ scale.xml。

```
<?xml version="1.0" encoding="utf-8"?>
<set xmlns:android="http://schemas.android.com/apk/res/android" >
<scale
android:duration="4000"
android:fromXScale="0"
android:fromYScale="0"
```

```
android:pivotX="50%"
android:pivotY="100%"
android:toXScale="1"
android:toYScale="1" />
</set>
```

✧ translate.xml。

```
<?xml version="1.0" encoding="utf-8"?>
<set xmlns:android="http://schemas.android.com/apk/res/android" >
<translate
android:duration="4000"
android:fromXDelta="-100"
android:fromYDelta="-100"
android:toXDelta="100"
android:toYDelta="100" />
</set>
```

✧ rotate.xml。

```
<?xml version="1.0" encoding="utf-8"?>
<set xmlns:android="http://schemas.android.com/apk/res/android" >
<rotate
android:duration="5000"
android:fromDegrees="0"
android:pivotX="50%"
android:pivotY="50%"
android:toDegrees="360" />
</set>
```

上述四个补间动画文件分别使用<alpha>、<scale>、<translate>、<rotate>元素定义了四种变化效果，并设置了对应的控制参数。

(2) 添加逐帧动画文件 robat.xml，代码如下：

```
<?xml version="1.0" encoding="utf-8"?>
<animation-list xmlns:android="http://schemas.android.com/apk/res/android"
    android:oneshot="false">
    <item android:drawable="@mipmap/r0" android:duration="100" />
    <item android:drawable="@mipmap//r1" android:duration="100" />
    <item android:drawable="@mipmap//r2" android:duration="100" />
    <item android:drawable="@mipmap//r3" android:duration="100" />
    <item android:drawable="@mipmap//r4" android:duration="100" />
    <item android:drawable="@mipmap//r5" android:duration="100" />
    <item android:drawable="@mipmap//r6" android:duration="100" />
    <item android:drawable="@mipmap//r7" android:duration="100" />
```

```
</animation-list>
```

上述逐帧动画文件指定了需要顺序播放的图片资源。

(3) 编写布局文件，代码如下：

```
<?xml version="1.0" encoding="utf-8"?>
<LinearLayout xmlns:android="http://schemas.android.com/apk/res/android"
    android:orientation="vertical"
    android:background="#ededed"
    android:layout_width="match_parent"
    android:layout_height="match_parent">
    <TextView
        android:text="补间动画"
        android:textSize="20sp"
        android:layout_width="match_parent"
        android:layout_height="wrap_content"/>
    <LinearLayout
        android:layout_width="match_parent"
        android:layout_height="wrap_content">
        <Button
            android:id="@+id/alpha"
            android:text="透明"
            android:layout_weight="1"
            android:layout_width="match_parent"
            android:layout_height="wrap_content"/>
        <Button
            android:id="@+id/scale"
            android:text="缩放"
            android:layout_weight="1"
            android:layout_width="match_parent"
            android:layout_height="wrap_content"/>
        <Button
            android:id="@+id/translate"
            android:text="移动"
            android:layout_weight="1"
            android:layout_width="match_parent"
            android:layout_height="wrap_content"/>
        <Button
            android:id="@+id/rotate"
            android:text="旋转"
            android:layout_weight="1"
            android:layout_width="match_parent"
```

```
                android:layout_height="wrap_content"/>
    </LinearLayout>
    <ImageView
        android:id="@+id/tween"
        android:src="@drawable/android"
        android:layout_width="wrap_content"
        android:layout_height="wrap_content"/>
    <TextView
        android:text="逐帧动画"
        android:textSize="20sp"
        android:layout_width="match_parent"
        android:layout_height="wrap_content"/>
    <ImageView
        android:id="@+id/frame"
        android:background="@anim/robat"
        android:layout_width="wrap_content"
        android:layout_height="wrap_content"/>
</LinearLayout>
```

上述文件定义了四个按钮控件和两个 ImageView 控件。单击不同的按钮，将对第一个 ImageView 执行相应的补间动画效果，而第二个 ImageView 则将循环播放逐帧动画。

(4) 编写 Activity，代码如下：

```
public class MainActivity extends AppCompatActivity {

    privateButton alpha;
    privateButton scale;
    privateButton translate;
    privateButton rotate;
    private ImageView tween; // 补间动画的 ImageView

    private ImageView frame; // 逐帧动画的 ImageView
    private AnimationDrawable drawable; // 逐帧动画使用的 AnimationDrawable

    @Override
    public void onCreate(Bundle savedInstanceState) {
        super.onCreate(savedInstanceState);
        setContentView(R.layout.activity_main);

        alpha = (Button) findViewById(R.id.alpha);
        scale = (Button) findViewById(R.id.scale);
        translate = (Button) findViewById(R.id.translate);
```

```
        rotate = (Button) findViewById(R.id.rotate);
        tween = (ImageView) findViewById(R.id.tween);

        frame = (ImageView) findViewById(R.id.frame);
        drawable = (AnimationDrawable) frame.getBackground();

        alpha.setOnClickListener(new OnClickListener() {
            @Override
            public void onClick(View v) {
                startTween(R.anim.alpha);
            }
        });
        scale.setOnClickListener(new OnClickListener() {
            @Override
            public void onClick(View v) {
                startTween(R.anim.scale);
            }
        });
        translate.setOnClickListener(new OnClickListener() {
            @Override
            public void onClick(View v) {
                startTween(R.anim.translate);
            }
        });
        rotate.setOnClickListener(new OnClickListener() {
            @Override
            public void onClick(View v) {
                startTween(R.anim.rotate);
            }
        });
    }
    // 执行某个补间动画，animId 代表 res/anim 下的动画资源名称
    void startTween(int animId) {
        Animation anim = AnimationUtils.loadAnimation(this, animId);
        tween.startAnimation(anim);
    }
    @Override
    public void onWindowFocusChanged(boolean hasFocus) {
        super.onWindowFocusChanged(hasFocus);
        drawable.start(); // 开始逐帧动画
```

```
    }
}
```

运行项目，可以看到逐帧动画已开始运行，而单击界面顶部的不同按钮，将显示对应的补间动画效果，如图 S7-3 所示。

图 S7-3　演示动画效果

2．绘制图形

绘制图形主要用到的是 android.graphics.Canvas 和 android.graphics.Paint 类。Canvas 类提供了绘制各种常见图形的方法(非常类似于 Java 中的 Graphics)，Paint 类则用于设置绘制的风格，如颜色、阴影、字体等。

Canvas 类的常用方法如表 S7-2 所示。

表 S7-2　Canvas 类的常用方法

方　法	说　明
drawText(String text, float x, float y, Paint paint)	绘制文字
drawPoint(float x, float y, Paint paint)	画点
drawLine(float x1, float y1, float x2, float y2, Paint paint)	画线
drawRect(RectF rect, Paint paint)	画矩形
drawCircle(float x, float y, float r, Paint paint)	画圆
drawOval(RectF rect, Paint paint)	画椭圆
drawRoundRect(RectF rect, float x, float y, Paint paint)	画圆角矩形
clipRect(float left, float top, float right, float bottom)	剪切矩形
clipRefion(Region region)	剪切区域

Paint 类的常用方法如表 S7-3 所示。

表 S7-3　Paint 类的常用方法

方　　法	说　　明
setColor(int color)	设置颜色
setStrokeWidth(float width)	设置线宽
setTextAlign(Paint.Align align)	设置文字对齐方式
setTextSize(float size)	设置文字大小
setShader(Shader shader)	设置渲染方式
setAlpha(int alpha)	设置透明度
setAntiAlias(boolean b)	抗锯齿
reset()	重置为默认值

在 Android 界面中绘制图形时，通常需要重写 View 的 onDraw()方法。onDraw()方法具有参数 Canvas，利用此参数可以绘制各种图形。

下面是基本图形绘制功能的实现示例。编写 Activity，代码如下：

```
public class MainActiivty extends AppCompatActivity {
    @Override
    public void onCreate(Bundle savedInstanceState) {
        super.onCreate(savedInstanceState);
        // 不显示标题
        requestWindowFeature(Window.FEATURE_NO_TITLE);
        // 全屏
        getWindow().setFlags(WindowManager.LayoutParams.FLAG_FULLSCREEN,
                    WindowManager.LayoutParams.FLAG_FULLSCREEN);
        // 使用自定义的 ExampleView
        setContentView(new ExampleView(this));
    }
    static class ExampleView extends View {
        public ExampleView(Context context) {
            super(context);
        }
        @Override
        protected void onDraw(Canvas c) {
            super.onDraw(c);
            c.drawColor(Color.parseColor("#ededed"));//设置背景色
            Paint p = new Paint();
            p.setAntiAlias(true); // 抗锯齿
```

```
                p.setColor(Color.GREEN); // 颜色
                p.setStyle(Paint.Style.STROKE); // 实线
                p.setStrokeWidth(4); // 线粗
                c.drawLine(0, 0, 320, 480, p); // 画直线

                p.setColor(Color.BLACK);
                p.setTextSize(40); // 文字大小
                c.drawText("Example", 20, 60, p); // 绘制文字

                p.setColor(Color.CYAN);
                c.drawCircle(160, 240, 160, p); // 画圆
                // 渲染效果
                Shader s = new LinearGradient(0, 0, 20, 20,
                        new int[] { Color.RED,
                                Color.MAGENTA }, null, TileMode.REPEAT);
                p.setShader(s); // 设置渲染效果
                p.setStyle(Paint.Style.FILL); // 填充
                // 画圆角矩形
                c.drawRoundRect(new RectF(80, 120, 240, 240), 20, 20, p);
        }
    }
}
```

运行项目，效果如图 S7-4 所示。

图 S7-4 测试图形绘制功能

注意 图形绘制涉及大量的类和方法，本实践只给出了一个简单的示例，如需了解更详细的信息，可以查阅 Android SDK API。

拓展练习

准备扑克牌的正面和背面图片各一张，制作补间动画，模拟翻转扑克牌的效果。

实践 8　片段(Fragment)

实践指导

实践　查桌功能

在点餐系统中实现查桌功能。

【分析】

(1) 使用在 Activity 中添加片段 Fragment 的方式实现查桌功能。

(2) 单击程序主菜单界面中的【查桌】图标，跳转到查桌界面。

(3) 新建 TableActivity 和 TableFragment，然后使用动态添加的方式，将 TableFragment 添加到 TableActivity 中。

(4) 在 TableFragment 中搭建界面，获取查桌数据。

注 意　获取查桌数据的操作需要连接点餐系统后台服务器，本章实践篇只将功能的演示界面展示出来，后续对网络操作进行介绍后将完善此功能。

【参考解决方案】

(1) 修改主菜单界面 MainActivity，实现在单击【查桌】图标时跳转到 TableActivity 的功能，代码如下：

```
public class MainActivity extends BasicActivity {
    …… // 省略其他代码

    @Override
    public void onCreate(Bundle savedInstanceState) {
        super.onCreate(savedInstanceState);

        …… // 省略其他代码
        gdv.setOnItemClickListener(
            new AdapterView.OnItemClickListener() {
                @Override
                public void onItemClick(AdapterView<?> arg0, View arg1,
                  int idx, long arg3) {
                    switch (idx) {
                        case 2:
                            //showMessageDialog("查桌", R.drawable.info,null);
                            showActivity(MainActivity.this,TableActivity.class);
```

```
                        break;
                        ……// 省略其他代码
                    }
                }
            });
        }
}
```

(2) 新建 TableFragment，编写其对应的布局文件 fragment_table.xml，代码如下：

```
<?xml version="1.0" encoding="utf-8"?>
<LinearLayout xmlns:android="http://schemas.android.com/apk/res/android"
    android:layout_width="match_parent"
    android:layout_height="match_parent"
    android:background="#ededed"
    android:orientation="vertical"
    android:padding="10dp" >

    <TableLayout
        android:layout_width="match_parent"
        android:layout_height="wrap_content"
        android:stretchColumns="0">

        <TableRow android:layout_width="match_parent">

            <EditText
                android:id="@+id/et_seats"
                android:layout_width="match_parent"
                android:layout_height="wrap_content"
                android:hint="需要座位数"
                android:inputType="number"
                android:singleLine="true" />

            <CheckBox
                android:id="@+id/cb_needEmptyCkb"
                android:layout_width="match_parent"
                android:layout_height="wrap_content"
                android:layout_gravity="right"
                android:checked="true"
                android:text="空桌" />
        </TableRow>
```

```
        <LinearLayout android:layout_width="match_parent">

            <Button
                android:id="@+id/btn_query"
                android:layout_width="match_parent"
                android:layout_height="wrap_content"
                android:layout_weight="1"
                android:drawableLeft="@mipmap/search"
                android:text="查询" />

            <Button
                android:id="@+id/btn_cansel"
                android:layout_width="match_parent"
                android:layout_height="wrap_content"
                android:layout_weight="1"
                android:drawableLeft="@mipmap/back"
                android:text="取消" />
        </LinearLayout>
    </TableLayout>

    <ListView
        android:id="@+id/lv_fragment"
        android:layout_width="wrap_content"
        android:layout_height="wrap_content"
        android:layout_marginTop="5dp"
        android:background="@drawable/bg1"
        android:visibility="gone" />

</LinearLayout>
```

(3) 编写 TableFragment，代码如下：

```
public class TableFragment extends Fragment {


    @Override
    public View onCreateView(LayoutInflater inflater, @Nullable ViewGroup container, @Nullable Bundle
savedInstanceState) {
        View view = inflater.inflate(R.layout.fragment_table, null);
        return view;
    }
```

```
    @Override
    public void onViewCreated(View view, @Nullable Bundle savedInstanceState) {
        super.onViewCreated(view, savedInstanceState);
    }

}
```

(4) 编写 TableActivity，代码如下：

```
public class TableActivity extends BasicActivity {

    @Override
    protected void onCreate(Bundle savedInstanceState) {
        super.onCreate(savedInstanceState);
        setContentView(R.layout.table);
        initView();
    }
    private void initView() {
        TableFragment tableFragment = new TableFragment();
        //得到 fragmentManager，把 Activity 和 Fragment 联系在一起
        FragmentManager fragmentManager = getSupportFragmentManager();
        //得到事务
        FragmentTransaction transaction = fragmentManager.beginTransaction();
        //添加
        transaction.add(R.id.fl_container,tableFragment);
        //提交事务
        transaction.commit();
    }

    @Override
    protected String getName() {
        return "查桌";
    }
}
```

(5) 编写 table.xml，代码如下：

```
<?xml version="1.0" encoding="utf-8"?>
<LinearLayout xmlns:android="http://schemas.android.com/apk/res/android"
    xmlns:tools="http://schemas.android.com/tools"
    android:layout_width="match_parent"
    android:layout_height="match_parent"
```

```
    android:orientation="vertical"
    tools:context="com.ugrow.repast_ph07.ui.TableActivity">

    <FrameLayout
        android:id="@+id/fl_container"
        android:layout_width="match_parent"
        android:layout_height="match_parent"/>

</LinearLayout>
```

上述代码在 Activity 中使用 transaction.add(R.id.fl_container,tableFragment)方法，将对应的 Fragment 动态添加到帧布局 FrameLayout 中，其中第一个参数为帧布局 FrameLayout 的 ID，第二个参数为添加的 Fragment 实例的名称。

编写完毕，在 Android Studio 中运行项目，单击【查桌】图标，会跳转到查桌界面，如图 S8-1 所示。

图 S8-1　跳转到查桌界面

知识拓展

在 Android 应用程序中，轮播图是常见的控件，通常 App 的首页上方和电商类 App 的商品详情图片都会使用轮播图。轮播图可以手动左右滚动，也可以自动轮播，本知识拓展将介绍如何使用 ViewPager 实现自动轮播。

在 Android Studio 中新建项目，实现轮播图的自动轮播。

在 activity_main.xml 中编写以下代码：

```
<?xml version="1.0" encoding="utf-8"?>
<RelativeLayout xmlns:android="http://schemas.android.com/apk/res/android"
    xmlns:tools="http://schemas.android.com/tools"
    android:layout_width="match_parent"
    android:layout_height="match_parent"
    tools:context="com.ugrow.ph08ex1_viewpager.MainActivity">


    <android.support.v4.view.ViewPager
        android:id="@+id/vp"
        android:layout_width="match_parent"
        android:layout_height="200dp"/>
    <LinearLayout
        android:layout_width="match_parent"
        android:layout_height="wrap_content"
        android:layout_alignBottom="@+id/vp"
        android:layout_marginBottom="20dp"
        android:orientation="vertical" >
        <LinearLayout
            android:id="@+id/ll_points"
            android:layout_width="match_parent"
            android:layout_height="wrap_content"
            android:gravity="center_horizontal"
            android:layout_marginTop="20dp"
            android:orientation="horizontal" >
        </LinearLayout>
    </LinearLayout>
</RelativeLayout>
```

上述代码中，ViewPager 是 v4 包中的一个类，继承自 ViewGroup，是一个容器，用来存放轮播的图片资源；线性布局 LinearLayout 则用来显示轮播时图片下方的小圆点，该小圆点用于标示图片的选择状态。

编写 MainActivity，代码如下：

```
public class MainActivity extends AppCompatActivity {

    private ViewPager mVp;
    private LinearLayout llPoints;
    private int[] images;
    private List<ImageView> mList;
    //默认起始位置
    private int prePostion = 0;
```

```
private boolean isLoop = true;

private Handler maHandler = new Handler() {
    public void handleMessage(android.os.Message msg) {
        switch (msg.what) {
            case 1:
                int currentItem = mVp.getCurrentItem();
                if (currentItem < 4) {
                    mVp.setCurrentItem(currentItem + 1);
                } else {
                    mVp.setCurrentItem(0);
                }
                break;
            default:
                break;
        }
    }
};

@Override
protected void onCreate(Bundle savedInstanceState) {
    super.onCreate(savedInstanceState);
    setContentView(R.layout.activity_main);
    initView();
    images = getImages();
    mList = new ArrayList<ImageView>();
    for (int i = 0; i < images.length; i++) {
        ImageView imageView = new ImageView(this);
        imageView.setBackgroundResource(images[i]);
        mList.add(imageView);
        View view = new View(this);
        view.setBackgroundResource(R.mipmap.dot_normal);
        LinearLayout.LayoutParams layoutParams = new LinearLayout.LayoutParams(25, 25);
        layoutParams.leftMargin = 15;
        view.setLayoutParams(layoutParams);
        llPoints.addView(view);
    }
    MyAdapter myAdapter = new MyAdapter();
    mVp.setAdapter(myAdapter);
    //显示首张图片时，当前小圆点默认为红色
```

```
llPoints.getChildAt(0).setBackgroundResource(R.mipmap.dot_enable);

mVp.setOnPageChangeListener(new ViewPager.OnPageChangeListener() {

    @Override
    public void onPageSelected(int position) {
        //选中某张图片时，将当前小圆点的颜色由灰色变为红色
        llPoints.getChildAt(position).setBackgroundResource(R.mipmap.dot_enable);
        //选中某张图片时，将其他小圆点的颜色变为灰色
        llPoints.getChildAt(prePostion).setBackgroundResource(R.mipmap.dot_normal);
        prePostion = position;

    }

    @Override
    public void onPageScrolled(int arg0, float arg1, int arg2) {

    }

    @Override
    public void onPageScrollStateChanged(int arg0) {

    }
});
new Thread(new Runnable() {

    @Override
    public void run() {
        while (isLoop) {
            try {
                Thread.sleep(2000);
                maHandler.sendEmptyMessage(1);

            } catch (InterruptedException e) {
                // TODO Auto-generated catch block
                e.printStackTrace();
            }
        }
    }
}).start();
```

```
    }

    private void initView() {
        mVp = (ViewPager) findViewById(R.id.vp);
        llPoints = (LinearLayout) findViewById(R.id.ll_points);
    }

    private int[] getImages() {
        return new int[]{R.mipmap.a, R.mipmap.b, R.mipmap.c, R.mipmap.d, R.mipmap.e};
    }

    class MyAdapter extends PagerAdapter {

        @Override
        public int getCount() {
            return mList != null ? mList.size() : 0;
        }

        @Override
        public boolean isViewFromObject(View arg0, Object arg1) {
            return arg0 == arg1;
        }

        @Override
        public Object instantiateItem(ViewGroup container, int position) {
            container.addView(mList.get(position));
            return mList.get(position);
        }

        @Override
        public void destroyItem(ViewGroup container, int position, Object object) {
            container.removeView(mList.get(position));
        }
    }
}
```

上述代码中，首先，将本地图片资源加载到程序当中，并根据图片的张数，动态增加小圆点的个数；然后，建立适配器控件 ViewPager，向该控件中添加图片；最后，使用 Handler 实现 ViewPager 的自动轮播。

编写完毕，在 Android Studio 中运行程序，效果如图 S8-2 所示。

图 S8-2　使用 ViewPager 实现图片自动轮播效果

拓展练习

参考本章知识拓展内容，编写应用程序，使用 Fragment 方式实现图片自动轮播。

实践 9 网络通信

实践指导

实践 9.1 服务器端程序

编写点餐系统的服务器端程序。

【分析】

(1) 点餐系统服务器端使用 Java 技术，采用 Servlet 实现，并使用 HTTP 协议与 Android 客户端进行通信，通信内容将通过 JSON 方式传输。

(2) 可以使用谷歌提供的 gson-2.2.2.jar 第三方开发包简单快捷地进行 JSON 数据的创建与解析。

(3) 新建 Web 项目，将客户端的实体类复制到项目中，并在服务器的数据库中给每个实体类建立对应的表。

(4) 编写用户业务类 UserService，实现根据用户编号和密码登录的功能。

(5) 编写餐品业务类 FoodService，实现获取所有餐品和餐品类型的功能。

(6) 编写餐桌业务类 TableService，实现以下功能：

✧ 获取所有餐桌。

✧ 根据需要的座位数和是否要求空桌来查询餐桌。

(7) 编写订单业务类 OrderService，实现以下功能：

✧ 添加订单，需要更新对应餐桌的就餐人数、插入订单及明细。

✧ 根据编号查询订单。

✧ 订单结账，需要更改订单的状态，并修改对应餐桌的就餐人数为 0。

(8) 编写 Servlet，实现以下功能：

✧ 用户登录。

✧ 更新数据。

✧ 查询餐桌。

✧ 查询订单。

✧ 添加订单。

✧ 结账。

【参考解决方案】

(1) 新建 Web 项目 repastWeb，将客户端项目中的实体类复制到 repastWeb 项目的

com.dh.repastweb.entity 包下，在服务器端的实体类中取消 get()和 set()方法。

(2) 在服务器端的数据库中分别创建用户、餐品、餐品类型、餐桌、订单、订单明细等信息对应的表。本例中使用 Oracle 数据库，创建表的 SQL 语句如下：

```
create table USERS
(
  ID        NUMBER not null,
  CODE      NVARCHAR2(10) not null,
  PASSWORD NVARCHAR2(10) not null,
  NAME      NVARCHAR2(10) not null
);
create table FOOD
(
  ID           NUMBER not null,
  CODE         NVARCHAR2(5) not null,
  NAME         NVARCHAR2(20) not null,
  TYPE_ID      NUMBER not null,
  PRICE        NUMBER not null,
  DESCRIPTION NVARCHAR2(100)
);
create table FOOD_TYPE
(
  ID   NUMBER not null,
  NAME NVARCHAR2(10) not null
);
create table TABLES
(
  ID           NUMBER not null,
  CODE         NVARCHAR2(5) not null,
  SEATS        NUMBER not null,
  CUSTOMERS    NUMBER,
  DESCRIPTION NVARCHAR2(50)
);
create table ORDERS
(
  ID           NUMBER not null,
  CODE         NVARCHAR2(10) not null,
  TABLE_ID     NUMBER not null,
  WAITER_ID    NUMBER not null,
  ORDER_TIME   DATE not null,
```

```
  CUSTOMERS    NUMBER not null,
  STATUS       NUMBER not null,
  DESCRIPTION NVARCHAR2(50)
);
create table ORDER_DETAIL
(
  ID           NUMBER not null,
  ORDER_ID     NUMBER not null,
  FOOD_ID      NUMBER not null,
  NUM          NUMBER not null,
  DESCRIPTION NVARCHAR2(50)
);
```

(3) 编写用于数据库操作的工具类 com.dh.repastweb.db.Db，代码如下：

```
public class Db {

	public static final String URL = "jdbc:oracle:thin:@localhost:1521:orcl";
	public static final String NAME = "admin";
	public static final String PASSWORD = "admin";

	private Db() {
		// 不可实例化
	}
	static {
		try {
			Class.forName("oracle.jdbc.driver.OracleDriver");
		} catch (Exception e) {
			e.printStackTrace();
		}
	}
	/**
	 * 获取 Connection，<strong>Connection 用后需要关闭。</strong>
	 *
	 * @return Connection
	 */
	public static Connection getConnection() throws SQLException {
		return DriverManager.getConnection(URL, NAME, PASSWORD);
	}

	/**
```

```
     * 执行查询，返回多个对象
     *
     * @param connConnection
     * @param adapter 将 ResultSet 转化为 List
     * @param sqlSQL 语句
     * @param paramsSQL 语句参数值
     * @return List 结果
     */
    public static <T> List<T> query(Connection conn,
                ResultSetAdapter<T> adapter, String sql, Object... params)
                throws SQLException {
        List<T> list = new ArrayList<T>();
        PreparedStatement ps = null;
        ResultSet rs = null;
        int i = 1;
        try {
            ps = conn.prepareStatement(sql);
            for (Object p : params)
                if (p instanceof java.util.Date)
                    ps.setTimestamp(i++, new java.sql.Timestamp(
                            ((java.util.Date) p).getTime()));
                else
                    ps.setObject(i++, p);
            rs = ps.executeQuery();
            while (rs.next())
                list.add(adapter.convert(rs));
        } finally {
            if (rs != null)
                rs.close();
            if (ps != null)
                ps.close();
        }
        return list;
    }

    /**
     * 执行查询，返回一个对象
     *
     * @param connConnection
```

```
 * @param adapter 将 ResultSet 转化为 List
 * @param sqlSQL 语句
 * @param paramsSQL 语句参数值
 * @return 结果
 */
public static <T> T get(Connection conn, ResultSetAdapter<T> adapter,
            String sql, Object... params) throws SQLException {
      List<T> list = query(conn, adapter, sql, params);
      if (list.isEmpty())
            return null;
      return list.get(0);
}

/**
 * 执行更新
 *
 * @param connConnection
 * @param sqlSQL 语句
 * @param paramsSQL 语句参数值
 * @return 受影响的行数
 */
public static int update(Connection conn, String sql,
       Object... params)throws SQLException {
      PreparedStatement ps = null;
      int i = 1;
      try {
            ps = conn.prepareStatement(sql);
            for (Object p : params)
                  if (p instanceof java.util.Date)
                        ps.setTimestamp(i++, new java.sql.Timestamp(
                                    ((java.util.Date) p).getTime()));
                  else
                        ps.setObject(i++, p);
            return ps.executeUpdate();
      } finally {
            if (ps != null)
                  ps.close();
      }
}
```

```
/**
 * 执行更新，并获取生成的值
 *
 * @param connConnection
 * @param sqlSQL 语句
 * @param paramsSQL 语句参数值
 * @return 结构为[受影响的行数，生成的值 1，生成的值 2，生成的值 3...]
 */
public static List<Object> updateAndGetId(Connection conn, String sql,
                int[] keyColumns, Object... params) throws SQLException {
        PreparedStatement ps = null;
        ResultSet rs = null;
        List<Object> r = new ArrayList<Object>();
        int i = 1;
        try {
                ps = conn.prepareStatement(sql, keyColumns);
                for (Object p : params)
                        if (p instanceof java.util.Date)
                                ps.setTimestamp(i++, new java.sql.Timestamp(
                                                ((java.util.Date) p).getTime()));
                        else
                                ps.setObject(i++, p);
                r.add(ps.executeUpdate());
                rs = ps.getGeneratedKeys();
                while (rs.next())
                        for (int k : keyColumns)
                                r.add(rs.getObject(k));
        } finally {
                if (rs != null)
                        rs.close();
                if (ps != null)
                        ps.close();
        }
        return r;
}

/**
 * 执行批量更新
```

```
 *
 * @param connConnection
 * @param sqlSQL 语句
 * @param paramsSQL 语句参数值
 * @return 受影响的行数
 */
public static int[] updateBatch(Connection conn, String sql,
            Object[]... params) throws SQLException {
    PreparedStatement ps = null;
    try {
        ps = conn.prepareStatement(sql);
        for (Object[] pp : params) {
            int i = 1;
            for (Object p : pp)
                if (p instanceof java.util.Date)
                    ps.setObject(i++, new java.sql.Date(
                            ((java.util.Date) p).getTime()));
                else
                    ps.setObject(i++, p);
            ps.addBatch();
        }
        return ps.executeBatch();
    } finally {
        if (ps != null)
            ps.close();
    }
}

/**
 * 将 ResultSet 中的一条记录转化为{@link T}类型
 *
 * @param <T>需要转化成的类型
 */
public static interface ResultSetAdapter<T> {
    /**
     *将 ResultSet 中的一条记录转化为{@link T}类型。
     *aRowInResultSet 的 next()方法
     *会被调用，不要再次调用。
     *
```

```
            * @param aRowInResultSet 需要转化的 ResultSet
            * @return T 类型的对象
            */
        T convert(ResultSet aRowInResultSet) throws SQLException;
    }

    /**
     * 获取指定表中最后一条记录的 ID
     *
     * @param conn
     * @param table
     * @return
     * @throws SQLException
     */
    public static int getLastId(Connection conn, String table)
                throws SQLException {
        PreparedStatement ps = null;
        ResultSet rs = null;
        try {
                ps = conn.prepareStatement("select max(id) from " + table);
                rs = ps.executeQuery();
                if (rs.next()) {
                        return rs.getInt(1);
                }
        } finally {
                if (rs != null)
                        rs.close();
                if (ps != null)
                        ps.close();
        }
        return 0;
    }
}
```

上述代码中的 Db 类定义了执行查询和更新语句的一些辅助方法，还定义了接口 ResultSetAdapter<T>，用于将 ResultSet 中的一条记录转化为一个实体的实例。

(4) 针对每个实体类编写 ResultSetAdapter<T>接口的实现类，代码如下：

```
public enum UserAdapter implements Db.ResultSetAdapter<User> {
    SINGLETON;
```

```
    @Override
    public User convert(ResultSet rs) throws SQLException {
        User user = new User();
        user.id = rs.getInt("id");
        user.code = rs.getString("code");
        user.password = rs.getString("password");
        user.name = rs.getString("name");
        return user;
    }
}

public enum FoodAdapter implements Db.ResultSetAdapter<Food> {
    SINGLETON;

    @Override
    public Food convert(ResultSet rs) throws SQLException {
        Food t = new Food();
        t.id = rs.getInt("id");
        t.code = rs.getString("code");
        t.typeId = rs.getInt("type_id");
        t.price = rs.getInt("price");
        t.name = rs.getString("name");
        t.description = rs.getString("description");
        return t;
    }
}

public enum FoodTypeAdapter implements Db.ResultSetAdapter<FoodType> {
    SINGLETON;

    @Override
    public FoodType convert(ResultSet rs) throws SQLException {
        FoodType t = new FoodType();
        t.id = rs.getInt("id");
        t.name = rs.getString("name");
        return t;
    }
}
```

```
public enum TableAdapter implements Db.ResultSetAdapter<Table> {
	SINGLETON;

	@Override
	public Table convert(ResultSet rs) throws SQLException {
		Table t = new Table();
		t.id = rs.getInt("id");
		t.code = rs.getString("code");
		t.seats = rs.getInt("seats");
		t.customers = rs.getInt("customers");
		t.description = rs.getString("description");
		return t;
	}
}

public enum OrderAdapter implements Db.ResultSetAdapter<Order> {
	SINGLETON;

	@Override
	public Order convert(ResultSet rs) throws SQLException {
		Order t = new Order();
		t.id = rs.getInt("id");
		t.code = rs.getString("code");
		t.tableId = rs.getInt("table_id");
		t.waiterId = rs.getInt("waiter_id");
		t.orderTime = rs.getTimestamp("order_time");
		t.customers = rs.getInt("customers");
		t.status = rs.getInt("status");
		t.description = rs.getString("description");
		return t;
	}
}

public enum OrderDetailAdapter implements Db.ResultSetAdapter<OrderDetail> {
	SINGLETON;
	@Override
	public OrderDetail convert(ResultSet rs) throws SQLException {
		OrderDetail t = new OrderDetail();
		t.id = rs.getInt("id");
```

```
            t.foodId = rs.getInt("food_id");
            t.num = rs.getInt("num");
            t.description = rs.getString("description");
            return t;
        }
}
```

(5) 编写用户业务类 com.dh.repastweb.service.UserService，代码如下：

```
public class UserService {
        /**
         * 登录
         *
         * @param code
         * 用户编号
         * @param password
         * 密码
         * @return 登录成功后，返回 User 实例，否则返回 null
         */
        public User login(String code, String password) {
                String sql = "select * from users where code = ?
                        and password = ?";
                Connection conn = null;
                try {
                        conn = Db.getConnection();
                        return Db.get(conn, UserAdapter.SINGLETON, sql, code, password);
                } catch (SQLException e) {
                        e.printStackTrace();
                } finally {
                        if (conn != null)
                                try {
                                        conn.close();
                                } catch (SQLException e) {
                                        e.printStackTrace();
                                }
                }
                return null;
        }
}
```

(6) 编写餐品业务类 com.dh.repastweb.service.FoodService，代码如下：

```
public class FoodService {
```

```
/**
 * 获取所有食品
 *
 * @return 食品
 */
public List<Food> getFoods() {
        String sql = "select * from food";
        Connection conn = null;
        try {
                conn = Db.getConnection();
                return Db.query(conn, FoodAdapter.SINGLETON, sql);
        } catch (SQLException e) {
                e.printStackTrace();
        } finally {
                if (conn != null)
                        try {
                                conn.close();
                        } catch (SQLException e) {
                                e.printStackTrace();
                        }
        }
        return new ArrayList<Food>(0);
}

/**
 * 获取所有食品类型
 *
 * @return 食品类型
 */
public List<FoodType> getFoodTypes() {
        String sql = "select * from food_type";
        Connection conn = null;
        try {
                conn = Db.getConnection();
                return Db.query(conn, FoodTypeAdapter.SINGLETON, sql);
        } catch (SQLException e) {
                e.printStackTrace();
        } finally {
                if (conn != null)
```

```
                    try {
                        conn.close();
                    } catch (SQLException e) {
                        e.printStackTrace();
                    }
            }
            return new ArrayList<FoodType>(0);
    }
}
```

(7) 编写餐桌业务类 com.dh.repastweb.service.TableService，代码如下：

```
public class TableService {
    /**
     * 获取所有餐桌
     *
     * @return 餐桌
     */
    public List<Table> getTables() {
        String sql = "select * from tables";
        Connection conn = null;
        try {
            conn = Db.getConnection();
            return Db.query(conn, TableAdapter.SINGLETON, sql);
        } catch (SQLException e) {
            e.printStackTrace();
        } finally {
            if (conn != null)
                try {
                    conn.close();
                } catch (SQLException e) {
                    e.printStackTrace();
                }
        }
        return new ArrayList<Table>(0);
    }

    /**
     * 查询餐桌
     *
     * @param seats
```

```
     * 最少需要的座位数
     * @param needEmpty
     * 是否必须空桌
     * @return 餐桌
     */
    public List<Table> query(int seats, boolean needEmpty)
                throws RuntimeException {
        String sql="select * from tables where seats- customers >= ? ";
        if (needEmpty)
            sql += "and customers = 0";
        Connection conn = null;
        try {
            conn = Db.getConnection();
            return Db.query(conn, TableAdapter.SINGLETON, sql, seats);
        } catch (SQLException e) {
            e.printStackTrace();
            throw new RuntimeException(e.getMessage());
        } finally {
            if (conn != null)
                try {
                    conn.close();
                } catch (SQLException e) {
                    e.printStackTrace();
                }
        }
    }
}
```

(8) 编写订单业务类 com.dh.repastweb.service.OrderService，代码如下：

```
public class OrderService {
    /**
     * 添加订单
     *
     * @param order
     * 订单
     */
    public int addOrder(Order order) throws RuntimeException {
        String sql1 = "update tables set customers = ? where id = ?";
        String sql2 = "insert into orders(id,table_id,waiter_id,"
                + "order_time,customers,status,description)"
```

```
            + "values (od_seq.nextval,?,?,?,?,?,?)";
    String sql3 = "insert into order_detail "
            + "(id,order_id,food_id,num,description) "
            + "values (od_d_seq.nextval,?,?,?,?)";
    Connection conn = null;
    try {
        conn = Db.getConnection();
        conn.setAutoCommit(false);

        // 修改餐桌就餐人数
        Db.update(conn, sql1, order.customers, order.tableId);
        // 添加订单
        List<Object> l = Db.updateAndGetId(conn,sql2,new int[] {1},
                order.tableId, order.waiterId, new Date(),
                    order.customers,0, order.description);
        // 添加订单明细
        List<Object[]> params = new ArrayList<Object[]>();
        for (OrderDetail od : order.orderDetails) {
            Object[] p = new Object[4];
            p[0] = l.get(1);
            p[1] = od.foodId;
            p[2] = od.num;
            p[3] = od.description;
            params.add(p);
        }
        Db.updateBatch(conn, sql3, params.toArray(new Object[params.size()][]));
        conn.commit();
        int lastId = Db.getLastId(conn, "orders");
        return lastId;
    } catch (SQLException e) {
        e.printStackTrace();
        try {
            if (conn != null)
                conn.rollback();
        } catch (SQLException e1) {
            e1.printStackTrace();
        }
        throw new RuntimeException("查询失败！ ");
    } finally {
```

```
                if (conn != null)
                    try {
                        conn.close();
                    } catch (SQLException e) {
                        e.printStackTrace();
                    }
        }
    }

    /**
     * 根据 ID 获取订单信息
     *
     * @param orderId
     * 订单 ID
     * @return 订单
     */
    public Order getOrder(String code) throws RuntimeException {
        String sql1 = "select * from orders where code = ?";
        String sql2 = "select * from order_detail where order_id = ?";
        Connection conn = null;
        try {
            conn = Db.getConnection();
            Order order = Db.get(conn, OrderAdapter.SINGLETON, sql1,
                code);
            if (order == null)
                return null;
            List<OrderDetail> orderDetails = Db.query(conn,
                    OrderDetailAdapter.SINGLETON, sql2, order.id);
            order.orderDetails = orderDetails;
            return order;
        } catch (SQLException e) {
            e.printStackTrace();
            throw new RuntimeException(e.getMessage());
        } finally {
            if (conn != null)
                try {
                    conn.close();
                } catch (SQLException e) {
                    e.printStackTrace();
```

```
                }
        }
}

/**
 * 订单结账
 *
 * @param orderId
 * 订单 ID
 */
public boolean pay(String orderId) {
        // 更新 order
        String sql1 = "update orders set status = 1 where id = ?";
        // 更新 table
        String sql2 = "update tables set customers = 0 where id =
                (select table_id from orders where orders.id = ?)";
        Connection conn = null;
        try {
                conn = Db.getConnection();
                conn.setAutoCommit(false);
                Db.update(conn, sql1, orderId);
                Db.update(conn, sql2, orderId);
                conn.commit();
                return true;
        } catch (SQLException e) {
                e.printStackTrace();
                try {
                        if (conn != null)
                                conn.rollback();
                } catch (SQLException e1) {
                        e1.printStackTrace();
                }
                return false;
        } finally {
                if (conn != null)
                        try {
                                conn.close();
                        } catch (SQLException e) {
                                e.printStackTrace();
```

```
                    }
            }
    }

}
```

(9) 编写 com.dh.repastweb.servlet.OnlyServlet，用于处理客户端请求，代码如下：

```
public class OnlyServlet extends HttpServlet {

    private UserService userService = new UserService();
    private FoodService foodService = new FoodService();
    private TableService tableService = new TableService();
    private OrderService orderService = new OrderService();

    @Override
    protected void service(HttpServletRequest req, HttpServletResponse
            resp)   throws ServletException, IOException {
            req.setCharacterEncoding("UTF-8");
            resp.setContentType("text/html");
            resp.setHeader("content-type", "text/html;charset=UTF-8");

            // 获取 to 参数值
            String to = req.getParameter("to");
            if ("login".equals(to)) {
                    login(req, resp);
            } else if ("update".equals(to)) {
                    update(req, resp);
            } else if ("queryTable".equals(to)) {
                    queryTable(req, resp);
            } else if ("getOrder".equals(to)) {
                    getOrderInfo(req, resp);
            } else if ("addOrder".equals(to)) {
                    addOrder(req, resp);
            } else if ("pay".equals(to)) {
                    pay(req, resp);
            }
    }

    /**
     * 用户登录
```

```
 *
 * @param req
 * @param resp
 * @throws IOException
 */
private void login(HttpServletRequest req, HttpServletResponse resp) throws IOException {
        String code = req.getParameter("code");
        String password = req.getParameter("password");
        User user = userService.login(code, password);

        boolean isSuccess = true;
        String errMsg = "";
        // 查询失败
        if (user == null) {
                isSuccess = false;
                errMsg = "用户名或密码错误!";
        }
        sendResultToClient(resp, Tools.createJson(isSuccess, errMsg, user));
}

/**
 * 数据更新
 *
 * @param req
 * @param resp
 * @throws IOException
 */
private void update(HttpServletRequest req, HttpServletResponse resp)
                throws IOException {

        UpdateData data = new UpdateData();

        data.setFoodTypes(foodService.getFoodTypes());
        data.setFoods(foodService.getFoods());
        data.setTables(tableService.getTables());

        sendResultToClient(resp, Tools.createJson(true, "", data));
}
```

```
/**
 * 添加菜单
 *
 * @param req
 * @param resp
 * @throws IOException
 */
void addOrder(HttpServletRequest req, HttpServletResponse resp) throws IOException {

        String data = req.getParameter("order");
        Order o = new Gson().fromJson(data, Order.class);

        boolean isSuccess = true;
        String errMsg = "";
        int lastId = 0;
        try {
                lastId = orderService.addOrder(o);
        } catch (RuntimeException e) {
                // 添加菜单失败
                isSuccess = false;
                errMsg = "添加菜单失败!";
        }
        sendResultToClient(resp, Tools.createJson(isSuccess, errMsg,   lastId));
}

/**
 * 获取订单信息
 *
 * @param req
 * @param resp
 * @throws IOException
 */
private void getOrderInfo(HttpServletRequest req, HttpServletResponse resp)
                throws IOException {
        String code = req.getParameter("code");
        try {
                Order order = orderService.getOrder(code);
                // 这里不考虑 Order 是否为空，如果为空，说明查询无记录
                sendResultToClient(resp, Tools.createJson(true, "", order));
```

```
        } catch (RuntimeException e) {
            // 这里捕获查询异常
            sendResultToClient(resp, Tools.createJson(false, "查询异常！", null));
        }

}

/**
 * 结账
 *
 * @param req
 * @param resp
 * @throws IOException
 */
private void pay(HttpServletRequest req, HttpServletResponse resp)
            throws IOException {
        String orderId = req.getParameter("orderId");

        boolean isSuccess = orderService.pay(orderId);
        String errMsg = "结账成功！";
        // 结账失败
        if (!isSuccess) {
            errMsg = "结账失败!";
        }
        sendResultToClient(resp, Tools.createJson(isSuccess, errMsg,    null));
}

/**
 * 查桌
 *
 * @param req
 * @param resp
 * @throws IOException
 */
private void queryTable(HttpServletRequest req, HttpServletResponse resp)
            throws IOException {
        String s = req.getParameter("seats");
        String n = req.getParameter("needEmpty");
        int seats = 0;
        boolean needEmpty = false;
```

```
        try {
            if (s != null)
                seats = Integer.parseInt(s);
        } catch (NumberFormatException e) {
            e.printStackTrace();
        }
        if (n != null)
            needEmpty = true;

        try {
            List<Table> tables = tableService.query(seats, needEmpty);
            // 这里不考虑 tables 是否为空，如果为空，说明查询无记录
            sendResultToClient(resp, Tools.createJson(true, "", tables));
        } catch (RuntimeException e) {
            // 这里捕获查询异常
            sendResultToClient(resp, Tools.createJson(false, "查询异常！ ", null));
        }
    }
    /**
     * 发送服务器响应信息到客户端
     *
     * @param resp
     * @param json
     * @throws IOException
     */
    private void sendResultToClient(HttpServletResponse resp, String json)
            throws IOException {
        PrintWriter out = resp.getWriter();
        out.print(json);
    }

}
```

上述代码中的 OnlyServlet 是唯一的 Servlet，负责处理所有的客户端请求，不同的请求采用不同的方法处理。

(10) 在 web.xml 中配置 OnlyServlet，代码如下：

```
<?xml version="1.0" encoding="UTF-8"?>
<web-app xmlns:xsi="http://www.w3.org/2001/XMLSchema-instance"
xmlns="http://java.sun.com/xml/ns/javaee" xmlns:web="http://java.sun.com/xml/ns/javaee/web-app_2_5.xsd"
```

```
xsi:schemaLocation="http://java.sun.com/xml/ns/javaee http://java.sun.com/xml/ns/javaee/web-app_3_0.xsd"
version="3.0">
<servlet>
<servlet-name>only</servlet-name>
<servlet-class>com.dh.repastweb.servlet.OnlyServlet</servlet-class>
</servlet>
<servlet-mapping>
<servlet-name>only</servlet-name>
<url-pattern>/</url-pattern>
</servlet-mapping>
</web-app>
```

实践 9.2　与服务器通信

在 Android 客户端编写工具类，实现以下功能：

(1) 使用 POST 方式提交数据。

(2) 向服务器发送数据。

【分析】

(1) 点餐系统中，登录、查询餐桌、结账等操作都需要向服务器提交特定数据，因此需要实现使用 HTTP 向服务器提交数据的功能。Android 从 API 23 开始停止对 HttpClient 的支持，如需使用，需要在 gradle 文件中 HttpClient 依赖，本书就是在 gradle 中添加了依赖。

(2) 发送和接收的数据均为 JSON 字符串，JSON 字符串与对应实体类可以通过 GSON 类方便地进行互相转换，提高了客户端与服务器的数据解析效率。

(3) 与服务器通信需要使用统一的 JSON 格式传输协议，所以需要创建一个实体类来统一数据传输格式。

【参考解决方案】

(1) 在 app 包下的 build.guild 中添加 com.google.code.gson.gson:2.8.1。

(2) 创建 com.ugrow.repast_ph09.entity.JsonData 实体类，用于统一客户端与服务器通信的数据格式，代码如下：

```
public class JsonData implements Serializable {
	private static final long serialVersionUID = 3223903391563380140L;
	private boolean success;
	private String msg;
	private String data;


	public boolean isSuccess() {
		return success;
	}
	public void setSuccess(boolean success) {
```

```
        this.success = success;
    }
    public String getData() {
        return data;
    }
    public void setData(String data) {
        this.data = data;
    }
    public String getMsg() {
        return msg;
    }
    public void setMsg(String msg) {
        this.msg = msg;
    }
    @Override
    public String toString() {
        return "JsonData [success=" + success + ", msg=" + msg
                + ", data=" + data + "]";
    }
}
```

上述代码创建了供客户端与服务器通信使用的基础类，主要用于封装由服务器向客户端返回的数据。

JSON 数据格式为：{"success":true|false,"msg":"信息", "data":"返回的数据"}，其中：

✧ success：表示操作是否成功，取值为 true 或 false。

✧ msg：表示返回的提示性信息，比如“数据查询失败”。

✧ data：表示返回的数据，比如查询结果，这些数据也是 JSON 格式。

(3) 编写 HTTP 辅助类 com.ugrow.repast_ph09.utils.HttpUtils，同时在 bulid.gradle 中添加依赖包，代码如下：

```
public class HttpUtils {
    private static final int TIMEOUT = 2000; // 超时时间
    public static final String SERVICE_ERR = "连接服务器失败";

    /**
     * POST 方式向服务器发出请求
     *
     * @param url
     * @param params
     * @return JsonData
     */
```

```
public static JsonData post(String url, Map<String, String> params) {
        HttpClient client = new DefaultHttpClient();
        HttpParams p = client.getParams();
        // 设置超时
        p.setParameter(CoreConnectionPNames.CONNECTION_TIMEOUT, TIMEOUT);
        p.setParameter(CoreConnectionPNames.SO_TIMEOUT, TIMEOUT);
        try {
                // 定义 POST 对象
                HttpPost post = new HttpPost(url);

                // 设置参数
                if (params != null && !params.isEmpty()) {
                        List<NameValuePair> formParams = new
                                ArrayList<NameValuePair>();
                        for (Map.Entry<String, String> e:params.entrySet()) {
                                BasicNameValuePair pair=new BasicNameValuePair(
                                        e.getKey(), e.getValue());
                                formParams.add(pair);
                        }
                        UrlEncodedFormEntity formEntity;
                        formEntity =new UrlEncodedFormEntity(formParams, "UTF-8");
                        post.setEntity(formEntity);
                }
                HttpResponse resp = client.execute(post);
                // 获取响应码
                int status = resp.getStatusLine().getStatusCode();
                if (status == 200) {
                        HttpEntity entity = resp.getEntity();
                        if (entity.getContentLength() == 0)
                                return null;
                        String respStr = EntityUtils.toString(resp.getEntity());
                        System.out.println("HttpUtils.post(): -" + respStr + "-");
                        client.getConnectionManager().shutdown();
                        JsonData jd = new Gson().fromJson(respStr, JsonData.class);
                        return jd;
                }
        } catch (Exception e) {
                e.printStackTrace();
        }
```

```
            return null;
        }
}
```

上述代码中，通过 post()方法以 POST 方式向指定 URL 上的服务器提交数据，并将服务器返回的 JSON 数据通过 GSON 类转换为 JsonData 对象，并返回给调用者。

可以通过非常简单的方式将 JSON 字符串数据解析为 JsonData 对象，代码如下：

```
JsonData jd = new Gson().fromJson(respStr, JsonData.class);
```

后续实践还将介绍如何将 JSON 字符串解析为 List 对象和将对象转换为 JSON 字符串的操作。

实践 9.3 登录验证

在 Android 客户端上连接服务器，并完成登录验证。

【分析】

(1) UserService 可使用 HttpUtils.post()方法向服务器提交用户编号和密码，并获取登录成功的 User 实例。

(2) 用户第一次使用点餐系统时，需要录入服务器地址后再登录，再次使用时，登录界面应显示已保存的地址，并允许用户修改。为此，需要在登录界面 LoginActivity 中添加输入服务器地址的 EditText，并在登录成功时保存用户录入的服务器地址。

(3) 需要在 AndroidMenifest.xml 文件中开通网络访问的权限。

【参考解决方案】

(1) 修改 com.ugrow.repast_ph09.service.UserService，代码如下：

```
public class UserService {
    private App app;
    public UserService(Context context) {
        app = (App) context.getApplicationContext();
    }
    /**
     * 登录
     *
     * @param code
     * 用户编号
     * @param password
     * 密码
     * @return 用户对象
     * @throws Exception
     * 服务器返回的错误信息
     */
    public User login(String code, String password)
```

```
            throws RuntimeException{
            String url = app.getServerUrl() + Constants.USER_LOGIN_URL;
            Map<String, String> params = new HashMap<String, String>();
            params.put("code", code);
            params.put("password", password);
            // 请求服务器，获取数据
            JsonData data = HttpUtils.post(url, params);
            // 如果服务器返回数据为 null，抛出异常
            if (data == null) {
                throw new RuntimeException(HttpUtils.SERVICE_ERR);
            }
            // 如果获取数据成功，返回 User 对象，
            //否则抛出异常，异常内容为服务器返回的内容
            if (data.isSuccess()) {
                //在这里将 data.getData()中的 Json 字符串转换为 User 对象
                return new Gson().fromJson(data.getData(), User.class);
            } else {
                throw new RuntimeException(data.getErrMsg());
            }
    }
}
```

上述代码中，在 login()方法里调用 HttpUtils.post()，从而向服务器提交用户编号和密码，并将返回结果解析为 JsonData 对象，如果 JsonData 不为 null，说明访问服务器成功。然后判断 data.isSuccess()，如果它为 false，表示登录失败，将 data.getErrMsg()中服务器返回的错误信息抛出；如果 data.isSuccess()为 true，表示登录成功，将 data.getData()中的 JSON 字符串转换为 User 对象并返回。将 JSON 字符串转换为对象的方法如下：

```
User user = new Gson().fromJson(data.getData(), User.class);
```

(2) 修改用户界面布局文件 activity_login.xml，在其中添加录入服务器地址的 EditText，代码如下：

```
<?xml version="1.0" encoding="utf-8"?>
<LinearLayout xmlns:android="http://schemas.android.com/apk/res/android"
    android:layout_width="match_parent"
    android:layout_height=" match__parent"
    android:background="#ededed"
    android:orientation="vertical"
    android:padding="10dp" >
<TableLayout
        android:layout_width=" match__parent"
        android:layout_height="fill_parent"
```

```
        android:stretchColumns="1" >
<EditText
            android:id="@+id/serverUrlEdt"
            android:layout_width=" match__parent"
            android:layout_height="wrap_content"
            android:hint="请输入服务器地址" />
<EditText
            android:id="@+id/codeEdt"
            android:layout_width=" match__parent"
            android:layout_height="wrap_content"
            android:hint="请输入编号"
            android:singleLine="true"
            android:text="fwy1" />
<EditText
            android:id="@+id/passwordEdt"
            android:layout_width=" match__parent"
            android:layout_height="wrap_content"
            android:hint="请输入密码"
            android:inputType="textPassword"
            android:singleLine="true"
            android:text="000000" />
<LinearLayout android:layout_width=" match__parent" >
<Button
                android:id="@+id/loginBtn"
                android:layout_width=" match__parent"
                android:layout_height="wrap_content"
                android:layout_weight="1"
                android:drawableLeft="@drawable/key"
                android:text="登录" />
<Button
                android:id="@+id/cancelBtn"
                android:layout_width=" match__parent"
                android:layout_height="wrap_content"
                android:layout_weight="1"
                android:drawableLeft="@drawable/back"
                android:text="取消" />
</LinearLayout>
</TableLayout>
</LinearLayout>
```

(3) 修改 com.ugrow.repast_ph09.ui.LoginActivity，代码如下：

```
public class LoginActivity extends BasicActivity {
	private EditText serverUrlEdt;
	private EditText codeEdt;
	private EditText passwordEdt;
	private Button loginBtn;
	private Button cancelBtn;

	private UserService userService;
	// 定义常量—当服务器返回异常数据(比如用户名密码错误)
	private static final int MSG_ERR = 0;
	// 定义常量—服务器返回数据正常
	private static final int MSG_SUCCESS = 1;

	@Override
	protected void onCreate(Bundle savedInstanceState) {
		super.onCreate(savedInstanceState);
		setContentView(R.layout.activity_login);

		String title = getString(R.string.app_name) + " " + getName();
		setTitle(title);

		userService = new UserService(this);

		serverUrlEdt = (EditText) findViewById(R.id.serverUrlEdt);
		codeEdt = (EditText) findViewById(R.id.codeEdt);
		passwordEdt = (EditText) findViewById(R.id.passwordEdt);
		loginBtn = (Button) findViewById(R.id.loginBtn);
		cancelBtn = (Button) findViewById(R.id.cancelBtn);

		serverUrlEdt.setText(Tools.getStringPre(this, "serverUrl"));

		loginBtn.setOnClickListener(new OnClickListener() {
			@Override
			public void onClick(View v) {
				String serverUrl = serverUrlEdt.getText().toString();
				String code = codeEdt.getText().toString();
				String password = passwordEdt.getText().toString();
				if (serverUrl.length() == 0) {
					showMessageDialog("请输入服务器地址",
							R.mipmap.warning, null);
					return;
				}
```

```
                if (code.length() == 0 || password.length() == 0) {
                    showMessageDialog("请输入登录名和密码",
                            R.mipmap.warning, null);
                    return;
                }
                Tools.putStringPre(LoginActivity.this, "serverUrl", serverUrl);
                login(code, password);
            }
        });
        cancelBtn.setOnClickListener(new OnClickListener() {
            @Override
            public void onClick(View v) {
                LoginActivity.this.finish();
            }
        });
    }
    /**
     * 用户登录
     *
     * @param code
     * @param password
     */
    private void login(final String code, final String password) {
        new Thread(new Runnable() {
            @Override
            public void run() {
                try {
                    User user = userService.login(code, password);
                    App app = (App) getApplicationContext();
                    app.user = user;
                    handler.sendMessage(Tools.createMsg(MSG_SUCCESS, null));
                } catch (Exception e) {
                    e.printStackTrace();
                    handler.sendMessage(Tools.createMsg(MSG_ERR,
                            e.getMessage()));
                }
            }
        }).start();
```

```
        }
        /**
         * 定义 Handler，用于更新 UI 主线程
         */
        Handler handler = new Handler() {
                @Override
                public void handleMessage(Message msg) {
                        switch (msg.what) {
                        case MSG_ERR:
                                showMessageDialog((String) msg.obj,
                                        R.mipmap.warning, null);
                                break;
                        case MSG_SUCCESS:
                                showActivity(LoginActivity.this, MainActivity.class);
                                finish();
                                break;
                        }
                }
        };
        @Override
        protected String getName() {
                return "登录";
        }
}
```

上述代码中，首先使用 onCreate()方法获取已保存的服务器地址，并允许用户修改此地址；在单击【登录】按钮时，会自动使用界面中输入的地址连接服务器，如果连接成功，则保存新地址，这里需要注意访问网络必须启动新的线程，不能在主线程中直接访问网络；最后通过 Handler 更新 UI。

(4) 修改 AndroidManifest.xml，开通网络访问权限，代码如下：

```
<uses-permission
android:name="android.permission.INTERNET" />
```

运行服务器端项目 repastWeb，然后运行客户端项目 repast_ph07。第一次登录时需要输入服务器地址，如果未输入，则单击【登录】按钮时会显示提示对话框，如图 S9-1 所示；输入正确的地址并登录成功后，再次登录将显示上次登录成功时使用的地址。

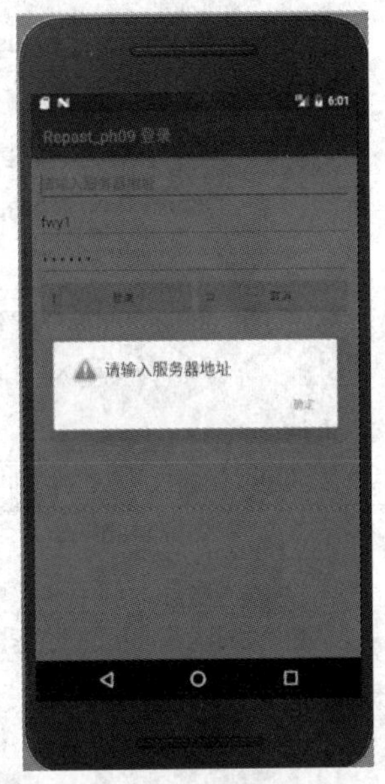

图 S9-1　登录界面

实践 9.4　更新数据

完善 Android 客户端的更新数据功能。

【分析】

(1) 更新数据需要同时更新食品类型、食品以及餐桌的信息，为了更方便地获取并解析 JSON 数据，需要创建一个实体类，这个实体类需要包含 List<FoodType>、List<Food>和 List<Table>。

(2) 修改更新数据的 UpdateDataService，连接服务器下载数据并将数据更新到本地数据库。

【参考解决方案】

(1) 创建实体类 com.ugrow.repast_ph09.entity.UpdateData.java，代码如下：

```
public class UpdateData implements Serializable {
    private static final long serialVersionUID = -5373726425663464772L;
    private List<Food> foods = null;
    private List<FoodType> foodTypes = null;
    private List<Table> tables = null;

    public List<Food> getFoods() {
        return foods;
    }
    public void setFoods(List<Food> foods) {
        this.foods = foods;
    }
    public List<FoodType> getFoodTypes() {
        return foodTypes;
    }
    public void setFoodTypes(List<FoodType> foodTypes) {
        this.foodTypes = foodTypes;
    }
    public List<Table> getTables() {
        return tables;
    }
    public void setTables(List<Table> tables) {
        this.tables = tables;
    }
}
```

上述代码中创建了 foods、foodTypes、tables 三个对象。可以使用 GSON 将 JSON 格式的更新数据转换为 UpdateData 对象，然后直接通过 UpdateData 对象获取 foods、

foodTypes 和 tables，转换方式参见实践 9.3。

(2) 修改 com.ugrow.repast_ph09.service.UpdateDataService，代码如下：

```
public class UpdateDataService extends Service {
    ...... // 省略已有代码
    /**
     * 从服务器下载数据，并更新到本地数据库
     *
     * @throws Exception
     */
    private void updateData() throws RuntimeException {
        App app = (App) getApplicationContext();
        DbHelper dbHelper = new DbHelper(app);
        // 删除已有数据
        dbHelper.deleteDb();
        String url = app.getServerUrl() + Constants.UPDATE_URL;
        // 请求服务器获取数据
        JsonData jd = HttpUtils.post(url, null);
        // 如果服务器返回数据为 null，抛出异常
        if (jd == null) {
            throw new RuntimeException(HttpUtils.SERVICE_ERR);
        }
        UpdateData ud = null;
        // 如果请求成功，解析返回 json 数据，否则将服务器返回的异常信息抛出
        if (jd.isSuccess())
        {
            ud = new Gson().fromJson(jd.getData(), UpdateData.class);
        } else
        {
            throw new RuntimeException(jd.getMsg());
        }
        List<Food> foods = ud.getFoods();
        List<FoodType> foodTypes = ud.getFoodTypes();
        List<Table> tables = ud.getTables();
        // 插入到本地数据库
        ...... // 省略已有代码
    }
}
```

上述代码将 JsonData 对象中的 JSON 字符串数据转换为 UpdateData 对象，并通过 UpdateData 对象的 getFoods()、getFoodTypes()、getTables()方法获取 Food、FoodType、

Table 三个列表对象。

实践 9.5　查桌功能

在 Android 客户端上实现查桌功能。

【分析】

(1) 顾客进入饭店时，服务员需要根据顾客人数查询可用的餐桌，查询时需要判断每个餐桌当前的就餐人数，而这些信息只能从服务器获取。因此，需要提供从服务器查询餐桌的功能。

(2) 修改餐桌业务类 TableService，添加所需的座位数，并检查是否空桌。

(3) 编写查桌界面 TableActivity，使单击主菜单界面中的【查桌】图标时显示此界面。

(4) 查桌界面使用 ListView 显示查询结果，需要新建一个 ListView 的 Item 布局文件。

【参考解决方案】

(1) 修改 com.ugrow.repast_ph09.service.TableService，代码如下：

```
public class TableService {
	App app;
	DbHelper dbHelper; // 本地数据库操作工具类

	public TableService(Context context) {
		dbHelper = new DbHelper(context);
		app = (App) context.getApplicationContext();
	}

	/**
	 * 查询餐桌。从服务器查询
	 *
	 * @param seats 最少需要的座位数
	 * @param needEmpty 是否必须空桌
	 * @return 餐桌
	 * @throws Exception
	 */
	public List<Table> query(int seats, boolean needEmpty)
		throws RuntimeException {
		String url = app.getServerUrl() + Constants.TABLE_SEARCH_URL;
		Map<String, String> params = new HashMap<String, String>();
		params.put("seats", seats + "");
		if (needEmpty)
			params.put("needEmpty", "true");
```

```
            // 请求服务器获取数据
            JsonData jd = HttpUtils.post(url, params);
            // 如果服务器返回数据为 null，抛出异常
            if (jd == null) {
                throw new RuntimeException(HttpUtils.SERVICE_ERR);
            }
            List<Table> tables = null;
            // 如果请求成功，解析返回 JSON 数据，否则将服务器返回的异常信息抛出
            if (jd.isSuccess()) {
                tables = new Gson().fromJson(jd.getData(),
                            new TypeToken<List<Table>>() {
                            }.getType());
                return tables;
            } else {
                throw new RuntimeException(jd.getMsg());
            }
        }

        /**
         * 获取所有餐桌。本地查询，查询结果中的当前就餐人数不可用
         *
         * @return 餐桌
         */
        public List<Table> getTables() {
            ...... // 省略原有代码
        }
}
```

上述代码在 query()方法内连接服务器，并调用 HttpUtils.post()方法提交需要的座位数和是否空桌的信息，然后从服务器获取查询结果。这里使用了通过 GSON 将 JSON 字符串转换为 List 对象的方法，具体代码如下：

```
List<Table> tables = new Gson().fromJson(jd.getData(),
                            new TypeToken<List<Table>>() {
                            }.getType());
```

通过上述代码，就可以方便地将 JSON 字符串转换为对应的 List 对象了。

(2) 编写查桌界面中显示查询结果的 ListView 使用的布局文件 table_item.xml，代码如下：

```
<?xml version="1.0" encoding="utf-8"?>
<TableLayout xmlns:android="http://schemas.android.com/apk/res/android"
    android:layout_width="match_parent"
```

```
    android:layout_height="wrap_content"
    android:stretchColumns="4" >

<TableRow
        android:layout_width="match_parent"
        android:textColor="#80FFFF" >
<TextView
            android:id="@+id/noTxv"
            android:layout_height="wrap_content"
            android:gravity="right"
            android:textColor="#fff"
            android:width="20dp" />
<TextView
            android:id="@+id/codeTxv"
            android:layout_width="wrap_content"
            android:layout_height="wrap_content"
            android:gravity="left"
            android:paddingLeft="10dp"
            android:textColor="#00FFFF"
            android:width="100dp" />
<TextView
            android:id="@+id/seatsTxv"
            android:layout_height="wrap_content"
            android:gravity="right"
            android:paddingRight="3dp"
            android:textColor="#A0FF42"
            android:width="40dp" />
<TextView
            android:id="@+id/customersTxv"
            android:layout_width="wrap_content"
            android:layout_height="wrap_content"
            android:gravity="left"
            android:paddingLeft="3dp"
            android:textColor="#FF8080" />
<TextView
            android:id="@+id/descriptionTxv"
            android:layout_width="wrap_content"
            android:layout_height="wrap_content"
            android:gravity="right"
```

```
            android:paddingRight="5dp"
            android:textColor="#00FFFF" />
</TableRow>

</TableLayout>
```

(3) 修改查桌界面 com.ugrow.repast_ph09.ui.TableFragment，代码如下：

```
public class TableFragment extends Fragment {

    private EditText seatsEdt;
    private Button mBtnQuery;
    private Button mBtnCansel;
    private ListView tablesLtv;
    private CheckBox needEmptyCkb;
    // 查询结果列表
    private List<Map<String, Object>> tableList = new ArrayList<Map<String, Object>>();
    private String[] tablesLtvKeys = new String[] { "no", "code", "seats",
            "customers", "description" };
    private int[] tablesLtvIds = new int[] { R.id.noTxv, R.id.codeTxv,
            R.id.seatsTxv, R.id.customersTxv, R.id.descriptionTxv };

    private TableService tableService;
    private List<Table> tables = null;

    // 定义常量—查询订单成功
    private static final int MSG_SUCCESS = 1;
    // 定义常量—查询订单失败
    private static final int MSG_ERR = 0;


    @Override
    public View onCreateView(LayoutInflater inflater, @Nullable ViewGroup container, @Nullable Bundle
savedInstanceState) {
        View view = inflater.inflate(R.layout.fragment_table, null);
        return view;
    }

    @Override
    public void onViewCreated(View view, @Nullable Bundle savedInstanceState) {
        super.onViewCreated(view, savedInstanceState);
        seatsEdt = (EditText) view.findViewById(R.id.et_seats);
```

```
    needEmptyCkb = (CheckBox) view.findViewById(R.id.cb_needEmptyCkb);
    mBtnQuery = (Button) view.findViewById(R.id.btn_query);
    mBtnCansel = (Button) view.findViewById(R.id.btn_cansel);
    tablesLtv = (ListView) view.findViewById(R.id.lv_fragment);
    initTablesLtv();
    initData();
}

private void initData() {
    tableService = new TableService(getActivity());
    mBtnQuery.setOnClickListener(new View.OnClickListener() {
        @Override
        public void onClick(View v) {
            // 隐藏键盘
            Tools.closeInput(getActivity(), seatsEdt);

            int seats = 0;
            String s = seatsEdt.getText().toString();
            if (s.length() > 0)
                seats = Integer.parseInt(s);
            boolean needEmpty = needEmptyCkb.isChecked();

            searchTables(seats, needEmpty);
        }
    });
    mBtnCansel.setOnClickListener(new View.OnClickListener() {
        @Override
        public void onClick(View v) {
            getActivity().finish();
        }
    });
}
void initTablesLtv() {
    SimpleAdapter sa = new SimpleAdapter(getActivity(), tableList, R.layout.table_item,
            tablesLtvKeys, tablesLtvIds);
    tablesLtv.setAdapter(sa);

}
/**
 * 定义 Handler，用于更新 UI 主线程
```

```
    */
    Handler handler = new Handler() {
        @Override
        public void handleMessage(Message msg) {

            switch (msg.what) {
                case MSG_ERR:
                    //showMessageDialog((String) msg.obj, R.mipmap.warning, null);
                    Toast.makeText(getActivity(),"查询失败",Toast.LENGTH_SHORT).show();
                    break;
                case MSG_SUCCESS:
                    showTables();
            }
        }
    };
    /**
     * 桌位查询
     *
     * @param seats
     * @param needEmpty
     */
    protected void searchTables(final int seats, final boolean needEmpty) {

        new Thread(new Runnable() {

            @Override
            public void run() {
                try {
                    tables = tableService.query(seats, needEmpty);
                    if (tables == null) {
                        handler.sendMessage(Tools.createMsg(MSG_ERR,
                                Constants.TABLE_NO_NEED_TABLES));

                    }else {
                        handler.sendMessage(Tools.createMsg(MSG_SUCCESS, null));
                    }
                } catch (Exception e) {
                    e.printStackTrace();
                    handler.sendMessage(Tools.createMsg(MSG_ERR, e.getMessage()));
                    return;
```

```
                }
            }
        }).start();

    }
    /**
     * 显示桌位
     */
    private void showTables() {
        tableList.clear();
        for (Table t : tables) {
            Map<String, Object> line = new HashMap<String, Object>();
            line.put("tableId", t.getId());
            line.put("no", tableList.size() + 1);
            line.put("code", t.getCode());
            line.put("seats", t.getSeats() + "座");
            if (t.getCustomers() == 0)
                line.put("customers", "");
            else
                line.put("customers", t.getCustomers() + "位就餐");
            line.put("description", t.getDescription());
            tableList.add(line);
        }
        ((SimpleAdapter) tablesLtv.getAdapter()).notifyDataSetChanged();
        if (tables.size() == 0){
            tablesLtv.setVisibility(View.GONE);
        }else{
            tablesLtv.setVisibility(View.VISIBLE);
        }

    }
}
```

根据上述代码，在单击【查询】按钮时，程序会调用 TableService 的 query()方法从服务器获取餐桌数据，并显示在 ListView 中。

(4) 修改 AndroidManifest.xml，在其中配置 TableActivity，代码如下：

```
<activity android:name=".ui.TableActivity" />
```

运行 Repast_ph09 项目，登录后在主菜单界面中单击【查桌】图标，将显示查桌界面，录入所需座位数以及是否必须空桌后，就可查询对应的餐桌信息，如图 S9-2 所示。

图 S9-2　测试查桌功能

实践 9.6　下单功能

在 Android 客户端上实现下单功能。

【分析】

在点餐界面中单击【下单】按钮时，需要向服务器提交订单数据，提交成功后，需返回订单 ID，用于订单查询与结账。

【参考解决方案】

(1) 修改 com.ugrow.repast_ph09.service.OrderService，代码如下：

```
public class OrderService {

    private App app;
    private FoodService foodService;

    public OrderService(Context context) {
        foodService = new FoodService(context);
        app = (App) context.getApplicationContext();
    }

    /**
     * 添加订单
     *
     * @param order 订单
```

```
     * @throws Exception
     */
    public int addOrder(Order order) throws RuntimeException{
        String url = app.getServerUrl() + Constants.ORDER_ADD_URL;
        Map<String, String> params = new HashMap<String, String>();
        params.put("order", new Gson().toJson(order));
        // 请求服务器，获取数据
        JsonData data = HttpUtils.post(url, params);

        // 如果服务器返回数据为 null，抛出异常
        if (data == null) {
            throw new RuntimeException(HttpUtils.SERVICE_ERR);
        }
        // 如果获取数据成功，返回 true，否则抛出异常，异常内容为服务器返回的内容
        if (data.isSuccess()) {
            int lastId = 0;
            try {
                lastId = Integer.parseInt(data.getData());
                return lastId;
            } catch (NumberFormatException e) {
                throw new RuntimeException(e.getMessage());
            }

        } else {
            throw new RuntimeException(data.getMsg());
        }
    }
    /**
     * Order 数据传输对象
     */
    public static class OrderDto {
        ...... // 省略
    }
}
```

上述代码中，addOrder(Order order)方法可实现添加订单功能，如果添加成功，则返回新增订单的 ID，以供账单查询和结账使用；而通过 new Gson().toJson(order)语句可以将 Order 对象通过 GSON 转换为 JSON 字符串，并作为参数提交到服务器。

(2) 修改 com.ugrow.repast_ph09.ui.OrderActivity，代码如下：

```
public class OrderActivity extends BasicActivity {
```

```
...... // 省略已有代码

@Override
protected void onCreate(Bundle savedInstanceState) {
        super.onCreate(savedInstanceState);

        tableService = new TableService(this);
        foodService = new FoodService(this);
        orderService = new OrderService(this);

        ...... // 省略已有代码

        orderBtn.setOnClickListener(new OnClickListener() {
                @Override
                public void onClick(View v) {
                        int customers = 0;
                        String c = customersEdt.getText().toString();
                        if (c.length() > 0)
                                customers = Integer.parseInt(c);
                        if (customers <= 0) {
                                showMessageDialog("请输入顾客数量",
                                        R.drawable.warning, null);
                                return;
                        }
                        int sum = Integer.parseInt(
                                sumTxv.getText().toString());
                        if (sum <= 0) {
                                showMessageDialog("尚未选择任何餐品",
                                        R.drawable.warning, null);
                                return;
                        }

                        Table table = (Table) tableSpn.getSelectedItem();
                        Order order = new Order();
                        User waiter = ((App) getApplication()).user;

                        order.setTableId(table.getId());
                        order.setWaiterId(waiter.getId());
                        order.setCustomers(Integer.parseInt(customersEdt.getText().toString()));
                        order.setDescription(descriptionEdt.getText().toString());
```

```
                order.setOrderDetails(new ArrayList<OrderDetail>());

                for (Map<String, Object> line : orderedList) {
                    boolean checked = (Boolean) line.get("checked");
                    if (!checked)
                        continue;
                    OrderDetail od = new OrderDetail();
                    od.setFoodId((Integer) line.get("foodId"));
                    od.setNum((Integer) line.get("num"));
                    od.setDescription((String) line.get("description"));
                    order.getOrderDetails().add(od);
                }
                addOrder(order);
            }
        });
    }
    private void addOrder(final Order order) {
        new Thread(new Runnable() {
            @Override
            public void run() {
                try {
                    int lastId = orderService.addOrder(order);
                    handler.sendMessage(Tools.createMsg(MSG_SUCCESS,
                        lastId));
                } catch (Exception e) {
                    e.printStackTrace();
                    System.out.println(e.getMessage());
                    handler.sendMessage(Tools.createMsg(MSG_ERR,
                        e.getMessage()));
                }
            }
        }).start();
    }
    /**
     * 定义 Handler，用于更新 UI 主线程
     */
    private Handler handler = new Handler() {
        @Override
        public void handleMessage(Message msg) {
            switch (msg.what) {
```

```
                case MSG_ERR:
                    showMessageDialog((String) msg.obj, R.drawable.not, null);
                    break;
                case MSG_SUCCESS:
                    showMessageDialog(Constants.ORDER_ADD_SUCCESS
                            + " 订单编号: "
                                + msg.obj, R.drawable.info,
                                new DialogInterface.OnClickListener() {
                                    @Override
                                    public void onClick(DialogInterface
                                            arg0, int arg1) {
                                                finish();
                                    }
                                });
                    break;
                }
            }
        };
        ......// 省略已有代码
}
```

运行 Repast_ph09 项目，进入点餐界面，点餐后单击【下单】按钮，将提交数据到服务器，如图 S9-3 所示。

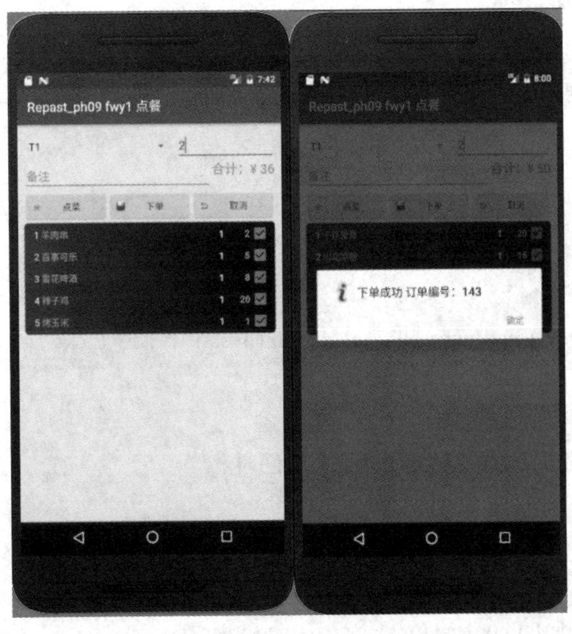

图 S9-3　测试下单功能

实践 9.7 结账功能

在 Android 客户端上实现结账功能。

【分析】

结账时，需要从服务器查询某个编号的订单，在用户确认后向服务器提交结账信息。

【参考解决方案】

修改 com.ugrow.repast_ph09.service.OrderService，在其中添加查询订单方法 getOrder() 和结账方法 pay()，代码如下：

```
public class OrderService {

    ...... // 省略已有代码

    /**
     * 根据 code 获取订单
     *
     * @param code
     *  订单 code
     * @return 订单
     */
    public OrderInfo getOrder(String code) {
        String url = app.getServerUrl() + Constants.ORDER_GET_URL;

        Map<String, String> params = new HashMap<String, String>();
        params.put("code", code);
        // 请求服务器，获取数据
        JsonData data = HttpUtils.post(url, params);

        // 如果服务器返回数据为 null，抛出异常
        if (data == null) {
            throw new RuntimeException(HttpUtils.SERVICE_ERR);
        }
        // 如果获取数据成功，返回 Order 对象，
        //否则抛出异常，异常内容为服务器返回的内容
        if (data.isSuccess()) {
            Order order = new Gson().fromJson(data.getData(), Order.class);
            // 对 Order 进行处理计算，得到最终的订单信息并返回
            OrderInfo orderInfo = handlerOrder(order);
            return orderInfo;
```

```
        } else {
                throw new RuntimeException(data.getMsg());
        }

}

/**
 * 结账
 *
 * @param orderId
 * 对应订单 ID
 */
public boolean pay(int orderId) {
        String url = app.getServerUrl() + Constants.ORDER_PAY_URL;

        Map<String, String> params = new HashMap<String, String>();
        params.put("orderId", orderId + "");
        // 请求服务器，获取数据
        JsonData data = HttpUtils.post(url, params);

        // 如果服务器返回数据为 null，抛出异常
        if (data == null) {
                throw new RuntimeException(HttpUtils.SERVICE_ERR);
        }
        // 如果获取数据成功，返回 true，否则抛出异常，异常内容为服务器返回的内容
        if (data.isSuccess()) {
                return true;
        } else {
                throw new RuntimeException(data.getMsg());
        }
}

/**
 * 对 Order 进行处理计算，得到最终的订单信息并返回
 *
 * @param order
 * @return
 */
private OrderInfo handlerOrder(Order order) {
```

```
            if (order == null) {
                return null;
            }
            OrderInfo info = new OrderInfo();
            info.setOrder(order);

            for (OrderDetail od : order.getOrderDetails()) {
                Food food = foodService.getFood(od.getFoodId());
                Map<String, Object> line = new HashMap<String, Object>();
                line.put("no", info.getOrderedList().size() + 1);
                line.put("name", food.getName());
                line.put("description", od.getDescription());
                line.put("num", od.getNum());
                line.put("price", food.getPrice());
                info.getOrderedList().add(line);
                info.setSum(info.getSum() + od.getNum() * food.getPrice());
            }
            return info;
        }
}
```

运行 Repast_ph09 项目，登录后在主菜单单击【结账】图标，进入结账界面，输入订单编号后单击【查询】按钮，将显示对应订单的详细信息，此时单击【结账】按钮，将完成结账操作，并输出提示信息，在结账界面再次查询此订单，将显示已结账，如图 S9-4 所示。

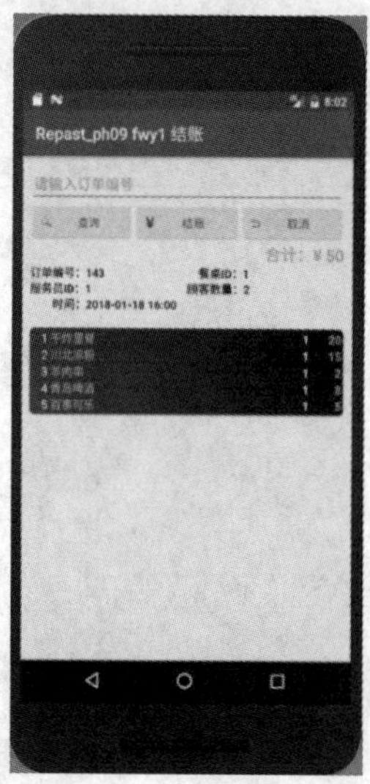

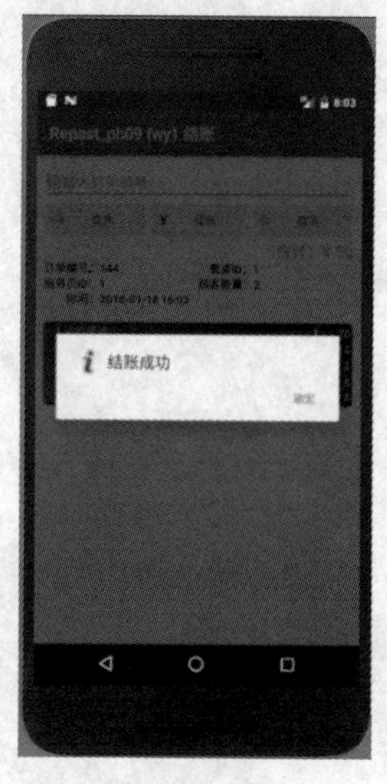

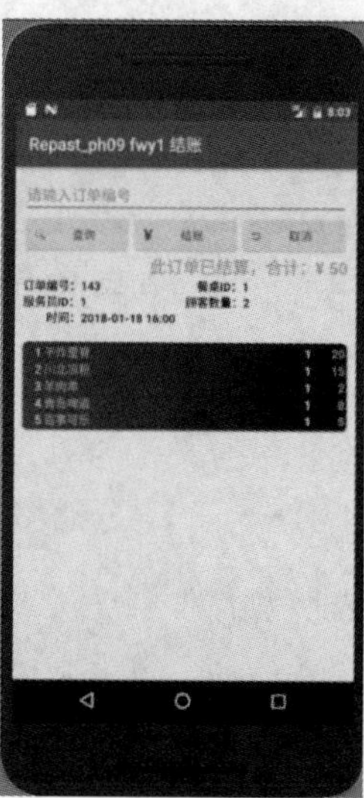

图 S9-4　测试结账功能

知识拓展

1. Wi-Fi 简介

Wi-Fi 全称 Wireless Fidelity，是一种无线网络通信技术，具有传输速度高和有效距离长的优点，并与各种 802.11DSSS 设备兼容。

Android 在 android.net.wifi 包中提供了许多用于操作 Wi-Fi 的类，这些类主要有：

(1) ScanResult：用于描述已检测到的接入点，包括接入点的地址、名称、身份认证、频率、信号强度等。

(2) WifiConfiguration：用于 Wi-Fi 网络的配置，包括安全配置等。

(3) WifiInfo：用于描述 Wi-Fi 无线连接的基本信息，包括接入点、连接状态、隐藏的接入点、IP 地址、连接速度、MAC 地址、网络 ID、信号强度等。

(4) WifiManager：Wi-Fi 连接的大部分管理功能都由 WifiManager 类提供，其常用方法如表 S9-1 所示。

表 S9-1　WifiManager 类的常用方法

方　　法	说　　明
addNetwork()	添加已配制的网络连接
calculateSignalLevel()	计算信号强度
compareSignalLevel()	比较信号强度
createWifiLock()	创建 Wi-Fi 锁
disableNetwork()	取消一个已配制的网络
disconnect()	断开接入点
enableNetwork()	启用网络
getConfiguredNetworks()	获取客户端所有已配制的网络列表
getConnectionInfo()	获取正在使用的连接的信息
getDhcpInfo()	获取最后一次成功的 DHCP 请求的地址
getScanResults()	获取扫描的接入点
getWifiState()	获取可用 Wi-Fi 的状态
pingSupplicant()	检查客户端对请求的反应
reassociate()	从当前接入点重新连接
removeNetwork()	从已配制的网络列表中删除指定 ID 的网络
saveConfiguration()	保存配置好的网络列表
setWifiEnabled()	设置 Wi-Fi 是否可用
startScan()	开始扫描存在的接入点
updateNetwork()	更新配置好的网络

2. 蓝牙

蓝牙是一种短距离(一般为 10 米以内)的无线通信技术标准，能在移动电话、PDA、无线耳机、笔记本电脑等众多设备之间传输数据。利用蓝牙技术，可以有效地简化移动设备之间、移动设备与 Internet 之间的通信过程。

Android 支持对蓝牙进行编程，与蓝牙有关的类位于 android.bluetooth 包下，其主要的类及功能如表 S9-2 所示。

表 S9-2 蓝牙主要相关类

类	说　明
BluetoothAdapter	本地蓝牙适配器
BluetoothClass	蓝牙类
BluetoothClass.Device	蓝牙设备
BluetoothClass.Device.Major	蓝牙设备管理
BluetoothClass.Service	蓝牙服务
BluetoothDevice	远程蓝牙设备
BluetoothSocket	蓝牙服务器端 Socket
BluetoothServerSocket	蓝牙 Socket

其中，BluetoothAdapter 是最重要的一个类，其常用方法如表 S9-3 所示。

表 S9-3 BluetoothAdapter 的常用方法

类	说　明
cancelDiscovery()	取消当前搜索过程
checkBluetoothAddress()	检查蓝牙地址是否正确
disable()	关闭蓝牙适配器
enable()	打开蓝牙适配器
getAddress()	获取本地蓝牙的硬件适配器地址
getDefaultAdapter()	获取默认的蓝牙适配器
getName()	获取蓝牙名称
getRemoteDevice()	获取指定地址的 BluetoothDevice 对象
getScanMode()	获取扫描模式
getState()	获取状态
isDiscovering()	设置是否允许被搜索
isEnabled()	设置是否打开
setName()	设置名称
startDiscovery()	开始搜索

注 意

Android 模拟器不支持 Wi-Fi 和蓝牙，必须连接实际的终端才可以编程测试。

拓展练习

修改实践 9.2，将其中的 HttpUtils.sendObject()方法改为使用 HttpClient 实现。

实践 10　第三方框架

实践指导

实践 10.1　ButterKnife

在 Android 开发中，使用 findViewById()方法初始化控件是基本的操作，为解决反复使用该方法造成的代码重复问题，Square 公司的 Android 工程师 JakeWharton 开发了一款依赖注入的开源框架 ButterKnife，中文译为“黄油刀”或“匕首”。

ButterKnife 是专门为 Android View 设计的绑定注解工具， 其主要优势如下：

(1) 强大的 View 绑定和 Click 事件处理功能，有效简化代码，提升应用开发效率。

(2) 方便地处理 Adapter 中 ViewHolder 类的绑定问题。

(3) 运行时不会影响应用程序效率，使用时配置方便。

(4) 代码清晰，可读性强。

在初始化成员变量时使用 ButterKnife 的注解方法@BindView()，向其中传入某个 View 的 ID，ButterKnife 就能自动找到该 ID 对应的 View，并对其进行转换(将 View 转换为特定的子类)，示例代码如下：

```
public class ExampleActivity extends Activity {
    @BindView(R.id.title)
    TextView title;
    @BindView(R.id.subtitle)
    TextView subtitle;
    @BindView(R.id.footer)
    TextView footer;

    @Override
    public void onCreate(Bundle savedInstanceState) {
        super.onCreate(savedInstanceState);
        setContentView(R.layout.simple_activity);
        ButterKnife.bind(this);
        // TODO Use fields...
    }
}
```

上述代码调用了 ButterKnife 中的 bind()方法初始化了文本控件 TextView，该方法中的代码如下：

```
public void bind(ExampleActivity activity) {
    activity.subtitle = (android.widget.TextView) activity.findViewById(2130968578);
    activity.footer = (android.widget.TextView) activity.findViewById(2130968579);
    activity.title = (android.widget.TextView) activity.findViewById(2130968577);
```

1．ButterKnife 注册与绑定

ButterKnife 的绑定方法分为以下三种：

(1) 在 Activity 中绑定 ButterKnife。

在 Activity 中绑定 ButterKnife，必须在 setContentView() 方法之后，使用 ButterKnife.bind(this)方法进行绑定，代码如下：

```
public class MainActivity extends AppCompatActivity{
    @Override
    protected void onCreate(Bundle savedInstanceState) {
        super.onCreate(savedInstanceState);
        setContentView(R.layout.activity_main);
        //绑定初始化ButterKnife
        ButterKnife.bind(this);
    }
}
```

由于每次都要在 onCreate()方法中将 Activity 与 ButterKnife 绑定，所以建议写一个基类 BaseActivity 来完成绑定操作，每次绑定时都新建一个子类继承这个基类即可。

(2) 在 Fragment 中绑定 ButterKnife。

在 onCreateView() 中使用 ButterKnife.bind(this, view) 方法，可以将 Fragment 与 ButterKnife 进行绑定，之后 ButterKnife 会返回一个 Unbinder 的实例，完成相关操作以后，需在生命周期回调方法 onDestroyView()中调用 ButterKnife.unbind()方法进行解绑操作，将 View 设置为 NULL，代码如下：

```
public class ButterknifeFragment extends Fragment{
    private Unbinder unbinder;
    @Override
    public View onCreateView(LayoutInflater inflater, ViewGroup container,
                              Bundle savedInstanceState) {
        View view = inflater.inflate(R.layout.fragment, container, false);
        // ButterKnife.bind(this, view)
        unbinder = ButterKnife.bind(this, view);
        return view;
    }

    /**
     * 使用 onDestroyView()方法进行解绑操作
     */
    @Override
```

```
    public void onDestroyView() {
        super.onDestroyView();
        unbinder.unbind();
    }
}
```

(3) 在适配器 Adapter 中绑定 ButterKnife。

在适配器 Adapter 的 ViewHolder 中绑定 ButterKnife，需要在 ViewHolder 中加一个构造方法，将 View 传递进去，然后使用 ButterKnife.bind(this, view)方法进行绑定，代码如下：

```
public class MyAdapter extends BaseAdapter {

  @Override
  public View getView(int position, View view, ViewGroup parent) {
    ViewHolder holder;
    if (view != null) {
      holder = (ViewHolder) view.getTag();
    } else {
      view = inflater.inflate(R.layout.testlayout, parent, false);
      holder = new ViewHolder(view);
      view.setTag(holder);
    }

    holder.name.setText("Donkor");
    holder.job.setText("Android");
    // etc...
    return view;
  }

  static class ViewHolder {
    @BindView(R.id.title) TextView name;
    @BindView(R.id.job) TextView job;

    public ViewHolder(View view) {
      ButterKnife.bind(this, view);
    }
  }
}
```

ButterKnife 使用时的注意事项如下：

(1) 在 Activity 类中绑定时，必须通过在 setContentView()方法之后调用 ButterKnife.bind(this)方法来绑定 ButterKnife，且父 Activity 类绑定后，子 Activity 类不需要再绑定。

(2) 在非 Activity 类(例如 Fragment 或 ViewHolder)中绑定时：需通过 ButterKnife.bind(this，view)方法绑定 ButterKnife，注意其中的参数 this 不能替换成 getActivity()。

(3) 在 Activity 中绑定 ButterKnife 之后不需进行解绑操作，但在 Fragment 中绑定

ButterKnife，使用完毕后必须在 onDestroyView()方法中进行解绑操作。

(4) 使用 ButterKnife 绑定的方法和控件不能用 private or static 关键字修饰，否则程序会报错。错误信息为“@BindView fields must not be private or static”。

(5) 使用 ButterKnife 注解框架时不能使用 setContentView()方法加载布局(但某些其他注解框架可以，如 Xutils 框架)。

(6) 使用 Activity 为根视图绑定任意视图(View)对象时，如果使用类似 MVC 的设计模式，可以在 Activity 中调用 ButterKnife.bind(this, activity)方法来绑定视图对象。

(7) 如果在 View 子节点的布局里或者自定义 View 的构造方法里使用了 inflate()方法，则可以调用 ButterKnife.bind(this，view)方法。

2. ButterKnife 常用功能

(1) 绑定资源。

可以使用@BindBool()、@BindColor()、@BindDimen()、@BindDrawable()、@BindInt()与@BindString()注解方法，将特定的资源绑定到某个类的成员上。使用时，需将对应资源的 ID 传入对应的注解方法中。

例如，使用@BindString()方法绑定字符串资源时，需要传入资源的 ID(如 R.string.id_string)，示例代码如下：

```
public class ExampleActivity extends Activity {
        @BindString(R.string.title)
        String title;
        @BindDrawable(R.drawable.graphic)
        Drawable graphic;
        @BindColor(R.color.red)
        int red; // int or ColorStateList field
        @BindDimen(R.dimen.spacer)
        Float spacer; // int (for pixel size) or float (for exact value) field
  // ...
}
```

(2) 绑定单个 View。

在 Activity 或 Fragment 中使用@BindView()方法，可以将需要的 View 控件与之绑定，传入的参数为该 View 控件的 ID，示例代码如下：

```
        @BindView( R2.id.button)
        public Button button;
```

(3) 绑定多个 View。

在 Activity 或 Fragment 中使用@BindViews()方法，可以一次性绑定多个 View 控件，生成一个集合或数组，传入的参数为各个 View 控件的 ID，示例代码如下：

```
public class MainActivity extends AppCompatActivity {

    @BindViews({ R2.id.button1, R2.id.button2,   R2.id.button3})
    public List<Button> buttonList ;
```

```
    @Override
    protected void onCreate(Bundle savedInstanceState) {
        super.onCreate(savedInstanceState);
        setContentView(R.layout.activity_main);
        ButterKnife.bind(this);
        buttonList.get( 0 ).setText( "hello 1 ");
        buttonList.get( 1 ).setText( "hello 2 ");
        buttonList.get( 2 ).setText( "hello 3 ");
    }
}
```

当使用@BindViews()方法获取 View 控件的集合之后，可以使用 ButterKnife.apply()方法，一次性地在集合中的所有 View 控件上执行一个相同的动作，示例代码如下：

```
ButterKnife.apply(buttonList, DISABLE);
ButterKnife.apply(buttonList, ENABLED, false);
```

集合中 View 控件的 Property 属性也可以使用 apply()方法进行设置，示例代码如下：

```
ButterKnife.apply(nameViews, View.ALPHA, 0.0f);
```

除使用 ButterKnife.apply()方法指定 View 控件执行动作以外，还可以使用 Action 和 Setter 接口，指定 View 控件执行一些简单的动作，示例代码如下：

```
static final ButterKnife.Action<View> DISABLE = new ButterKnife.Action<View>() {
    @Override public void apply(View view, int index) {
        view.setEnabled(false);
    }
};
static final ButterKnife.Setter<View, Boolean> ENABLED = new ButterKnife.Setter<View, Boolean>() {
    @Override public void set(View view, Boolean value, int index) {
        view.setEnabled(value);
    }
};
```

(4) 绑定事件。

在 Android 开发中，控件的事件可分为两种：单击事件(OnClick())和长按事件(OnLongClick())。使用 ButterKnife 框架可以将这两种事件绑定在相应的控件上，示例代码如下：

```
public class MainActivity extends AppCompatActivity {

    @OnClick(R.id.button1 )    //给button1设置一个单击事件
    public void showToast(){
        Toast.makeText(this, "单击事件", Toast.LENGTH_SHORT).show();
    }

    @OnLongClick( R.id.button2 )    //给button2设置一个长按事件
    public boolean showToast2(){
```

```
        Toast.makeText(this, "长按事件", Toast.LENGTH_SHORT).show();
        return true ;
    }

    @Override
    protected void onCreate(Bundle savedInstanceState) {
        super.onCreate(savedInstanceState);
        setContentView(R.layout.activity_main);
        //绑定Activity
        ButterKnife.bind( this ) ;
    }
}
```

上述代码分别使用@OnClick()方法和@OnLongClick()方法在单个 View 控件上绑定了监听事件，传入的参数 ID 只有一个。

ButterKnife 也可以同时在多个 View 控件上绑定事件，此时需要在注解方法中传入多个 View 控件的 ID，示例代码如下：

```
public class MainActivity extends AppCompatActivity {

    @Override
    protected void onCreate(Bundle savedInstanceState) {
        super.onCreate(savedInstanceState);
        setContentView(R.layout.activity_main);
        //绑定Activity
        ButterKnife.bind( this ) ;
    }
    @OnClick({R.id.ll_product_name, R.id.ll_product_lilv, R.id.ll_product_qixian,
R.id.ll_product_repayment_methods})
    public void onViewClicked(View view) {
        switch (view.getId()) {
            case R.id.ll_product_name:
                System.out.print("我是点击事件1");
                break;
            case R.id.ll_product_lilv:
                System.out.print("我是点击事件2");
                break;
            case R.id.ll_product_qixian:
                System.out.print("我是点击事件3");
                break;
            case R.id.ll_product_repayment_methods:
                System.out.print("我是点击事件4");
                break;
```

```
        }
    }
}
```

(5) 绑定自定义 View。

在 Android 开发中，有时会使用自定义的 View 控件，可使用 ButterKnife 框架为其绑定事件。绑定时直接使用注解方法@OnClick()即可，不需要指定控件的 ID，示例代码如下：

```
public class FancyButton extends Button {
    @OnClick()
    public void onClick() {
        // TODO do something!
    }
}
```

(6) 可选绑定。

在默认情况下，@bind()和监听器都必须要与 View 进行绑定，如果没有进行绑定，ButterKnife 就会抛出异常。

如果不想使用 ButterKnife 的默认设置，而是想创建一个自定义的绑定，那么只需在变量上使用@Nullable 注解或在方法上使用@Option 注解即可，示例代码如下：

```
    @Nullable
    @BindView(R.id.might_not_be_there)
    TextView mightNotBeThere;

    @Optional
    @OnClick(R.id.maybe_missing)
    void onMaybeMissingClicked() {
     // TODO ...
    }
```

注意：任何被命名为@Nullable 的注解都可以在成员变量上使用，建议使用 Android 注解库中的@Nullable 注解。

3．ButterKnife 其他功能

上一小节讲解了 ButterKnife 最常用的几种功能，此外 ButterKnife 还有以下功能：

(1) @OnCheckedChanged ：选中或取消选中 View。

(2) @OnEditorAction ：给软键盘的功能键绑定监听事件。

(3) @OnFocusChange ：改变 View 控件的焦点。

(4) @OnItemClickItem：单击 item(注意：如果 item 布局中有 Button 等有单击功能的控件，需将这些控件的 focusable 属性设置为 false)。

(5) @OnItemLongClickItem：长按 item(该方法返回 true 时，onItemClick()方法的执行会被拦截)。

(6) @OnItemSelected：选择 item。

(7) @OnPageChange：改变页面。

(8) @OnTextChanged：改变 EditText 中的文本。

【示例 10.1】 在 Android Studio 中新建项目 Ph10ex1_Butterknife，使用 ButterKnife 对 View 进行注解，并在其上绑定监听事件。

在使用 ButterKnife 之前，要先安装 Android Butterknife Zelezny 插件：单击 Android Studio 工具栏中的【File】/【Settings】命令，或者按快捷键【Ctrl】+【Alt】+【S】打开【Settings】设置菜单。

在弹出的界面左侧的项目列表中选择【Plugins】，然后在右边出现的搜索框中搜索 Zelezny 插件并下载安装，如图 S10-1 所示。安装完毕，重启 Android Studio。

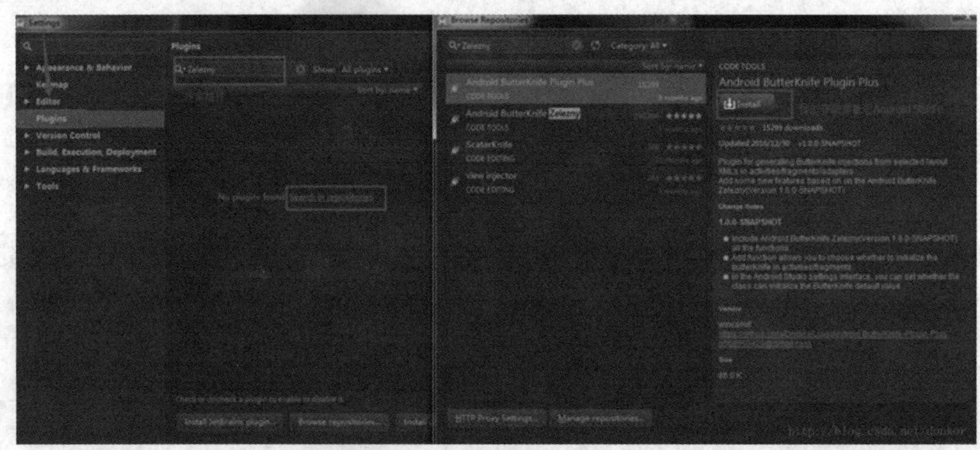

图 S10-1　安装 Zelezny 插件

Android Butterknife Zelezny 插件安装完成后，将项目切换到 Project 模式下，单击项目包的名称，然后在项目包的名称上单击鼠标右键，在弹出的菜单中选择【Open Module Settings】命令，如图 S10-2 所示。

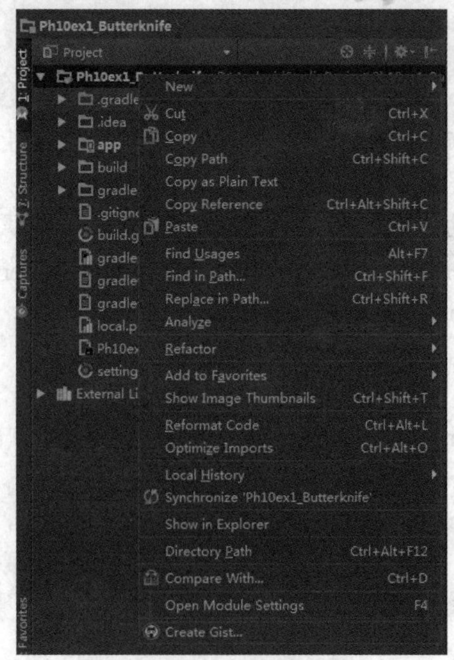

图 S10-2　进入项目设置

在弹出的界面【Project Structure】左侧项目列表中，选择【app】项，然后单击界面右侧标签【Dependencies】中的【+】号，在弹出的菜单中，选择【Library dependency】，如图 S10-3 所示。

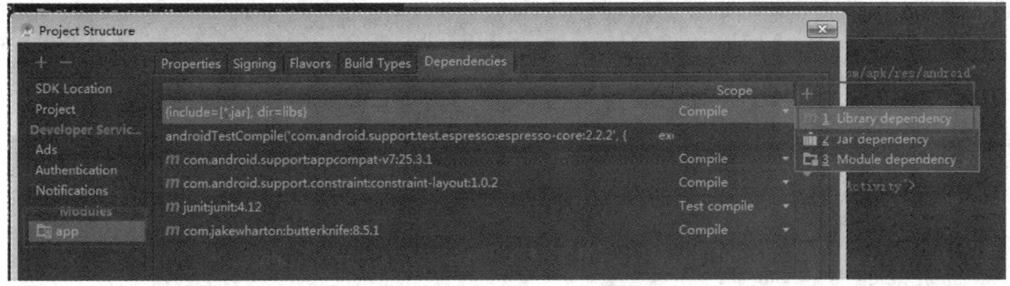

图 S10-3　选择添加依赖库

在弹出的界面【Choose Library Dependency 】顶端的搜索框中，输入依赖库的名称“com.jakewharton:butterknife:5.1.1”，然后单击【OK】按钮，关闭对话框，如图 S10-4 所示。

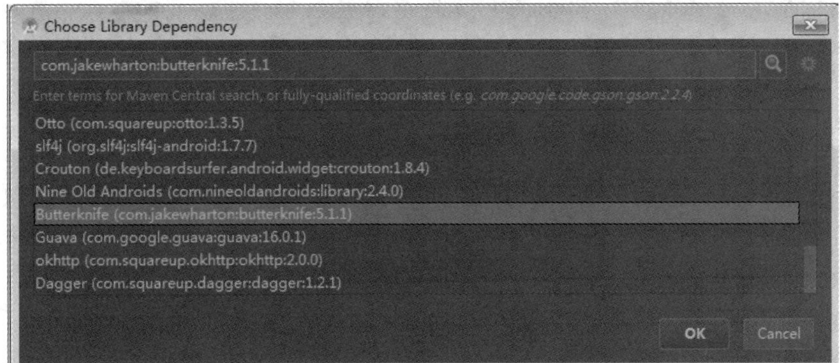

图 S10-4　查找并添加所需的 ButterKnife 依赖库

通过上述步骤将 ButterKnife 所需要的依赖库加入到 build.gradle 中，代码如下：

```
dependencies {
    compile fileTree(include: ['*.jar'], dir: 'libs')
    androidTestCompile('com.android.support.test.espresso:espresso-core:2.2.2', {
        exclude group: 'com.android.support', module: 'support-annotations'
    })
    compile 'com.android.support:appcompat-v7:25.3.1'
    compile 'com.android.support.constraint:constraint-layout:1.0.2'
    testCompile 'junit:junit:4.12'
    compile 'com.jakewharton:butterknife:5.1.1'
}
```

编写 activity_main.xml，代码如下：

```
<?xml version="1.0" encoding="utf-8"?>
<LinearLayout xmlns:android="http://schemas.android.com/apk/res/android"
    xmlns:tools="http://schemas.android.com/tools"
    android:layout_width="match_parent"
```

```
    android:layout_height="match_parent"
    android:orientation="vertical"
    tools:context="com.ugrow.ph10ex1_butterknife.MainActivity">

    <Button
        android:id="@+id/btn_one"
        android:layout_width="wrap_content"
        android:layout_height="wrap_content"
        android:text="按钮 one"/>
    <Button
        android:id="@+id/btn_sec"
        android:layout_width="wrap_content"
        android:layout_height="wrap_content"
        android:text="按钮 sec"/>

</LinearLayout>
```

编写 MainActivity，代码如下：

```
public class MainActivity extends AppCompatActivity {

    @Bind(R.id.btn_one)
    Button btnOne;
    @Bind(R.id.btn_sec)
    Button btnSec;

    @Override
    protected void onCreate(Bundle savedInstanceState) {
        super.onCreate(savedInstanceState);
        setContentView(R.layout.activity_main);
        ButterKnife.bind(this);
        btnOne.setText("第一个按钮");
        btnSec.setText("第二个按钮");
    }

    @OnClick({R.id.btn_one, R.id.btn_sec})
    public void onClick(View view) {
        switch (view.getId()) {
            case R.id.btn_one:
                Toast.makeText(MainActivity.this, btnOne.getText().toString(), Toast.LENGTH_SHORT).show();
                break;
            case R.id.btn_sec:
                Toast.makeText(MainActivity.this, btnSec.getText().toString(), Toast.LENGTH_SHORT).show();
```

```
                break;
            default:
                break;
        }
    }
}
```

在上述代码中右键单击布局文件的名称，在弹出的菜单中选择【Generate】命令，如图 S10-5 所示。

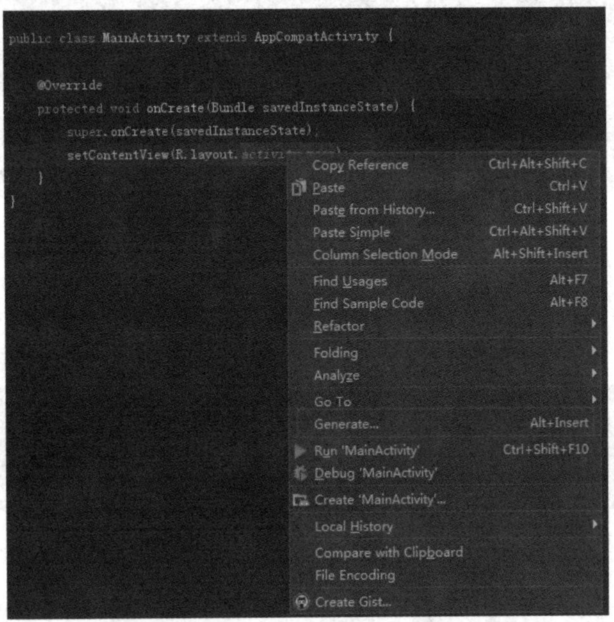

S10-5　开启控件初始化菜单

在弹出的【Generate】菜单中选择【Generate Butterknife Injections】命令，如图 S10-6 所示。

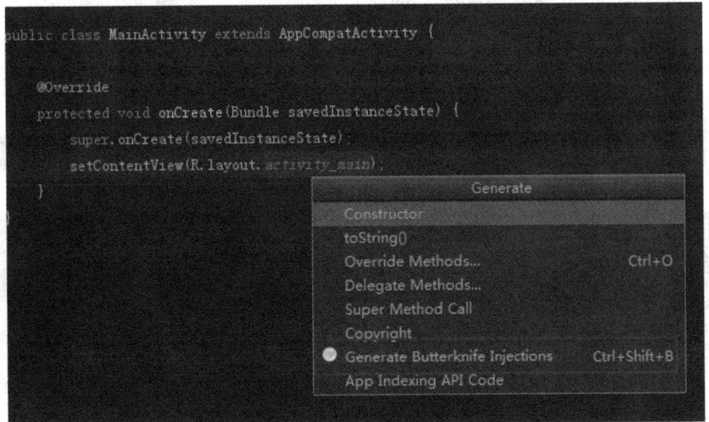

S10-6　将 Activity 与 ButterKnife 绑定

在弹出的界面中，选择要绑定的 View 控件，然后单击【Confirm】按钮，就会在 Activity 中绑定这些控件，如图 S10-7 所示。

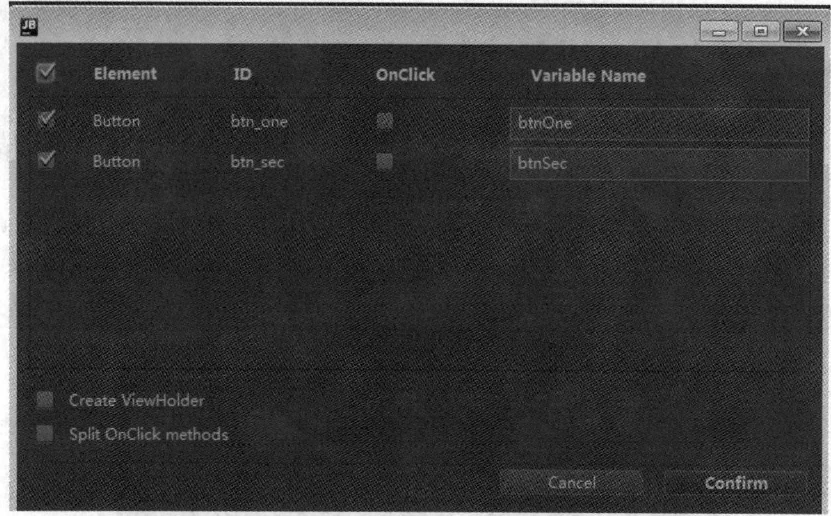

S10-7　将 Activity 与 View 控件绑定

编写完毕，在 Android Studio 中运行项目，效果如图 S10-8 所示。

分别单击应用界面上的两个按钮，会分别在模拟器界面中显示字符串“第一个按钮”和“第二个按钮”，如图 S10-9 所示。

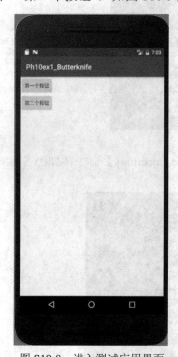

图 S10-8　进入测试应用界面

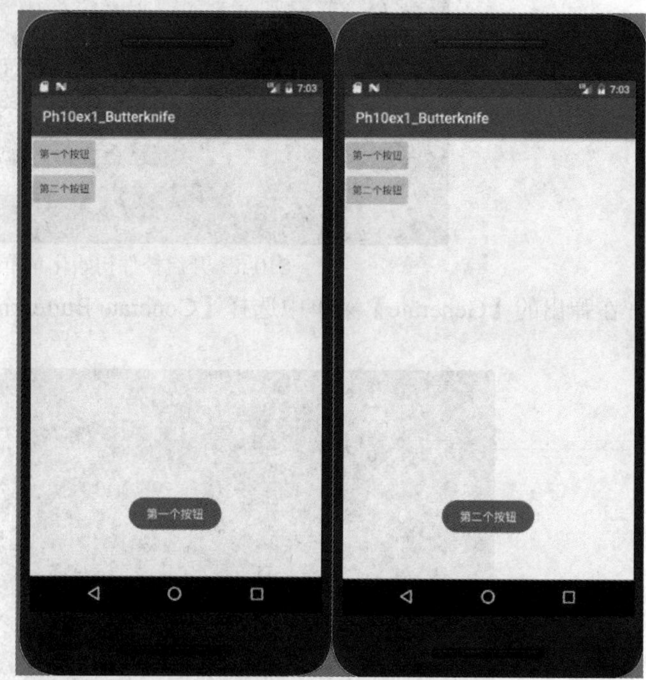

图 S10-9　测试 ButterKnife 监听功能

实践 10.2　Picasso

在 Android 应用中，加载图片是不可缺少的一项功能，而怎样才能更好地加载网络或本地图片就是 Android 应用开发中需要重点考虑的问题。

Picasso 是 Square 公司开发的一个开源 Android 图片加载缓存库，中文名译为“毕加索”，用于实现下载图片的功能。

Picasso 具有以下特性：

(1) 支持任务优先级，会优先加载优先级高的图片。

(2) 带有统计监控功能，可以统计缓存率，并实时监控已使用的内存。

(3) 能够根据当前网络状态自动调整并发线程数。

(4) 支持图片的延迟加载。

(5) 支持加载过重、错误处理、不同资源加载等。

(6) 在 Adapter 中回收不在视野中的 ImageView，并取消已回收 ImageView 的下载进程。

从上述特性可以看出，Picasso 不仅实现了图片异步加载的功能，同时也解决了 Android 加载图片时的一些常见问题。

Picasso 的主要应用如下。

1. 图片异步加载

Picasso 使用了流式接口的调用方式，其加载图片的核心实现类为 Picasso 类，加载图片的代码格式如下：

```
Picasso.with(context).load(Url).into(targetImageView);
```

由上述代码可知，Picasso 类实现图片加载功能至少需要三个参数：① Context，即上下文参数；② URL，即被加载图像的 URL 地址；③ targetImageView，即图片最终要在哪个 ImageView 控件上展示。

2. 图片转换

Picasso 可以使用最少的内存完成复杂的图片转换，转换后的图片会自动适应所要显示的 ImageView，以减少内存消耗或更适应于布局，示例代码如下：

```
Picasso.with(context)
  .load(url)
//裁剪图片尺寸
  .resize(50, 50)
//设置图片圆角
  .centerCrop()
  .into(imageView)
```

3. 加载过重、错误处理

使用 Picasso 加载图片时，可以使用 placeholder()方法显示加载过程中要显示的图片；使用 error()方法显示加载失败时要显示的图片，示例代码如下：

```
Picasso.with(context)
    .load(url)
    .placeholder(R.drawable.user_placeholder)
//如果重试3次(下载源代码可以根据需要修改)还是无法成功加载图片，则用错误占位符图片显示
    .error(R.drawable.user_placeholder_error)
```

```
    .into(imageView);
```

4．自动取消图片下载

当使用 Picasso 下载图片时，如果检测到 convertView 不为空(View 的重用)时，会自动取消之前的 convertView 下载任务，示例代码如下：

```
@Override
public void getView(intposition,View convertView,ViewGroup parent) {
  SquaredImageView view = (SquaredImageView) convertView;
  if(view ==null) {
    view =newSquaredImageView(context);
  }
  String url = getItem(position);
  Picasso.with(context).load(url).into(view);
}
```

5．加载不同资源

使用 Picasso 可以加载多种数据源，除了可以加载网络资源，还可以加载本地资源、Assets 资源等，示例代码如下：

```
//加载资源文件
Picasso.with(context).load(R.drawable.landing_screen).into(imageView1);
//加载本地文件
Picasso.with(context).load(new File("/images/oprah_bees.gif")).into(imageView2);
```

【示例 10.2】 在 Android Studio 中新建项目 Ph10ex2_Picasso，在其中使用 Picasso 加载网络图片。

首先在 build.gradle 中添加依赖库，代码如下：

```
dependencies {
    compile fileTree(include: ['*.jar'], dir: 'libs')
    androidTestCompile('com.android.support.test.espresso:espresso-core:2.2.2', {
        exclude group: 'com.android.support', module: 'support-annotations'
    })
    compile 'com.android.support:appcompat-v7:25.3.1'
    compile 'com.android.support.constraint:constraint-layout:1.0.2'
    testCompile 'junit:junit:4.12'
    compile 'com.jakewharton:butterknife:7.0.1'
    compile 'com.squareup.picasso:picasso:2.3.2'
}
```

在清单文件 AndroidMainfest.xml 中添加网络权限，代码如下：

```
<uses-permission android:name="android.permission.INTERNET"></uses-permission>
```

编写布局文件 activity_main.xml，代码如下：

```
<?xml version="1.0" encoding="utf-8"?>
<LinearLayout xmlns:android="http://schemas.android.com/apk/res/android"
```

```
    xmlns:tools="http://schemas.android.com/tools"
    android:layout_width="match_parent"
    android:layout_height="match_parent"
    android:orientation="vertical"
    tools:context="com.ugrow.ph10ex2_picasso.MainActivity">

   <ImageView
       android:id="@+id/iv"
       android:layout_width="match_parent"
       android:layout_height="200dp" />

    <Button
        android:id="@+id/btn"
        android:layout_width="wrap_content"
        android:layout_height="wrap_content"
        android:layout_marginTop="16dp"
        android:layout_gravity="center_horizontal"
        android:text="下载图片"/>

</LinearLayout>
```

编写 MainActivity，代码如下：

```
public class MainActivity extends AppCompatActivity {

    @Bind(R.id.iv)
    ImageView iv;
    @Bind(R.id.btn)
    Button btn;

    private String
url="http://f.hiphotos.baidu.com/image/h%3D300/sign=4a0a3dd10155b31983f9847573ab8286/503d269759ee3d
6db032f61b48166d224e4ade6e.jpg";

    @Override
    protected void onCreate(Bundle savedInstanceState) {
        super.onCreate(savedInstanceState);
        setContentView(R.layout.activity_main);
        ButterKnife.bind(this);
        btn.setOnClickListener(new View.OnClickListener() {
            @Override
            public void onClick(View view) {
```

```
Picasso.with(MainActivity.this).load(url).placeholder(R.mipmap.ic_launcher).error(R.mipmap.ic_launcher).fit().i
nto(iv);
            }
        });
    }
}
```

编写完毕，在 Android Studio 中运行项目，单击【下载图片】按钮，显示效果如图 S10-10 所示。

图片下载完成后的效果如图 S10-11 所示。

图 S10-10　Picasso 图片下载中

图 S10-11　图片下载完成

实践 10.3　XUtils

XUtils 是基于 Afinal 开发的一个功能较为完善的 Android 开源框架，其最新发布的版本为 XUtils3.0，在增加新功能的同时也提高了框架的性能。

XUtils 具有以下特点：

(1) XUtils 最低兼容 Android 4.0(API level 14)。

(2) XUtils 包含了很多实用的 Android 工具，如超大文件(超过 2G)上传支持，更全面的 http 请求协议支持(11 种谓词)，更加灵活的 ORM，更多的事件注解支持(且不受代码混淆影响)等。

(3) XUtils 中的网络模块 HttpUtils 将 HttpClient 替换为 UrlConnection，同时支持自动

解析回调方法的泛型，以实现更安全的断点续传策略。

(4) 支持标准的 Cookie 策略，并能区分 Domain 与 Path。

(5) 去除不常用的事件注解功能，提高了自身性能。

(6) XUtils 中的数据库模块 DbUtils 简化了操作数据库的 API，从而能达到与 greenDao 数据库一致的性能。

(7) 支持绑定 gif(受系统兼容性影响，部分 gif 文件只能静态显示)和 webp 格式的图片。

(8) 支持对图片进行圆角、圆形、方形等裁剪操作，支持图片的自动旋转操作。

XUtils 目前主要由四大模块组成，下面依次对各个模块进行介绍。

1. ViewUtils 模块

XUtils3 的注解模块，相当于 Android 中的 IOC 框架，可以使用完全注解方式进行 UI、资源和事件绑定。使用 ViewUtils 模块进行事件绑定的应用程序，在使用混淆工具混淆后仍可正常工作。目前，ViewUtils 模块已经支持常用的 20 种事件绑定方法，这些方法都被封装在 ViewCommonEventListener 类中。

ViewUtils 模块在实际开发中的使用方法如下：

(1) 在 Activity 中使用 ViewUtils 注解，示例代码如下：

```
public class MainActivity extends AppCompatActivity {
    @ViewInject(R.id.viewpager)
    ViewPager viewPager;
    @Override
    protected void onCreate(Bundle savedInstanceState) {
        super.onCreate(savedInstanceState);
        setContentView(R.layout.activity_main);
          x.view().inject(this);
        ...
    }
}
```

上述代码也可以修改为如下形式，区别在于对布局的加载方法不同。

```
@ContentView(R.layout.activity_main)
public class MainActivity extends AppCompatActivity {
    @ViewInject(R.id.viewpager)
    ViewPager viewPager;
    @Override
    protected void onCreate(Bundle savedInstanceState) {
        super.onCreate(savedInstanceState);
        x.view().inject(this);
        ...
    }
}
```

(2) 在 Fragment 中使用 ViewUtils 注解，示例代码如下：

```
@ContentView(R.layout.fragment_http)
public class HttpFragment extends Fragment {
    @Nullable
    @Override
    public View onCreateView(LayoutInflater inflater, @Nullable ViewGroup container, @Nullable Bundle
savedInstanceState) {
        return x.view().inject(this, inflater, container);
    }
    @Override
    public void onViewCreated(View v, @Nullable Bundle savedInstanceState) {
        super.onViewCreated(v, savedInstanceState);
    }
}
```

(3) 使用 ViewUtils 注解为按钮设置单击事件，示例代码如下：

```
@ViewInject(R.id.bt_main)
Button bt_main;
...
@Override
protected void onCreate(Bundle savedInstanceState) {
    ...
}
/**
* 单击事件
*/
@Event(type = View.OnClickListener.class,value = R.id.bt_main)
private void testInjectOnClick(View v){
    Toast.makeText(v,"OnClickListener", Toast.LENGTH_SHORT).show();
}
/**
* 长按事件
*/
@Event(type = View.OnLongClickListener.class,value = R.id.bt_main)
private boolean testOnLongClickListener(View v){
    Toast.makeText(v,"testOnLongClickListener", Toast.LENGTH_SHORT).show();
    return true;
}
```

注意：必须将所注解方法的修饰符声明为 private；type 默认为 View.OnClickListener.class，可以简化不写，如@Event(R.id.bt_main)。

2．DbUtils 模块

XUtils3 的数据库模块，相当于 Android 中的 ORM 框架，具有以下功能：① 可使用一行代码进行增删改查操作；② 支持事务处理功能，该功能默认为关闭；③ 通过注解方式对表名、列名、外键、唯一性约束、NOT NULL 约束、CHECK 约束等进行定义(需要对代码进行混淆时要注解表名和列名)；④ 支持绑定外键，且在保存实体时，与绑定外键相关联的实体会自动保存或更新；⑤ 自动加载外键关联实体，支持延时加载；⑥ 支持链式表达查询，具备更直观的查询语句。

DbUtils 模块在实际开发中的使用方法如下：

(1) 创建和删除数据库。

首先配置 DaoConfig，示例代码如下：

```
/**
* DaoConfig配置
* /
DbManager.DaoConfig daoConfig = new DbManager.DaoConfig()
        //设置数据库名称，默认为 xutils.db
        .setDbName("myapp.db")
        //设置创建表时的监听事件
        .setTableCreateListener(new DbManager.TableCreateListener() {
            @Override
            public void onTableCreated(DbManager db, TableEntity table){
                Log.i("JAVA", "onTableCreated：" + table.getName());
            }
        })
        //设置是否允许daoConfig 处理事务
        //.setAllowTransaction(true)
        //设置数据库路径，默认安装程序路径如下
        //.setDbDir(new File("/mnt/sdcard/"))
        //设置数据库的版本号
        //.setDbVersion(1)
        //设置数据库更新的监听事件
        .setDbUpgradeListener(new DbManager.DbUpgradeListener() {
            @Override
            public void onUpgrade(DbManager db, int oldVersion,
                    int newVersion) {
            }
        })
        //设置打开数据库时的监听事件
        .setDbOpenListener(new DbManager.DbOpenListener() {
            @Override
            public void onDbOpened(DbManager db) {
```

```
                //开启数据库的多线程操作，提升性能
                db.getDatabase().enableWriteAheadLogging();
            }
        });
DbManager db = x.getDb(daoConfig);
```

然后创建数据库表中的实体类 ChildInfo，示例代码如下：

```
/**
* onCreated = "sql"：当第一次创建表，需要插入数据时，在此编写 SQL 语句
*/
@Table(name = "child_info",onCreated = "")
public class ChildInfo {
    /**
     * name = "id"：指定数据库表中的一个字段
     * isId = true：指定是否为主键
     * autoGen = true：指定是否自动增长
     * property = "NOT NULL"：添加约束
     */
    @Column(name = "id",isId = true,autoGen = true,property = "NOT NULL")
    private int id;
    @Column(name = "c_name")
    private String cName;

    public ChildInfo(String cName) {
        this.cName = cName;
    }
    //默认的构造方法必须写出，如果没有，这张表是创建不成功的
    public ChildInfo() {
    }
    public int getId() {
        return id;
    }
    public void setId(int id) {
        this.id = id;
    }
    public String getcName() {
        return cName;
    }
    public void setcName(String cName) {
        this.cName = cName;
    }
```

```
    @Override
    public String toString() {
        return "ChildInfo{"+"id="+id+",cName='"+cName+'\''+'}';
    }
}
```

最后定义数据库的创建和删除方法，示例代码如下：

```
//创建数据库
@Event(R.id.create_db)
private void createDB(View v) throws DbException {
    //用集合向child_info表中插入多条数据
    ArrayList childInfos = new ArrayList<>();
    childInfos.add(new ChildInfo("zhangsan"));
    childInfos.add(new ChildInfo("lisi"));
    childInfos.add(new ChildInfo("wangwu"));
    childInfos.add(new ChildInfo("zhaoliu"));
    childInfos.add(new ChildInfo("qianqi"));
    childInfos.add(new ChildInfo("sunba"));
    //db.save()方法不仅可以插入单个对象，还能插入集合
    db.save(childInfos);
}

//删除数据库
@Event(R.id.del_db)
private void delDB(View v) throws DbException {
    db.dropDb();
}
```

(2) 删除表，示例代码如下：

```
//删除表
@Event(R.id.del_table)
private void delTable(View v) throws DbException {
    db.dropTable(ChildInfo.class);
}
```

(3) 查询表中的数据，示例代码如下：

```
//查询表中的数据
@Event(R.id.select_table)
private void selelctDB(View v) throws DbException {
    //查询数据库表中第一条数据
    ChildInfo first = db.findFirst(ChildInfo.class);
    Log.i("JAVA",first.toString());
    //添加查询条件进行查询
```

```
    //第一种写法：
    WhereBuilder b = WhereBuilder.b();
    b.and("id",">",2); //构造修改的条件
    b.and("id","<",4);
    List all = db.selector(ChildInfo.class).where(b).findAll();//findAll()：查询所有结果
    for(ChildInfo childInfo :all){
        Log.i("JAVA",childInfo.toString());
    }
    //第二种写法：
    List all = db.selector(ChildInfo.class).where("id",">",2).and("id","<",4).findAll();
    for(ChildInfo childInfo :all){
        Log.i("JAVA",childInfo.toString());
    }
}
```

(4) 修改表中的数据，示例代码如下：

```
//修改表中的一条数据
@Event(R.id.update_table)
private void updateTable(View v) throws DbException {
    //第一种写法：
    ChildInfo first = db.findFirst(ChildInfo.class);
    first.setcName("zhansan2");
    db.update(first,"c_name"); //c_name：表中的字段名
    //第二种写法：
    WhereBuilder b = WhereBuilder.b();
    b.and("id","=",first.getId()); //构造修改的条件
    KeyValue name = new KeyValue("c_name","zhansan3");
    db.update(ChildInfo.class,b,name);
    //第三种写法：
    first.setcName("zhansan4");
    db.saveOrUpdate(first);
}
```

(5) 删除表中的数据，示例代码如下：

```
@Event(R.id.del_table_data)
private void delTableData(View v) throws DbException {
    //第一种写法：
    db.delete(ChildInfo.class); //child_info表中数据将被全部删除
    //第二种写法，添加删除条件
    WhereBuilder b = WhereBuilder.b();
    b.and("id",">",2); //构造修改的条件
    b.and("id","<",4);
```

```
    db.delete(ChildInfo.class, b);
}
```

3. HttpUtils 模块

XUtils3 的网络模块，其主要功能包括：① 支持通过同步，异步方式向服务器发送请求；② 支持大文件上传，上传大文件不会发生 OOM(内存泄漏)；③ 支持 GET、POST、PUT、MOVE、COPY、DELETE、HEAD、OPTIONS、TRACE 与 CONNECT 请求方式；④ 支持缓存请求(默认为 GET 请求方式)返回的文本内容，可以设置缓存默认的过期时间和当前请求的过期时间。

HttpUtils 模块大大方便了网络功能的开发，其在实际开发中的应用主要包括 GET 请求、POST 请求、上传文件、下载文件、使用缓存等功能。下面依次进行介绍。

(1) 使用 GET 请求方式模拟登录网站，示例代码如下：

```
String url = "http://www.baidu.com";
@Event(R.id.get)
private void get(View v){
    final ProgressDialog progressDialog = new ProgressDialog(getActivity());
    progressDialog.setMessage("请稍候...");
    RequestParams params = new RequestParams(url);
    params.addQueryStringParameter("username","abc");
    params.addQueryStringParameter("password","123");
    Callback.Cancelable cancelable = x.http().get(params, new Callback.CommonCallback() {
        @Override
        public void onSuccess(String result) {
            Log.i("JAVA", "onSuccess result:" + result);
            progressDialog.cancel();
        }
        //请求异常后的回调方法
        @Override
        public void onError(Throwable ex, boolean isOnCallback) {
        }
        //主动调用取消请求的回调方法
        @Override
        public void onCancelled(CancelledException cex) {
        }
        @Override
        public void onFinished() {
            progressDialog.cancel();
        }
    });
    //主动调用取消请求
```

```
    cancelable.cancel();
}
```

(2) 使用 POST 请求方式模拟登录网站，示例代码如下：

```
String url = "http://www.baidu.com";
@Event(R.id.post)
private void post(View v){
    RequestParams params = new RequestParams(url);
    params.addBodyParameter("username","abc");
    params.addParameter("password","123");
    params.addHeader("head","android"); //为当前请求添加一个请求头
    x.http().post(params, new Callback.CommonCallback() {
        @Override
        public void onSuccess(String result) {
        }
        @Override
        public void onError(Throwable ex, boolean isOnCallback) {
        }
        @Override
        public void onCancelled(CancelledException cex) {
        }
        @Override
        public void onFinished() {
        }
    });
}
```

(3) 使用其他网络请求方式模拟登录网站，示例代码如下：

```
String url = "http://www.baidu.com";
@Event(R.id.other)
private void other(View v){
    RequestParams params = new RequestParams(url);
    params.addParameter("username","abc");
    x.http().request(HttpMethod.PUT, params, new Callback.CommonCallback() {
        @Override
        public void onSuccess(String result) {
        }
        @Override
        public void onError(Throwable ex, boolean isOnCallback) {
        }
        @Override
        public void onCancelled(CancelledException cex) {
```

```
        }
        @Override
        public void onFinished() {
        }
    });
}
```

(4) 上传文件，示例代码如下：

```
String url = "http://www.baidu.com";
@Event(R.id.upload)
private void upload(View v){
    String path="/mnt/sdcard/Download/icon.jpg";
    RequestParams params = new RequestParams(url);
    params.setMultipart(true);
    params.addBodyParameter("file",new File(path));
    x.http().post(params, new Callback.CommonCallback() {
        @Override
        public void onSuccess(String result) {
        }
        @Override
        public void onError(Throwable ex, boolean isOnCallback) {
        }
        @Override
        public void onCancelled(CancelledException cex) {
        }
        @Override
        public void onFinished() {
        }
    });
}
```

(5) 下载文件，示例代码如下：

```
String url = "http://www.baidu.com";
@Event(R.id.download)
private void download(View v){
    url = "http://127.0.0.1/server/ABC.apk";
    RequestParams params = new RequestParams(url);
     //自定义保存路径
     Environment.getExternalStorageDirectory()：SD卡的根目录
 params.setSaveFilePath(Environment.getExternalStorageDirectory()+"/myapp/");
    //自动为文件命名
    params.setAutoRename(true);
```

```
    x.http().post(params, new Callback.ProgressCallback() {
        @Override
        public void onSuccess(File result) {
            //apk下载完成后，调用系统的安装方法
            Intent intent = new Intent(Intent.ACTION_VIEW);
            intent.setDataAndType(Uri.fromFile(result), "application/vnd.android.package-archive");
            getActivity().startActivity(intent);
        }
        @Override
        public void onError(Throwable ex, boolean isOnCallback) {
        }
        @Override
        public void onCancelled(CancelledException cex) {
        }
        @Override
        public void onFinished() {
        }
        //网络请求之前回调
        @Override
        public void onWaiting() {
        }
        //网络请求开始时回调
        @Override
        public void onStarted() {
        }
        //下载时不断获取进度信息
        @Override
        public void onLoading(long total, long current, boolean isDownloading) {
            //当前进度和文件总大小
            Log.i("JAVA","current："+ current +"，total："+total);
        }
    });
}
```

(6) 请求网络时使用缓存，示例代码如下：

```
String url = "http://www.baidu.com";
@Event(R.id.cache)
private void cache(View v) {
    RequestParams params = new RequestParams(url);
    params.setCacheMaxAge(1000*60); //为请求添加缓存时间
    Callback.Cancelable cancelable = x.http().get(params, new Callback.CacheCallback() {
```

```
        @Override
        public void onSuccess(String result) {
            Log.i("JAVA","onSuccess："+result);
        }
        @Override
        public void onError(Throwable ex, boolean isOnCallback) {
        }
        @Override
        public void onCancelled(CancelledException cex) {
        }
        @Override
        public void onFinished() {
        }
        //result：缓存内容
        @Override
        public boolean onCache(String result) {
            //返回true：返回的缓存内容正确，使用本地缓存；返回false：返回的缓存内容不正确，不使
用本地缓存，重新请求网络
            Log.i("JAVA","cache："+result);
            return true;
        }
    });
}
```

4．BitmapUtils 模块

XUtils3 的图片模块，其主要功能包括：① 加载图片时，会自动纠正加载过程中出现的 OOM 错误和 Android 容器快速滑动时出现的图片错位等现象；② 支持加载网络图片和本地图片；③ 使用 LRU 算法进行内存管理，从而更好地管理图片内存；④ 配置加载线程数量、缓存大小、缓存路径、加载显示动画等。

BitmapUtils 模块的重点在于：四个加载图片的 bind()方法、loadDrawable()方法、loadFile()方法和 ImageOptions()方法。BitmapUtils 模块在实际开发中的使用方法如下所示。

(1) 首先获取 ImageView 控件，示例代码如下：

```
@ViewInject(R.id.image01)
ImageView image01;
@ViewInject(R.id.image02)
ImageView image02;
@ViewInject(R.id.image03)
ImageView image03;
```

(2) 然后获取网络图片的地址，示例代码如下：

```
String[] urls={
    "http://img.android.com/a.jpg",
    "http://img.android.com/b.jpg"
    "http://img.android.com/c.jpg"
    ...
};
```

(3) 最后声明 setPic()方法，编写图片下载操作的代码，相关代码如下：

```
private void setPic() {
    /**
     * 通过ImageOptions.Builder().set方法设置图片的属性
     */
    ImageOptions options = new ImageOptions.Builder().setFadeIn(true).build(); //淡入效果
    //ImageOptions.Builder()的一些其他属性：
    //.setCircular(true) //设置图片显示为圆形
    //.setSquare(true) //设置图片显示为正方形
    //setCrop(true).setSize(200,200) //设置大小
    //.setAnimation(animation) //设置动画
    //.setFailureDrawable(Drawable failureDrawable) //设置加载失败的动画
    //.setFailureDrawableId(int failureDrawable) //以资源ID 方式设置加载失败的动画
    //.setLoadingDrawable(Drawable loadingDrawable) //设置加载中的动画
    //.setLoadingDrawableId(int loadingDrawable) //以资源ID 方式设置加载中的动画
    //.setIgnoreGif(false) //忽略gif图片
    //.setParamsBuilder(ParamsBuilder paramsBuilder) //在网络请求中添加一些参数
    //.setRaduis(int raduis) //设置拐角弧度
    //.setUseMemCache(true) //设置使用MemCache，默认true

    /**
     * 加载图片的4个bind()方法
     */
    x.image().bind(image01, urls[0]);
    x.image().bind(image02, urls[1], options);
    x.image().bind(image03, urls[2], options, new Callback.CommonCallback() {
        @Override
        public void onSuccess(Drawable result) {
        }
        @Override
        public void onError(Throwable ex, boolean isOnCallback) {
        }
        @Override
        public void onCancelled(CancelledException cex) {
```

```
        }
        @Override
        public void onFinished() {
        }
    });
    x.image().bind(image04, urls[3], options, new Callback.CommonCallback() {
        @Override
        public void onSuccess(Drawable result) {
        }
        @Override
        public void onError(Throwable ex, boolean isOnCallback) {
        }
        @Override
        public void onCancelled(CancelledException cex) {
        }
        @Override
        public void onFinished() {
        }
    });

    /**
     * 使用 loadDrawable()方法加载图片
     */
    Callback.Cancelable cancelable = x.image().loadDrawable(urls[0], options, new Callback.CommonCallback()
{
        @Override
        public void onSuccess(Drawable result) {
            image03.setImageDrawable(result);
        }
        @Override
        public void onError(Throwable ex, boolean isOnCallback) {
        }
        @Override
        public void onCancelled(CancelledException cex) {
        }
        @Override
        public void onFinished() {
        }
    });
    //主动取消loadDrawable()方法
```

```
    //cancelable.cancel();

    /**
     * loadFile()方法
     * 当通过bind()或者 loadDrawable()方法加载了一张图片后，它会保存到本地文件中，当需要这张
图片时，可以通过 loadFile()方法进行查找。
     * urls[0]：网络地址
     */
    x.image().loadFile(urls[0],options,new Callback.CacheCallback(){
        @Override
        public boolean onCache(File result) {
            //在这里可以进行图片另存为等操作
            Log.i("JAVA","file："+result.getPath()+result.getName());
            return true; //使用本地缓存
        }
        @Override
        public void onSuccess(File result) {
            Log.i("JAVA","file");
        }
        @Override
        public void onError(Throwable ex, boolean isOnCallback) {
        }
        @Override
        public void onCancelled(CancelledException cex) {
        }
        @Override
        public void onFinished() {
        }
    });
}
```

【示例 10.3】 在 Android Studio 中新建项目 Ph10ex3_XUtils，逐一实现 XUtils 的四大模块。

使用 XUtils 之前，需要在 build.gradle 中编写以下代码，添加依赖库：

```
dependencies {
    compile fileTree(dir: 'libs', include: ['*.jar'])
    androidTestCompile('com.android.support.test.espresso:espresso-core:2.2.2', {
        exclude group: 'com.android.support', module: 'support-annotations'
    })
    compile 'com.android.support:appcompat-v7:26.+'
    compile 'com.android.support.constraint:constraint-layout:1.0.2'
```

```
    compile 'org.xutils:xutils:3.3.36'
    testCompile 'junit:junit:4.12'
}
```

同时，需要在清单文件 AndroidMainfest.xml 中添加权限，代码如下：

```
<uses-permission android:name="android.permission.INTERNET" />
<uses-permission android:name="android.permission.WRITE_EXTERNAL_STORAGE" />
```

然后新建 MyApp 类，并在其中初始化 XUtils：

```
// 在application的onCreate中初始化
/**
* 初始化XUtils3
*/
public class MyApp extends Application {

    @Override
    public void onCreate() {
        super.onCreate();
        //对XUtils进行初始化
        x.Ext.init(this);
        //设置是否为开发、调试模式
        x.Ext.setDebug(BuildConfig.DEBUG);//设置是否输出debug日志，开启debug会影响性能


    }
}
```

初始化完成后，需要在清单文件中注册 MyApp，代码如下：

```
 <application
        android:name=".MyApp"
        android:allowBackup="true"
        android:icon="@mipmap/ic_launcher"
        android:label="@string/app_name"
        android:roundIcon="@mipmap/ic_launcher_round"
        android:supportsRtl="true"
        android:theme="@style/AppTheme">


               .......
</application>
```

(1) 使用 XUtils 中的 ViewUtils 模块，实现对控件的初始化和事件绑定操作。

新建 ViewUtilsActivity 并编写其布局文件 activity_view_utils.xml，代码如下：

```
<?xml version="1.0" encoding="utf-8"?>
<LinearLayout xmlns:android="http://schemas.android.com/apk/res/android"
    xmlns:tools="http://schemas.android.com/tools"
```

```
    android:layout_width="match_parent"
    android:layout_height="match_parent"
    android:orientation="vertical"
    tools:context="com.ugrow.ph10ex3_xutils.ViewUtilsActivity">

    <Button
        android:id="@+id/btn_click"
        android:layout_width="wrap_content"
        android:layout_height="wrap_content"
        android:textColor="#000"
        android:text="点击" />

    <TextView
        android:id="@+id/tv"
        android:layout_width="wrap_content"
        android:layout_height="wrap_content"
        android:textColor="#000"/>

</LinearLayout>
```

然后编写 ViewUtilsActivity，代码如下：

```
@ContentView(R.layout.activity_view_utils)
public class ViewUtilsActivity extends AppCompatActivity {


    @ViewInject(R.id.tv)
    private TextView mTv;
    @ViewInject(R.id.btn_click)
    private Button mBtn;

    @Override
    protected void onCreate(Bundle savedInstanceState) {
        super.onCreate(savedInstanceState);
        x.view().inject(this);
    }
    @Event(type = View.OnClickListener.class,value = R.id.btn_click)
    private void onClick(View view){
        mTv.setText("按钮被点击了！");
    }
}
```

在清单文件中，将程序的入口修改为“ViewUtilsActivity”，在 Android Studio 中运行程序，单击【点击】按钮，效果如图 S10-12 所示。

图 S10-12　测试 ViewUtil 模块功能

(2) 使用 XUtils 中的 DbUtils 模块，实现对数据库中数据的增删改查。

新建 DbUtilsActivity 并编写其布局文件 activity_db_utils.xml，代码如下：

```
<?xml version="1.0" encoding="utf-8"?>
<LinearLayout xmlns:android="http://schemas.android.com/apk/res/android"
    xmlns:tools="http://schemas.android.com/tools"
    android:layout_width="match_parent"
    android:layout_height="match_parent"
    android:orientation="vertical"
    tools:context="com.ugrow.ph10ex3_xutils.DbUtilsActivity">

    <Button
        android:id="@+id/btn_insert"
        android:layout_width="wrap_content"
        android:layout_height="wrap_content"
        android:text="添加数据" />

    <Button
        android:id="@+id/btn_delete"
        android:layout_width="wrap_content"
        android:layout_height="wrap_content"
        android:text="删除数据" />

    <Button
```

```
        android:id="@+id/btn_update"
        android:layout_width="wrap_content"
        android:layout_height="wrap_content"
        android:text="修改数据" />

    <Button
        android:id="@+id/btn_select"
        android:layout_width="wrap_content"
        android:layout_height="wrap_content"
        android:text="查询数据" />

</LinearLayout>
```

新建 Person 实体类，在其中编写代码如下：

```
/**
 * onCreated = "sql"：当第一次创建表需要插入数据时在此写 SQL 语句
 */
//@Table(name = "person"):表名
@Table(name = "person",onCreated = "")
public class Person {
    /**
     * name = "id"：指定数据库表中的一个字段
     * isId = true：指定是否为主键
     * autoGen = true：指定是否自动增长
     * property = "NOT NULL"：添加约束
     */
    @Column(name = "_id",isId = true,autoGen = true,property = "NOT NULL")
    private int id;
    @Column(name = "name")
    private String name;
    @Column(name = "age")
    private int age;
    @Column(name = "sex")
    private String sex;

    public int getId() {
        return id;
    }

    public void setId(int id) {
        this.id = id;
```

```
    }

    public String getName() {
        return name;
    }

    public void setName(String name) {
        this.name = name;
    }

    public int getAge() {
        return age;
    }

    public void setAge(int age) {
        this.age = age;
    }

    public String getSex() {
        return sex;
    }

    public void setSex(String sex) {
        this.sex = sex;
    }

    @Override
    public String toString() {
        return "Person{" +
                "id=" + id +
                ", name='" + name + '\'' +
                ", age=" + age +
                ", sex='" + sex + '\'' +
                '}';
    }
}
```

编写 DbUtilsActivity，代码如下：

```
public class DbUtilsActivity extends AppCompatActivity {

    private DbManager db;
```

```
@Override
protected void onCreate(Bundle savedInstanceState) {
    super.onCreate(savedInstanceState);
    setContentView(R.layout.activity_db_utils);
    x.view().inject(this);
    initDb();
}

private void initDb() {
    DbManager.DaoConfig daoConfig = new DbManager.DaoConfig()
            //设置数据库名，默认为 xutils.db
            .setDbName("myapp.db")
            //设置创建表时的监听事件
            .setTableCreateListener(new DbManager.TableCreateListener() {
                @Override
                public void onTableCreated(DbManager db, TableEntity table){
                    Log.i("JAVA", "onTableCreated：" + table.getName());
                }
            })
            //设置是否允许 daoConfig 处理事务，默认为 true
            //.setAllowTransaction(true)
            //设置数据库路径，默认安装程序路径如下
            //.setDbDir(new File("/mnt/sdcard/"))
            //设置数据库的版本号
            //.setDbVersion(1)
            //设置更新数据库时的监听事件
            .setDbUpgradeListener(new DbManager.DbUpgradeListener() {
                @Override
                public void onUpgrade(DbManager db, int oldVersion,
                                      int newVersion) {
                }
            })
            //设置打开数据库时的监听事件
            .setDbOpenListener(new DbManager.DbOpenListener() {
                @Override
                public void onDbOpened(DbManager db) {
                    //开启数据库中的多线程操作，提升性能
                    db.getDatabase().enableWriteAheadLogging();
                }
            });
```

```
        db = x.getDb(daoConfig);
    }
    @Event({R.id.btn_insert,R.id.btn_delete,R.id.btn_update,R.id.btn_select})
    private void onClick(View view){
        switch (view.getId()){
            case R.id.btn_insert:
                Person person = new Person();
                person.setId(1);
                person.setName("张三");
                person.setAge(18);
                person.setSex("男");
                try {
                    db.save(person);
                    Toast.makeText(DbUtilsActivity.this,"添加成功！ ",Toast.LENGTH_SHORT).show();
                } catch (DbException e) {
                    e.printStackTrace();
                }
                break;
            case R.id.btn_delete:
                try {
                    db.dropTable(Person.class);
                    Toast.makeText(DbUtilsActivity.this,"删除表成功！ ",Toast.LENGTH_SHORT).show();
                } catch (DbException e) {
                    e.printStackTrace();
                }
                break;
            case R.id.btn_update:
                try {
                    Person first = db.findFirst(Person.class);
                    first.setName("李四");
                    db.update(first,"name");
                    Toast.makeText(DbUtilsActivity.this,"修改第一条数据成功！
",Toast.LENGTH_SHORT).show();
                } catch (DbException e) {
                    e.printStackTrace();
                }
                break;
            case R.id.btn_select:
                try {
                    Person first = db.findFirst(Person.class);
```

```
                if (first != null){
                    Toast.makeText(DbUtilsActivity.this,"查询第一条数据成功！",
Toast.LENGTH_SHORT).show();

Log.d("TAG","ID:"+first.getId()+",name:"+first.getName()+",age:"+first.getAge()+",sex:"+first.getSex());
                }else {
                    Toast.makeText(DbUtilsActivity.this,"暂无数据！
",Toast.LENGTH_SHORT).show();
                }

            } catch (DbException e) {
                e.printStackTrace();
            }
            break;
        default:
            break;
    }
  }

}
```

在清单文件中，将程序的入口修改为 DbUtilsActivity，在 Android Studio 中运行程序，效果如图 S10-13 所示。

图 S10-13　测试 DbUtils 模块功能

单击图 S10-13 中的【添加数据】按钮，会向数据库中添加一条数据；单击【删除数据】按钮，会将数据库中的 Person 表删除；单击【修改数据】按钮，会修改数据库中的第一条数据；单击“查询数据”按钮，会查询数据库中第一条数据，并将其打印如下：

```
02-01 02:05:13.679 4331-4331/com.ugrow.ph10ex3_xutils D/TAG: ID:1,name:李四,age:18,sex:男
```

(3) 使用 XUtils 中的 HttpUtils 模块，实现登录功能。

新建 HttpUtilsActivity 并编写其布局文件 activity_http_utils.xml，代码如下：

```
<?xml version="1.0" encoding="utf-8"?>
<LinearLayout xmlns:android="http://schemas.android.com/apk/res/android"
    xmlns:tools="http://schemas.android.com/tools"
    android:layout_width="match_parent"
    android:layout_height="match_parent"
    android:orientation="vertical"
    tools:context="com.ugrow.ph10ex3_xutils.HttpUtilsActivity">

    <EditText
        android:id="@+id/et_name"
        android:layout_width="match_parent"
        android:layout_height="wrap_content"
        android:hint="请输入用户名"/>
    <EditText
        android:id="@+id/et_pwd"
        android:layout_width="match_parent"
        android:layout_height="wrap_content"
        android:hint="请输入密码"/>

    <Button
        android:id="@+id/btn_login"
        android:layout_width="match_parent"
        android:layout_height="wrap_content"
        android:text="登录"
        android:textColor="#000"/>

</LinearLayout>
```

编写 HttpUtilsActivity，代码如下：

```
public class HttpUtilsActivity extends AppCompatActivity {
    @ViewInject(R.id.et_name)
    private EditText mEtName;
    @ViewInject(R.id.et_pwd)
    private EditText mEtPwd;
    @ViewInject(R.id.btn_login)
```

```
private Button mBtn;

private String url ="http://192.168.0.57:8080/repastWeb_V1/only/?to=login";
private ProgressDialog dialog;
@Override
protected void onCreate(Bundle savedInstanceState) {
    super.onCreate(savedInstanceState);
    setContentView(R.layout.activity_http_utils);
    x.view().inject(this);

}
@Event(R.id.btn_login)
private void onClick(View view){
    dialog = new ProgressDialog(HttpUtilsActivity.this);
    dialog.setMessage("正在登录，请稍后...");
    dialog.setCancelable(false);

    String name = mEtName.getText().toString();
    String pwd = mEtPwd.getText().toString();
    if (TextUtils.isEmpty(name)) {
        Toast.makeText(HttpUtilsActivity.this, "请输入用户名",
                Toast.LENGTH_SHORT).show();
        return;
    } else if (TextUtils.isEmpty(pwd)) {
        Toast.makeText(HttpUtilsActivity.this, "请输入密码",
                Toast.LENGTH_SHORT).show();
        return;
    }
    dialog.show();
    RequestParams params = new RequestParams(url);
    params.addQueryStringParameter("code",name);
    params.addQueryStringParameter("password",pwd);
    x.http().get(params, new Callback.CacheCallback<String>() {
        @Override
        public void onSuccess(String result) {
            Log.d("TAG",result);
        }

        @Override
        public void onError(Throwable ex, boolean isOnCallback) {
```

```
                Toast.makeText(HttpUtilsActivity.this,"登录失败！",Toast.LENGTH_LONG).show();
            }

            @Override
            public void onCancelled(CancelledException cex) {
                //主动调用取消请求的回调方法
            }

            @Override
            public void onFinished() {
                dialog.dismiss();
            }

            @Override
            public boolean onCache(String result) {
                return false;
            }
        });
    }
}
```

上述代码中，HttpUtils 模块调用的接口是点餐系统中的登录接口。

在清单文件中，将程序的入口修改为“HttpUtilsActivity”，在 Android Studio 中运行程序，在模拟器中输入用户名“fwy1”，密码“000000”，然后单击【登录】按钮，打开 LogCat 日志，将看到如下信息：

```
02-01 02:29:39.448 5994-5994/com.ugrow.ph10ex3_xutils D/TAG:
{"success":true,"msg":"","data":"{\"id\":1,\"code\":\"fwy1\",\"password\":\"000000\",\"name\":\"用户1\"}"}
```

(4) 使用 XUtils 中的 BitmapUtils 模块，实现网络图片的加载。

新建 BitmapUtilsActivity 并编写其布局文件 activity_bitmap_utils.xml，代码如下：

```
<?xml version="1.0" encoding="utf-8"?>
<LinearLayout xmlns:android="http://schemas.android.com/apk/res/android"
    xmlns:tools="http://schemas.android.com/tools"
    android:layout_width="match_parent"
    android:layout_height="match_parent"
    android:orientation="vertical"
    tools:context="com.ugrow.ph10ex3_xutils.BitmapUtilsActivity">

    <ImageView
        android:id="@+id/iv"
        android:layout_width="match_parent"
        android:layout_height="200dp" />
```

```
    <Button
        android:id="@+id/btn_download"
        android:layout_width="wrap_content"
        android:layout_height="wrap_content"
        android:layout_gravity="center_horizontal"
        android:layout_marginTop="16dp"
        android:text="下载图片"/>
</LinearLayout>
```

编写 BitmapUtilsActivity，代码如下：

```
public class BitmapUtilsActivity extends AppCompatActivity {
    @ViewInject(R.id.iv)
    private ImageView iv;
    @ViewInject(R.id.btn_download)
    private Button btn;
    private String mUrl = "http://img1.imgtn.bdimg.com/it/u=2740873643,4064044956&fm=27&gp=0.jpg";

    @Override
    protected void onCreate(Bundle savedInstanceState) {
        super.onCreate(savedInstanceState);
        setContentView(R.layout.activity_bitmap_utils);
        x.view().inject(this);
    }

    @Event(R.id.btn_download)
    private void onClick(View view){
        ImageOptions options = new ImageOptions.Builder().setFadeIn(true).build(); //淡入效果
        //ImageOptions.Builder()的一些其他属性：
        //.setCircular(true) //设置图片显示为圆形
        //.setSquare(true) //设置图片显示为正方形
        //setCrop(true).setSize(200,200) //设置大小
        //.setAnimation(animation) //设置动画
        //.setFailureDrawable(Drawable failureDrawable) //设置加载失败的动画
        //.setFailureDrawableId(int failureDrawable) //以资源ID 方式设置加载失败的动画
        //.setLoadingDrawable(Drawable loadingDrawable) //设置加载中的动画
        //.setLoadingDrawableId(int loadingDrawable) //以资源ID 方式设置加载中的动画
        //.setIgnoreGif(false) //忽略Gif图片
        //.setParamsBuilder(ParamsBuilder paramsBuilder) //在网络请求中添加一些参数
        //.setRaduis(int raduis) //设置拐角弧度
        //.setUseMemCache(true) //设置使用MemCache，默认true
```

```
        x.image().bind(iv,mUrl,options);
    }

}
```

在清单文件中，将程序入口修改为“BitmapUtilsActivity”，在 Android Studio 中运行程序，单击【下载图片】按钮，效果如图 S10-14 所示。

S10-14　测试 BitmapUtils 模块功能

拓展练习

结合本章知识，综合应用框架 1、2、3，对点餐系统 App 进行重构，要求如下：

✧ 使用 HttpUtils 处理网络请求。
✧ 使用 DbUtils 管理数据库。
✧ 使用 ButterKnife 进行注解。
✧ 使用 Picasso 加载图片。

附录 Widget 列表

本附录中列出了 android.widget 包下的所有 Widget，即所有的 UI 组件。

类	说　明
AbsListView	各种列表控件的抽象父类
AbsoluteLayout	已废弃
AbsSeekBar	拖动条的抽象父类
AbsSpinner	下拉菜单的抽象父类
AdapterView<T extends Adapter>	代表一个视图，其子视图由 Adapter 决定
AdapterViewAnimator	当 AdapterView 的子视图切换时提供动画
AlphabetIndexer	为实现 SectionIndexer 接口的 Adapters 提供的帮助类
AnalogClock	表状时钟
ArrayAdapter<T>	数组适配器
AutoCompleteTextView	具有自动完成功能的 TextView
BaseAdapter	可用于 ListView 和 Spinner 的通用的适配器父类
BaseExpandableListAdapter	ExpandableListView 的适配器父类
Button	按钮
CheckBox	多选框
CheckedTextView	实现 Checkable 接口的 TextView
Chronometer	一个简单的定时器
CompoundButton	具有选中和未选中状态的按钮
CursorAdapter	可将 Cursor 数据绑定到 ListView 的适配器
CursorTreeAdapter	可将多个 Cursor 数据绑定到 ExpandableListView 的适配器
DatePicker	日期选择控件
DialerFilter	用于支持各种电话键盘
DigitalClock	数字时钟
EditText	可编辑的文本框
ExpandableListView	二级列表视图
Filter	用于过滤数据
FrameLayout	帧布局，用于显示堆叠在一起的多个视图
Gallery	画廊控件，用于显示水平滚动的多个项目
GridLayout	网格布局
GridView	表格视图
HeaderViewListAdapter	用于具有表头的 ListView 的适配器
HorizontalScrollView	水平滚动视图
ImageButton	图片按钮
ImageSwitcher	图片切换控件
ImageView	图片视图

续表

类	说　　明
LinearLayout	线性布局
ListView	列表视图
MediaController	针对 MediaPlayer 的控制器
MultiAutoCompleteTextView	可连续提示的自动完成文本框
PopupWindow	弹出窗口
ProgressBar	进度条
QuickContactBadge	联系人快捷图标
RadioButton	单选按钮
RadioGroup	单选按钮组
RatingBar	评级控件
RelativeLayout	相对布局
RemoteViews	远程（跨进程）视图
ResourceCursorAdapter	根据 XML 文件创建视图的适配器
ResourceCursorTreeAdapter	根据 XML 文件创建视图的适配器，适合 ExpandableListView
Scroller	用于封装滚动操作
ScrollView	可滚动视图
SeekBar	可拖动的进度条
SimpleAdapter	简单适配器，可将静态数据映射到视图
SimpleCursorAdapter	用于将 Cursor 数据映射到视图
SimpleCursorTreeAdapter	用于将 Cursor 数据映射到视图，并支持 ExpandableListView
SimpleExpandableListAdapter	用于将静态数据映射到 ExpandableListView
SlidingDrawer	可以将内容隐藏，并提供一个可拖回屏幕显示的把手
Spinner	下拉菜单
TabHost	选项卡视图容器
TableLayout	表格布局
TableRow	表格布局中的行
TabWidget	用于显示选项卡列表
TextSwitcher	专用于 TextView 的 ViewSwitcher
TextView	文本框
TimePicker	时间选择控件
Toast	用于提示简短的信息
ToggleButton	具有选中状态和未选中状态的按钮，带有亮度提示
TwoLineListItem	用于 ListView，有两个子视图
VideoView	用于显示视频的视图
ViewAnimator	FrameLayout 的父类，用于在变换视图时显示动画效果
ViewFlipper	用于在添加到该类的多个视图之间显示动画
ViewSwitcher	用于切换视图
ZoomButton	缩放按钮
ZoomButtonsController	缩放按钮控制器
ZoomControls	用于控制缩放的控件